The Peptides *Volume 8*

THE PEPTIDES

Analysis, Synthesis, Biology

Treatise Editors

S. UDENFRIEND AND J. MEIENHOFER

Volume 1
Major Methods of Peptide Bond Formation

Volume 2
Special Methods in Peptide Synthesis, Part A

Volume 3
Protection of Functional Groups in Peptide Synthesis

Volume 4
Modern Techniques of Conformational, Structural, and Configurational Analysis

Volume 5
Special Methods in Peptide Synthesis, Part B

Volume 6
Opioid Peptides: Biology, Chemistry, and Genetics

Volume 7
Edited by Victor J. Hruby
Conformation in Biology and Drug Design

Volume 8
Edited by Clark W. Smith
Chemistry, Biology, and Medicine of Neurohypophyseal Hormones and Their Analogs

The Peptides

Analysis, Synthesis, Biology

Edited by

SIDNEY UDENFRIEND
Roche Institute of Molecular Biology
Nutley, New Jersey

JOHANNES MEIENHOFER
Berlex Laboratories, Inc.
Cedar Knolls, New Jersey

VOLUME 8 Chemistry, Biology, and Medicine of Neurohypophyseal Hormones and Their Analogs

Edited by

CLARK W. SMITH
Biopolymer Chemistry
The Upjohn Company
Kalamazoo, Michigan

1987

ACADEMIC PRESS, INC.
Harcourt Brace Jovanovich, Publishers
Orlando San Diego New York Austin
Boston London Sydney Tokyo Toronto

ACADEMIC PRESS, INC.
Orlando, Florida 32887

United Kingdom Edition published by
ACADEMIC PRESS INC. (LONDON) LTD.
24–28 Oval Road, London NW1 7DX

Library of Congress Cataloging in Publication Data
(Revised for vol. 8)

The Peptides : analysis, synthesis, biology.

Vol. 6- edited by Sidney Udenfriend and others.
Includes bibliographies and indexes.
Contents: v. 1. Major methods of peptide bond formation.—v. 2. Special methods in peptide synthesis, part A.—[etc.]—v. 8. Chemistry, biology, and medicine of neurohypophyseal hormones and their analogs.
1. Peptides—Collected works. 2. Peptides.
I. Gross, Erhard.
QP552.P4P47 574.1'92456 78-31958
ISBN 0–12–304208–9 (v. 8: alk. paper)

PRINTED IN THE UNITED STATES OF AMERICA

86 87 88 89 9 8 7 6 5 4 3 2 1

Contents

Chapter 3 *Biochemistry and Biology of Vasopressin, Oxytocin, and Their Corresponding Neurophysins*

Dietmar Richter

Chapter 4 *Structure–Activity Relationships of Neurohypophyseal Peptides*

Victor J. Hruby and Clark W. Smith

Chapter 5 *The Conformation of Neurohypophyseal Hormones*

Judith C. Hempel

Chapter 6 *Central Nervous System Effects of Neurohypophyseal Peptides*

Paula L. Hoffman

Chapter 7 *Renal Effects and Hormonal Interaction with the Neurohypophyseal Hormone, Vasopressin*

San-e Ishikawa, Jin K. Kim, and Robert W. Schrier

Chapter 8 *Cardiovascular Effects of Neurohypophyseal Peptides*

Phillip G. Schmid and Kaushik P. Patel

Preface

The last treatise that covered the chemistry, biology, and medicine of neurohypophyseal hormones was the excellent work edited by B. Berde in Volume 23 of the "Handbook of Experimental Pharmacology" in 1968. During the interim there have been a number of reviews that covered specific aspects of the chemistry or the biology or the medicine of neurohypophyseal hormones. In Volume 8 of "The Peptides" I have attempted again to bring together many of the advances in our knowledge about this important class of peptides in all three disciplines into one volume.

The emphasis of Volume 8 is certainly different from that of the "Handbook" and reflects the advances in methods to study peptide hormone action at the molecular level. Powerful new methods in physical chemistry and biophysics have allowed the study of specific conformational features to be related to biological activity. Recombinant molecular biology has allowed the elucidation of the biosynthetic pathway of neurohypophyseal hormones. New methods in cell biology have advanced our knowledge of peptide hormone–receptor interactions and the role of the second messenger molecules. New techniques in protein biochemistry have even made it necessary to expand our definition of neurohypophyseal peptides.

Because the neurohypophyseal peptides have been used for so many basic studies in the chemistry, biophysics, biology, and medicine of peptide hormones, these reviews are intended as a reference for the specialist in neurohypophyseal peptides and as a guide for the novice investigator interested more generally in peptide hormone research.

The order of the chapters in this volume begins with studies on the neurohypophysis, proceeds through the chemistry of the peptides, and finishes with the biological effects of the hormones at their target tissues. In Chapter 1 by Alan G. Robinson and Joseph G. Verbalis the anatomy, physiology, and clinical disorders of the neurohypophysis are discussed. In Chapter 2 by Thomas L. O'Donohue and Joszef Z. Kiss we are introduced to "the other neurohypophyseal peptides." In Chapter 3 by Dietmar Richter the biosynthesis of neurohypophyseal peptides is presented. In Chapter 4 by Victor J. Hruby and Clark W. Smith the structure–activity relationships of neurohypophyseal peptides are discussed with an emphasis on the role of conformational features. Except for purposes of continuity of discussion, the extensive tables in this chapter do not include those analogs already listed in the earlier "Handbook of Experimental Pharmacology." In Chapter 5 Judith C. Hempel presents conformational studies of neurohypophyseal hormones by physical chemical techniques such as X-ray crystallography, NMR, and other spectroscopic methods. In Chapter 6 by Paula Hoffman some of the central nervous system effects of neurohypophyseal peptides are given. The effects on memory and learning by neurohypophyseal hormones have been an entirely new area of study since the last treatise. Effects on one of the major target tissues of neurohypophyseal hormones, the kidney, are presented by S. Ishikawa, J. K. Kim, and R. W. Schrier in Chapter 7. An integrated discussion of the effects of neurohypophyseal hormones on the cardiovascular systems, including clinical implications, is clearly presented in the final chapter by P. G. Schmid and K. P. Patel.

I am greatly indebted to Johannes Meienhofer for his constant encouragement and everlasting patience with me during the creation of this volume. We are all indebted to Vincent du Vigneaud and Roderich Walter for their many insights upon which much of the work presented here was built.

Clark W. Smith

The Peptides *Volume 8*

Chapter **1**

Clinical Disorders of the Neurohypophysis

ALAN G. ROBINSON, JOSEPH G. VERBALIS

Department of Medicine
School of Medicine
University of Pittsburgh
Pittsburgh, Pennsylvania 15261

I. ANATOMY

The peptide hormones of the posterior pituitary, oxytocin and vasopressin, are synthesized in specialized magnocellular neurons of the hypothalamus (*9, 103*). The magnocellular neurons are grouped into two major nuclei, the supraoptic nucleus (SON) and the paraventricular nucleus (PVN) (*102, 103*). In most species, both hormones are synthesized in both nuclei, although there is frequently a predominance of vasopressin in the supraoptic nucleus (*20, 102–104*). Each hormone is synthesized together with its neurophysin (*56*) in individual neurons, which are hormone-specific (*104*). The vasopressinergic and oxytocinergic neurons are clustered into rather discrete areas within each nucleus (*20, 103*), and there is, in addition, some suggestion that the clustering within the nucleus is on the basis of the ultimate projection of the neurons (*20, 55, 76, 83, 101*). For example, vasopressinergic neurons that project to the spinal cord may be clustered in an area separate from vasopressinergic neurons that project to the posterior pituitary. The major projection of both nuclei is to the posterior pituitary via the supraopticohypophyseal tract. In the posterior pituitary there is a large store of vasopressin, oxytocin, and their associated neurophysins in neurosecretory granules within the axon terminals.

The first clear demonstration that neurohypophyseal neurons projected to areas other than the posterior pituitary was the description of neurophysin terminals in the zona externa of the median eminence (*50, 78, 100*). Since the terminals in the median eminence ended on capillaries of the hypothalamohypophyseal portal system, it was concluded that the vasopressin was secreted into the pituitary

THE PEPTIDES, VOLUME 8

portal circulation. Measurement of neurophysin and vasopression in pituitary stalk blood of the monkey confirmed markedly elevated concentrations consistent with direct secretion of vasopressin into portal blood (*100*). As it had long been known that vasopressin could stimulate the release of adrenocorticotropic hormone (ACTH), the content of neurophysin and vasopressin in the median eminence was studied during a state of increased release of ACTH. The content of vasopressin and neurophysin was determined by immunohistochemistry in the zona externa after adrenalectomy (*71*). A marked increase in the neurophysin and vasopressin at this level was noted, which could be specifically prevented by pretreatment with glucocorticoid (*50, 78*). Subsequently, since the discovery of corticotropin-releasing factor (CRF) (*87*), it has been amply demonstrated in many species that vasopressin does potentiate CRF-mediated release of ACTH (*57, 88*).

Interestingly, the vasopressinergic neurons that project to the median eminence originate in the PVN (*3*), but in parvocellular rather than magnocellular neurons (*97*). This is also the site of synthesis corticotropin-releasing factor. Indeed, some reports have colocalized vasopressin and CRF within the same neurons (*10, 84*). In addition to the projections to the median eminence, numerous hypothalmic projections of both vasopressin and oxytocin neurons to the spinal cord, brainstem, and forebrain have been described (*76, 82, 83*). Although the function of these projections has not been elucidated, it is likely that vasopressin and oxytocin will be found to be involved in a variety of diverse central nervous system (CNS) functions. Recent studies have confirmed that the paraventricular nucleus is the source of virtually all the extrahypothalamic projections of vasopressin, while the SON projects almost exclusively to the posterior pituitary.

II. SYNAPTIC INPUT

It is now certain the response of the vasopressinergic neurons to changes in osmolality is not due to osmoreceptivity of the magnocellular neurons themselves, but rather due to separate neurons that communicate synaptically with the SON and PVN (*74, 75*). The osmoresponsive neurons are close to the SON and PVN because hypothalamo–neurohypophyseal explants maintained *in vitro* in organ culture continue to be responsive to osmotic stimulation (*74*). In these explants, the osmotically stimulated release of vasopressin can be blocked by nicotinic cholinergic antagonists, suggesting that acetylcholine is the main neurotransmitter of the osmoreceptors (although angiotensin has also been implicated) (*74, 75*). Lesion studies in several species have provided evidence that the osmoregulation of vasopressin is disrupted by lesions in the anteroventral region of the third ventricle, either in the nucleus medianus (*27, 38*) and/or in the organum vasculosum terminalis (*54*).

A second major input to the vasopressinergic cells of the SON and PVN is from the subfornical organ (*42*). The subfornical organ is thought to be important as a central sensor of volume depletion, because destruction of the subfornical organ will abolish the drinking response produced by infusion of angiotensin II peripherally (*53*). This pathway provides a potential mechanism whereby volume depletion can stimulate secretion of vasopressin via a humoral (angiotensin II) stimulus independently of the classic brainstem-mediated volume-receptor pathway.

The third major afferent input to the vasopressinergic neurons in the SON and PVN is from the brainstem (*82*). These pathways are predominantly noradrenergic or adrenergic and are probably responsible for the dense catecholaminergic innervation that has been demonstrated in the SON and PVN (*34*). Since catacholamines have generally been found to inhibit the secretion of vasopressin (*4, 39*), the brainstem input is thought to chronically inhibit the release of vasopressin. Anything that interferes with the function of these inhibitory pathways can potentially stimulate the release of vasopressin (*5*).

The fourth area of afferent input is from the limbic system, including the lateral septum and amygdala (*73, 86*). This input is largely to the vasopressinergic neurons of the PVN. Since glucocorticord receptors have been localized in these areas (*38*), it is possible that this input is important in regulation of the vasopressin and CRF neurons that project from the PVN to the median eminence.

III. PHYSIOLOGY

A. Oxytocin

The only recognized function for oxytocin is in women (*13, 25*), although the levels of oxytocin in men may be similar in response to pharmacologic stimuli (*1*). In women, oxytocin is released in response to suckling and is associated with myocontraction of ducts within the breasts to cause milk letdown (*13*). Whether oxytocin release at the time of nursing is essential for milk release is not known; data in animals suggest that milk production and release can be ongoing in the absence of oxytocin. At parturition, oxytocin can stimulate the uterus to cause uterine contractions, and it is frequently used as a pharmacologic agent to stimulate uterine contractions and initiate labor (*2, 12, 70*). However, whether oxytocin plays an important role in initiating labor in normal pregnancy is not known. Stretching of the cervix that occurs during labor may be a stimulus to cause increased release of oxytocin (*24*). Oxytocin levels have been reported to be high at the time of parturition and immediately after delivery (*27*). Thus, oxytocin may be more important for maintenance of uterine tone after expulsion of the fetus than for initiation of labor. The fetus itself is also known to be a source of oxytocin at parturition, but the importance of this secretion of oxytocin is not known (*14*).

B. Vasopressin

As alluded to in the description of synaptic input, the two major regulatory systems involved in the control of the release of arginine vasopressin (AVP) are osmotic receptors and volume receptors (*59*). In recent years it has been recognized that nausea (*66*), glucopenia (*8, 95*) and certain stresses (*22*) may also stimulate the release of vasopressin (*61, 64*). As these various other inputs are thought to be predominantly similar to volume receptor afferent pathways, the concept of ''volume'' receptors has been expanded to include all ''nonosmotic'' regulation of release of vasopressin.

1. Osmoreceptors

Secretion of vasopressin is exquisitely sensitive to changes in osmolality. A 1% increase in osmolality of the extracellular fluid will cause a measureable increase in the secretion of vasopressin (*59, 61*). The relationship between plasma osmolality and levels of vasopressin is illustrated in Fig. 1. The osmotic

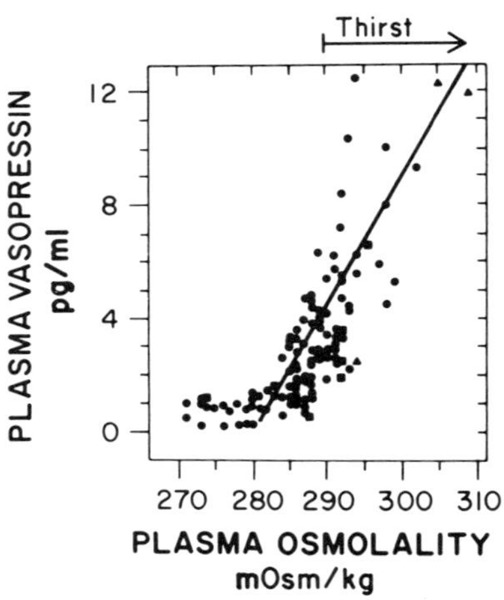

Figure 1. The relationship between plasma vasopressin and plasma osmolality for normal subjects. The thirst threshold is indicated at the top. [Reprinted with permission (*64*).]

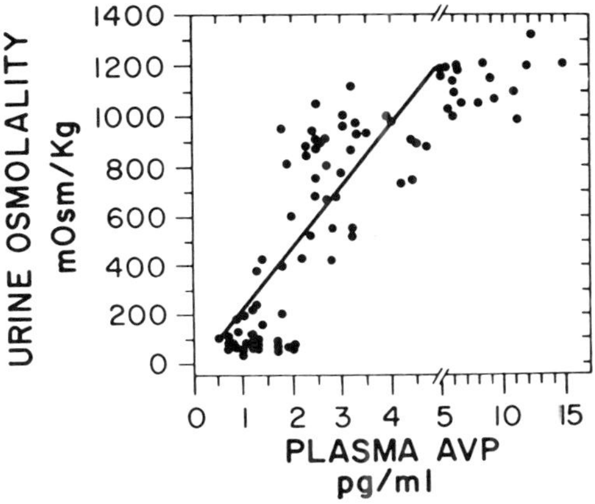

Figure 2. The relationship between urine osmolality and the level of AVP in plasma for normal subjects. [Reprinted with permission (*64*).]

threshold for release of vasopressin in normal subjects is approximately 280 mOsm/kg. While it is not possible to prove that no vasopressin is released below this threshold, it is certain that release of vasopressin is sufficiently suppressed to allow maximum water diuresis and no concentration of the urine. Between osmolalities of 280 and 295–300 there is a linear increase in the level of vasopressin with increases in osmolality. At a plasma osmolality of approximately 295–300, the level of vasopressin in plasma (5 pg/ml) is sufficient to cause maximum antidiuresis. This is demonstrated in Fig. 2, in which the relationship between plasma vasopressin and urine osmolality is illustrated. The entire physiologic range from maximum water diuresis to maximum antidiuresis occurs with vasopressin concentrations between 1 and 5 pg/ml (*62, 63*). While levels of vasopressin in plasma may be much greater than 5 pg/ml, these levels will not translate into further increases in antidiuresis. The sensitivity for the release of vasopressin on the one hand and the sensitivity of the renal response on the other is the explanation for the close correlation between plasma osmolality and urine osmolality (Fig. 3) (*41*); shifts in plasma osmolality between 280 and 295–300 will cause corresponding changes in urine osmolality from less than 50 to greater than 900 mOsm/kg.

In most situations, in the human the sensation of thirst is closely correlated with urinary concentration (*59*). The osmoreceptors for thirst are in the same area (or in fact may be localized in the same neurons) as those regulating the secretion of vasopressin (*80, 81*). There are data showing that in humans the threshold for thirst is somewhat higher than that for vasopressin, and that thirst is not stimulated until reaching a plasma osmolality of as high as 295 (*61*). Therefore, most

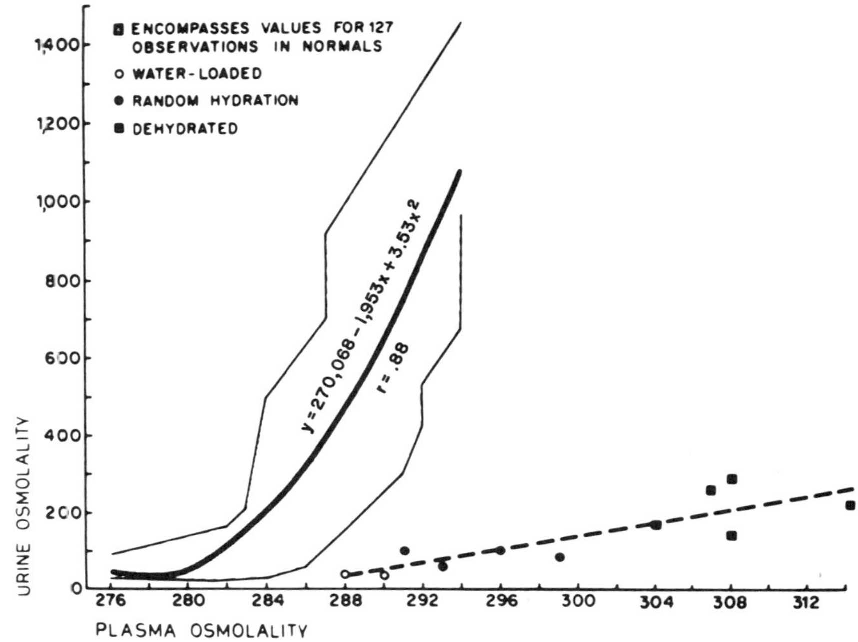

Figure 3. The relationship between urine osmolality and plasma osmolality in normal subjects (dark line). The individual data points and dotted line represent values from a patient with severe diabetes insipidus. [Reprinted with permission (*41*).]

humans may not sense thirst until vasopressin is at a sufficiently high level to produce near maximum urine osmolality. Most adults probably drink sufficient water by habit and by social custom such that urine osmolality goes through wide swings of dilution and concentration without the subject ever experiencing marked thirst. In addition, studies in animals indicate that the relationship between thirst and secretion of vasopressin may vary significantly between different species (*79, 93*).

2. Volume Receptor

The main volume receptors are probably the high-pressure baroreceptors in the carotid sinus and in the aortic arch, but there is also input from low-pressure volume receptors probably located in the left atrium (*23, 59, 93*). Afferent input from these receptors ascend to the brainstem, predominantly via cranial nerves 9 and 10, and synapse in the brainstem to relay their message to the supraoptic and paraventricular nuclei through the adrenergic group described above. This adrenergic input is predominantly inhibitory, and decreased pressure or volume results in a decreased level of inhibition and subsequently an increased secretion of vasopressin. The sensitivity of the baroreceptor system is much less than that for osmolality (*23*), as demonstrated in Fig. 4; whereas a 1% change in osmolality will cause an increase in the release of vasopressin, a 5–10% change in volume is necessary before vasopressin release is stimulated. However, once the threshold for release of vasopressin is reached, levels of vasopressin then increase exponentially with further decreased volume and finally reach levels which are much higher than those reached with osmolar stimulation. Thus, as described in Chapter 8, the levels of vasopressin in plasma after volume-receptor stimulation will

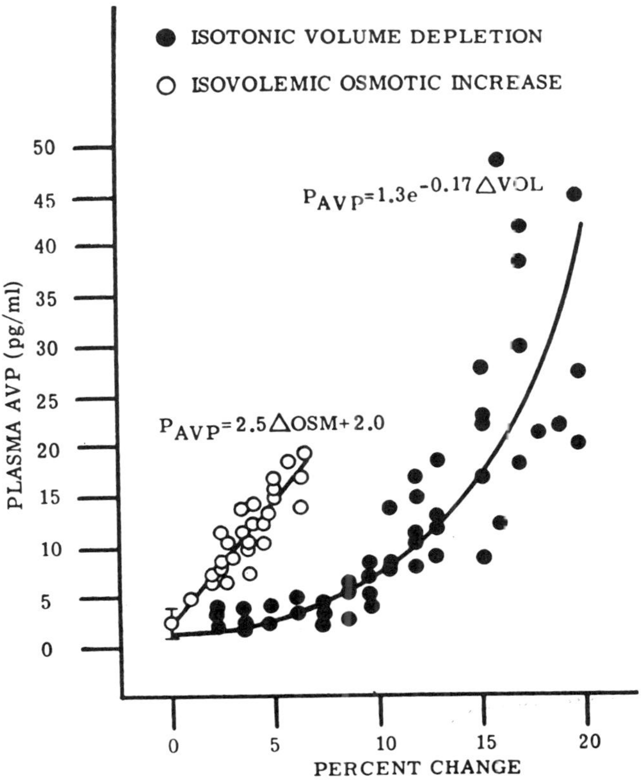

Figure 4. The change in plasma vasopressin (AVP) in response to a change in osmolality compared to blood pressure, both expressed as a percentage change from baseline. [Reprinted with permission (*23*).]

far exceed the levels needed for maximum concentration of the urine, and are in the range that may cause significant pressor effects.

There is also evidence that hemodynamic tone may influence the release of vasopressin by altering the response of osmotically stimulated release of vasopressin (*60, 71*). Both the osmotic threshold for release of vasopressin and the sensitivity of osmotically stimulated release of vasopressin can be affected by hemodynamic input. Decreases in blood volume or pressure that alone are not sufficient to stimulate the release of vasopressin may nonetheless cause a greater release of vasopressin in response to an osmotic stimulus. Interaction between osmotic and volume regulation of vasopressin secretion thereby allows for fine tuning of the renal concentrating and diluting ability.

Other nonosmotic stimuli that cause the release of vasopressin—for example, intestinal traction, surgery, smoking—may exert their effects through the same neural pathways as the baroreceptor system (*93*). Nausea (*66*) and glucopenia (*7, 85*), which stimulate the release of vasopressin via central receptors, may also stimulate secretion of vasopressin via suppression of tonic inhibitory input to the neurohypophysis by adreneric afferents from the brainstem.

IV. PATHOPHYSIOLOGIC MECHANISMS

The only known diseases of the neurohypophysis involve vasopressin (not oxytocin) and have opposite effects, diabetes insipidus (DI) causing hypernatremia, and the syndrome of inappropriate secretion of antidiuretic hormone (SIADH) causing hyponatremia. Although vasopressin is ultimately released from the posterior pituitary, the final common pathway for all of the pathogenetic mechanisms is at the level of the perikarya of the magnocellular neurons. Thus, diseases of the neurohypophyseal system are always manifestations of disorders that are anatomically above the level of the diaphragm sella.

A. Tumor

Tumors in the hypothalamic area may produce either diabetes insipidus or SIADH, depending on whether the tumors are locally invasive and destructive of the neurohypophysis or cause local pressure with interruption of inhibitory afferent inputs to the neurohypophysis, resulting in increased release of vasopressin (*95*). The most common tumors are pituitary ademonas, craniopharyngiomas, germinomas, metastatic tumors, and leukemias. Tumors elsewhere within the central nervous system may interfere with the inhibitory pathways described above and cause SIADH but are not likely to cause diabetes insipidus (*6, 58, 69*). Cancers elsewhere in the body (especially bronchogenic carcinoma,

carcinoma of the duodenum, and carcinoma of the pancreas) may cause SIADH by ectopic production of vasopressin (*50, 59*).

B. Trauma

Severe trauma to the head, especially when associated with skull fracture, may cause either diabetes insipidus or SIADH (*43, 44, 94*). In some cases this is associated with section of the pituitary stalk and/or hemorrhage in the hypothalamus, but in other cases there may be a general metabolic derangement. Clinically, this most often occurs after automobile or motorcycle accidents (*94*). One important clinical correlate is that patients with these disorders are likely to have coexistent panhypopituitarism and must be treated accordingly, e.g., with cortisol (*94*). Another major cause of "trauma" of the hypothalamus is pituitary surgery. During transsphenoidal hypophysectomy there is frequently transient diabetes insipidus, which occurs in 10–15% of the patients (*48*). Persistence of this disorder depends on the degree of damage to the hypothalamus (*29*). As with other endocrine systems, there is a considerable reserve of posterior pituitary tissue in excess of that necessary for water homeostasis. As shown in animals, decreased ability to concentrate the urine does not occur in dogs studies until 80–90% of the entire neurohypophysis has been destroyed (*29*). In fact, removal of all posterior pituitary tissue from the sella may not result in permanent diabetes insipidus in humans or experimental animals (*36, 51*). In studies of patients undergoing palliative hypophysectomy for malignancy or diabetic retinopathy, complete section of the pituitary stalk rarely produced permanent diabetes insipidus, but higher stalk section with more proximal damage of the supraopticohypophyseal tract regularly resulted in diabetes insipidus (*35*). When the stalk is sectioned, there is retrograde neuronal degeneration. As with nerves in general, the severity of damage to the perikaryion depends upon the location of the section of the axon. With lesions lower in the stalk, sufficient neurons survive to permit regeneration of neurosecretory terminals; with sections higher in the supraopticohypophyseal tract, extensive retrograde neuronal degeneration occurs and an insufficient number of magnocellular neurons may remain viable to produce vasopressin (*18, 32*). Neurohypophyseal injury that destroys 70–80% of the magnocellular neurons may produce "partial" diabetes insipidus (*40*). This disorder is described below.

C. Infiltrative Lesions of the Hypothalamus

Granulomatous disease such as tuberculosis, sarcoidosis, and histocytosis may infiltrate the hypothalamus to cause either SIADH or, more commonly, diabetes insipidus (*59, 95*).

D. Vascular Lesions

Cerebral thrombosis, cerebral hemorrhage, or aneurysm (with or without hemorrhage) may be associated with disorders of neurohypophyseal function (*59, 95*).

E. Pulmonary Diseases

In addition to bronchogenic carcinoma, pulmonary tuberculosis, pneumonia, aspergillosis, and a variety of other pulmonary diseases may be associated with SIADH (but not diabetes insipidus) (*59, 95*). Use of positive-pressure respirators has been associated with SIADH (*96*) and negative-pressure respirators with transient diabetes insipidus (*8*). Probably both effects are due to the effect of the respirators on the volume receptors on the lung and a consequence sensing of increased or decreased intravascular volume, possibly on the basis of the effect of the increased or decreased venous return on cardiac output and stimulation or inhibition of baroreceptors.

F. Infectious Diseases

Various infectious diseases have been associated with the syndrome of inappropriate secretion of antidiuretic hormone. This affect is predominantly through the volume-receptor system in the lung, as pulmonary disease, or as a central nervous system disorder that is probably associated with decreased conduction of inhibitory influences through the adrenergic and nonadrenergic pathways previously described.

G. Idiopathic

In some cases both the cause of SIADH and the cause of diabetes insipidus cannot be determined. For diabetes insipidus, recent evidence indicates that in some cases ''idiopathic'' diabetes insipidus may be due to autoimmune destruction of cells of the supraoptic and paraventricular nucleus. Scherbaum and Botazzo (*67*) found that as many as 30% of patients who have idiopathic diabetes insipidus had circulating antibodies that would attach to cells of the neurohypophysis in sections of fetal hypothalamus. While the role of these antibodies in the pathogenesis of the disease is not established, the data are suggestive that there is an autoimmune process in some cases of diabetes insipidus similar to other endocrine disorders.

Table I. Some Drugs Commonly Associated with Siadh or Diabetes Insipidus (DI)

Level of action	May cause SIADH with hyponatremia	May cause DI with hyponatremia
Hypothalamus	Vincristine	Vinblastin
	Nicotine	
	Carbamazepine	
	Clofibrate	
	Phenothiazine	
	Chlorpropamide	
Kidney	Vasopressin	Lithium
	Oxytocin	Demeclocycline
	DDAVP	Tolazamide
	Chlorpropamide	Propoxyphene
	Prostaglandin inhibitors	Acetohexamide
	Cyclophosphamide	

H. Drugs

A variety of drugs may cause SIADH or diabetes insipidus and may act either at the level of the hypothalamus or at the level of the kidney. Drugs that are used to treat one disease are likely to be causative of the other disorder—for example, drugs used to treat diabetes insipidus may cause SIADH in some patients, and drugs used to treat SIADH may cause diabetes insipidus. Table I includes a variety of drugs illustrative of those that have been reported to produce either SIADH or diabetes insipidus.

V. HYPOFUNCTION OF THE NEUROHYPOPHYSIS: DIABETES INSIPIDUS

In its classic form, diabetes insipidus is a disease in which there is no production of vasopressin. This causes an inability to concentrate the urine and causes extreme polyuria. With the diuresis of dilute urine there is an increase in plasma omsolality, which stimulates thirst, leading to progressive polydipsia. Therefore, the differential diagnosis is of disorders with polyuria and polydipsia (*59, 89, 95*). Once an osmotic diuresis is ruled out, the major differential is between nephrogenic diabetes insipidus (which has been discussed in Chapter 7) and primary polydipsia. With primary polydipsia, the initial stimulus is to the thirst center, and the increase in water intake causes hemodilution and inhibition of secretion of vasopressin. With absent vasopressin effect on the kidney, there is maximum diuresis and excretion of large volumes of unconcentrated urine. Thus both in primary polydipsia and in primary central diabetes insipidus the patient is extremely thirsty, with marked polydipsia and polyuria of a hypoosmolar urine. The diagnosis of nephrogenic diabetes insipidus can be established by measurement of plasma vasopressin, which will be elevated despite the lack of urinary concentration (*62*), whereas, for both primary polydipsia and central diabetes insipidus the vasopressin level will be low. To differentiate the latter cases, some further form of testing that involves the dynamic secretion of vasopressin is necessary. These tests can be divided into indirect tests, which involve measurement of plasma and urine osmolality and calculation of osmolal clearance and free water clearance, and direct tests, which involve measurement of the level of vasopressin in plasma.

A. Indirect Tests

As described above under normal physiology, there is a close link between plasma osmolality and plasma vasopressin, between plasma vasopressin and urine osmolality, and hence between plasma osmolality and urine osmolality. All of the indirect tests are based upon some correlation of plasma osmolality with urine osmolality: that is, the patient's own renal response is used as an internal bioassay for the level of vasopressin. The level of plasma osmolality is increased either by administration of a hypertonic solution of salt or by the excretion of a hypotonic urine, while the patient is not allowed to replace the urinary losses with water (dehydration test). The use of these two tests over years of clinical practice has recently been summarized by Moses (*44*). The infusion of hypertonic saline, the Hickey–Hare test, utilizes either 3% or 5% NaCl solution, which either is infused for a standard volume and time or is infused with careful monitoring of the serum sodium to a specific and elevated serum sodium. Regardless, urine osmolality and urine volume are measured, and osmolar clearance and free water clearance are calculated (*44*). When the serum sodium is elevated above the osmotic threshold for secretion of vasopressin, a normal subject will have a decrease in free water clearance. Patients with diabetes insipidus will have no response.

The infusion of hypertonic NaCl has now been largely supplanted by the dehydration test (*40*). Dehydration allows development of hypertonicity with some volume contraction, so that both receptor systems may be stimulated to release vasopressin to concentrate the urine. There are several variations on this test, which have been popularized in recent years by Miller and Moses (*40, 44*). The essence of the test is to withhold water until the patient achieves an elevated serum sodium and then to determine the osmolality of the urine. In the test as described by Miller and Moses, consecutive hourly urine samples were obtained and the osmolality measured. When at least two consecutive urines showed a difference in osmolality of less than 10%, it was assumed that this was the maximum urine osmolality which could be obtained by intrinsic secretion of vasopressin. Cases with complete hypothalamic diabetes insipidus showed little concentration of their urine, while patients with partial diabetes insipidus, nephrogenic diabetes insipidus, and primary polydipsia showed some concentration of urine with osmolalities greater than 200 but usually less than 500 mOsm/kg. Administration of 5 units of aqueous arginine vasopressin, given subcutaneously at the time of maximal osmolality, would cause a further increase in urine osmolality in patients with complete diabetes insipidus and partial diabetes insipidus, but no change in those with primary polydipsia. In primary polydipsia the lack of concentration is thought to be due to washout of the renal medulla due to the prolonged polyuria. (As noted in Ref. (*40*), patients with nephrogenic diabetes insipidus may also respond to administered vasopressin because the dose is pharmacologic and will cause further concentration of the urine, suggesting that the defect in these patients is only a partial inability to respond to vasopressin.)

Several workers have altered the standard dehydration test as originally described to include the requirement that the patient achieve an absolute elevation of serum sodium (*16, 17, 92*). The patient should be weighed before the test and weighed periodically during the test and have vital signs monitored. Ideally, the test is performed where serum sodium can be monitored continually. Under these conditions a patient can be safely dehydrated until the serum sodium is between 145 and 150. If this is done there is a better distinction between cases of primary polydipsia and partial diabetes insipidus. Furthermore, measured levels of vasopressin at the time of maximum dehydration will better distinguish between partial diabetes insipidus and nephrogenic diabetes insipidus (*99*).

Czernichow *et al.* (*16, 17*) have utilized in children a standard 14-hr dehydration test to establish the diagnosis of diabetes insipidus. In their test all patients with central diabetes insipidus became hypernatremic during the period of dehydration and maintained a hypoosmolar urine. Patients with partial diabetes insipidus had an intermediate elevation of serum sodium and an intermediate increase in urine osmolality, whereas normal subjects had virtually no increase in their plasma sodium at a time when their urine was markedly concentrated. Except for the most severe cases of diabetes insipidus, in whom this test may be too long, this has been an extremely sensitive and reliable test to differentiate partial from complete diabetes insipidus.

B. Direct Tests

The direct tests similarly utilize some method of increasing serum sodium but rely on measurement of levels of vasopressin in plasma in addition to the urine parameters outlined above. As mentioned previously, nephrogenic diabetes insipidus can always be distinguished from central diabetes insipidus by the measured levels of vasopressin in plasma. However, in the differentiation of partial diabetes insipidus from primary polydipsia, the measurement of vasopressin levels in plasma may also be helpful (*98, 99*). The difficulty in distinguishing these cases stems from the fact that the renal medulla may be washed out by the excessive polyuria that occurs in both disorders. Thus, some cases of primary polydipsia will have a low concentration of their urine with dehydration, and some cases of partial central diabetes insipidus will achieve the maximum renal response with the low levels of vasopressin that they are able to secrete and will not respond further to administered vasopressin. In this situation a sensitive radioimmunoassay of vasopressin can be used to distinguish the low levels of vasopressin present in partial central diabetes insipidus from the higher levels seen in primary polydipsia.

C. Postoperative Diabetes Insipidus

After surgery in the pituitary hypothalamic area there may be transient diabetes insipidus or permanent diabetes insipidus. In addition, a third pattern may be seen, which is the "triphasic response." This response is thought to be characteristic of complete stalk section. It is characterized by an abrupt onset of diabetes insipidus lasting several days, followed by apparent resolution and a "normal interphase" lasting 2–14 days, and then eventual return of diabetes insipidus (*31, 35, 49, 72*). The period of interphase has now been demonstrated to be a period of inappropriate release of antidiuretic hormone and, depending on the water intake, may be associated with hyponatremia (*30, 49, 72*). Several studies have demonstrated that the uncontrolled secretion of vasopressin is the result of necrosis of the disconnected but *in situ* posterior pituitary (*30, 32, 72*). However, it is now clear that the triphasic response may occur even after complete hypophysectomy if a sufficient amount of pituitary stalk and median eminence tissue is left intact and the neurohypophysis is severely damaged (*94*).

A typical patient's triphasic response is shown in Fig. 5. The initially severe DI was followed by inappropriate antidiuresis with serum sodium as low as 125 mEq/l. This required fluid restriction and hypertonic saline for therapy. Eventually the diabetes insipidus returned.

D. Diabetes Insipidus with Inadequate Thirst

Because of the proximity of the thirst center to the osmoreceptors and to the magnocellular neuron system in the anterior hypothalamus, it is possible that diseases in the hypothalamus will destroy the thirst center along with the vasopressinergic neurons. These patients are especially likely to develop severe hy-

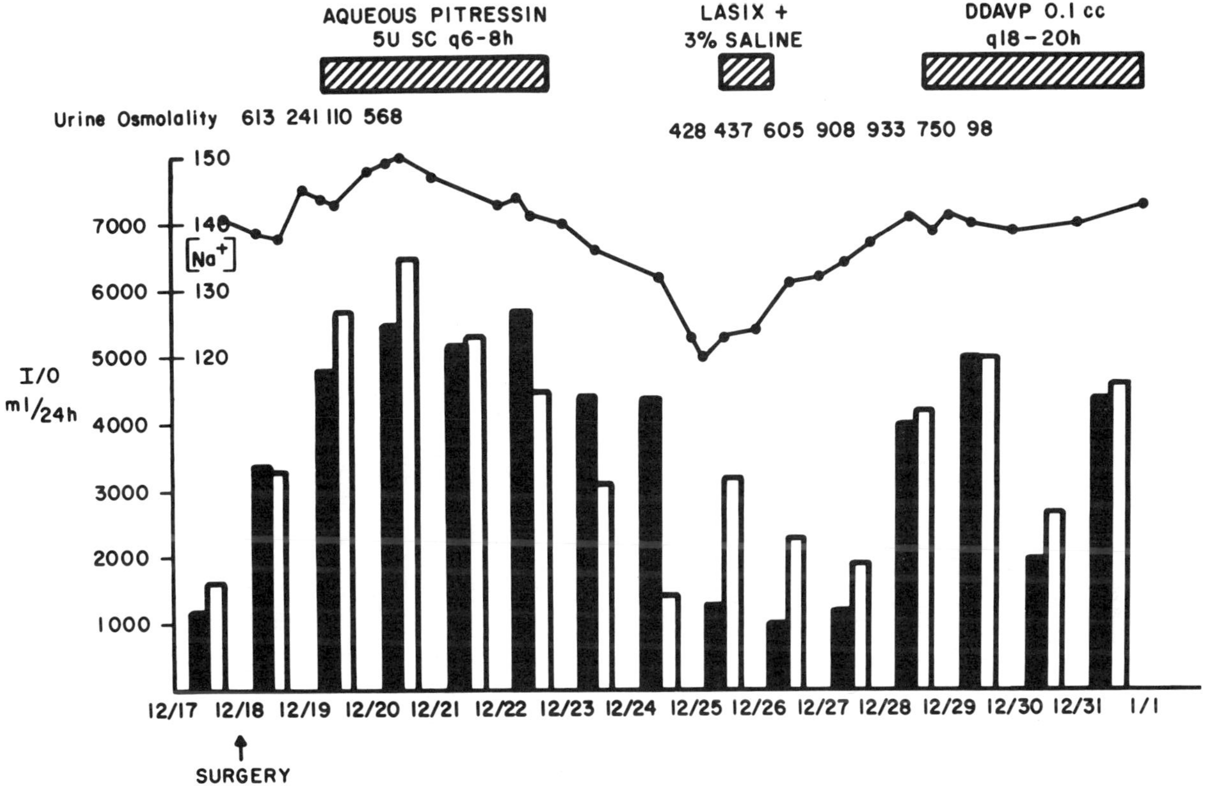

Figure 5. Clinical parameters of a patient with triphasic diabetes insipidus after pituitary surgery. The solid bars indicate the intake and the open bars the output, in milliliters per day. Surgery is indicated at 12/18. On the first day postoperatively the patient developed diabetes insipidus with large urine output and required intermittent treatment with aqueous pitressin for 4 days. Subsequently, pitressin was discontinued but the patient continued to drink large volumes of water in the second phase. It was demonstrated that the release of vasopressin was not under physiologic control because the sodium fell to 120. Urine osmolality noted at the top is inappropriately concentrated throughout the second phase of SIADH. On 12/28, diabetes insipidus recurred with a large urine output and urine osmolality of 98. Treatment with DDAVP was begun.

pernatremia and may become very hypernatremic without a large amount of polyuria. In such patients, a fixed amount of fluid must be prescribed in addition to an antidiuretic hormone (*59, 91*).

E. Absent Osmoreceptor with Intact Volume Receptor

This is probably the pathogenic mechanism of the previously described cases of "essential hypernatremia." These patients have chronic hypernatremia but lack the sensation of thirst; they have no increased secretion of vasopressin with hypertonicity but do respond with increased secretion of vasopressin with volume-receptor stimulation (*21, 28, 77*). The cases are distinct from primarx adipsia in that there is also inadequate secretion of vasopressin in response to hyperosmolality. The pathophysiology is that the patients do not drink and become dehydrated and hypernatremic. In response to the hypernatremia, they do not adequately concentrate their urine and vasopressin is not secreted until volume loss is sufficient to stimulate the release of vasopressin on a hemodynamic basis. Studies in rats have demonstrated that ablative lesions in the anterior hypothalamus will produce disorders of both regulation of thirst and regulation of osmotic secretion of vasopressin, without significantly affecting volume-mediated release (*11, 26, 37, 90*). These cases in humans can be defined by indirect testing, as first demonstrated by DeRubertis *et al.* (*21*). An infusion of saline, or of hypertonic saline, will turn off the baroreceptor stimulation and secretion of vasopressin, causing excretion of a dilute urine despite progressive hypernatremia. In a few reported cases with severe hypertonicity, some vasopressin was released, indicating osmoreceptor disregulation rather than a complete absence of response (*59*). Stimulation of volume receptors, either by tilt-table testing or by hypotension induced by pharmacologic agents, resulted in a prompt release of vasopressin and concentration of the urine. Patients with this disorder are able to appropriately dilute their urine and are not in danger of developing hyponatremia. Consequently, they are best treated with a fixed volume of water, which they must drink each day, to maintain serum sodium within the normal

F. Treatment of Diabetes Insipidus

Replacement of water is emphasized as a treatment of central diabetes insipidus. Water taken in sufficient quantities will correct any metabolic abnormality, and all pharmacologic therapies are designed to reduce the amount of water intake necessary to a tolerable level. The tolerable level will vary from patient to patient and from day to day in a given patient. If thirst is adequate, the physician should not be disturbed with an occasional lapse in pharmacologic treatment, which, in fact, may be beneficial to avoid overtreatment and water intoxication. The most physiologic replacement is with the missing hormone vasopressin, but because of its relatively short duration of action, treatment with this agent is unsatisfactory. Today the best therapy is with 1-desamino-D-arginine vasopressin (DDAVP) (*65*). This is absorbed from the nasal mucosa and has a long duration of action, 6–24 hr, without producing any noticable stimulation of vascular vasopressin receptors. This agent has recently been approved for parental administration in the United States (*46*) and now is also available for treatment of patients when intranasal is not feasible. The use of DDAVP has made the other vasopressinergic agents (e.g., aqueous vasopressin, lysine vasopressin, and pitressin tannate in oil) largely unnecessary (*65*).

There are pharmacologic agents that can be administered orally and that are useful in the treatment of partial diabetes insipidus. None of these agents is of value when vasopressin is completely absent. The agents act at the level of the hypothalamus to induce increased secretion of vasopressin and/or at the level of the renal tubule to increase the hydroosmotic action of vasopressin. Chlorpropamide, the best-studied agent, in doses of 100–500 mg/day will produce antidiuresis in patients with partial diabetes insipidus. The major action of chlorpropamide is to enhance the effect of small amounts of vasopressin to increase water transport in the renal tubule (*47, 52*). Carbamazepine and chlofibrate will also cause antidiuresis in patients with partial diabetes insipidus, primarily by increasing secretion of vasopressin (*45, 65*). Thiazide diuretics decrease extracellular fluid volume and through a decrease in glomerular filtration rate (GFR) may cause some decrease in urinary output in patients with both central and nephrogenic diabetes insipidus (*95*). Recently prostaglandin inhibitors have been reported to augment the effect of naturally secreted or administered vasopressinergic agents (*46*) and have been reported to be especially useful in nephrogenic diabetes insipidus. For a complete review of therapy, readers are referred to recent publications of Robinson and Verbalis (*65, 91*).

VI. HYPERFUNCTION OF THE POSTERIOR PITUITARY—SYNDROME OF INAPPROPRIATE SECRETION OF ANTIDIURETIC HORMONE

Hyperfunction of the posterior pituitary is only recognized for the hormone vasopressin and is known as the syndrome of inappropriate secretion of antidiuretic hormone (SIADH) (*6, 69*). With excess secretion of vasopressin there is an inability of the kidneys to maximally dilute urine, and subsequent water retention causes dilution of plasma with expansion of the extracellular fluid volume. Expansion of the extracellular fluid then causes an increased excretion of sodium. The water retention with dilution and the progressive natriuresis produce hyponatremia. The clinical parameters of SIADH are hyponatremia of the plasma and an inappropriately concentrated urine. The urine need not be markedly concentrated but only inappropriate for the degree of hyponatremia. There are two basic categories of excess secretion of vasopressin: excess secretion by the neurohypophysis, and excess secretion ectopically by nonneurohypophyseal tissue. The latter has been described most commonly with bronchogenic carcinoma of the lung, but has also been reported in pancreatic carcinoma, thymoma, hepatoma, lymphoma, carcinoid tumors, and several other disorders (*58*). Inflammatory diseases of the lung are also associated with SIADH and in most cases are probably due to some malfunction of volume receptors in the chest, but in some cases may be due to production of vasopressin by some pulmonary cells (*93*). As mentioned previously, various central nervous system disorders are associated with release of vasopressin from the neurohypophysis.

A. Diagnosis

In the past the diagnosis of SIADH has been based on indirect tests (*6, 69*). If the patient had a concentrated urine that was inappropriate to the degree of hyponatremia and there was no other disease such as adrenal insufficiency that might cause increased natriuresis, the diagnosis of SIADH was considered established. Recently, measurement of vasopressin in plasma has become available and a more direct diagnosis of SIADH can be made when the measured level of vasopressin is inappropriate for the degree of hyponatremia. When patients with hyponatremia and presumed diagnosis of SIADH are tested with water loading and/or administration of hypertonic NaCl, four subgroups of patients have been described by Robertson (*63*) (see Fig. 6): (a) patients with erratic secretion of vasopressin unresponsive to any stimulus, (b) patients with "reset osmostat" in which there is a normal slope of the plasma osmolality/plasma vasopressin (POsm/PAVP) curve but at lower levels of POsm, (c) patients in whom the basal level of secretion of vasopressin is fixed but in whom elevation of POsm is associated with further increases in the level of vasopressin, and (d) patients in whom antidiuresis is not accompanied by measured levels of vasopressin in plasma and in whom the antidiuresis may be due to some antidiuretic substance not recognized as vasopressin, or in whom there is a supersensitivity of the kidney to low levels of vasopressin. Of interest is the fact that this classification has not helped to distinguish among the various etiologies of SIADH, and in particular all patterns have been described with ectopic production of vasopressin.

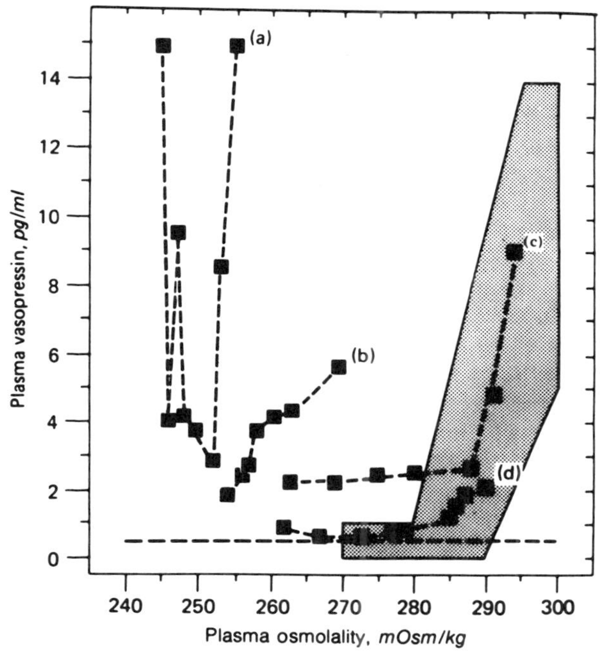

Figure 6. The four patterns of release of vasopressin, as described by Robertson (*63*), in patients with inappropriate secretion of antidiuretic hormone. [Reprinted with permission (63).]

B. Treatment

Just as administration of water is a primary form of therapy for diabetes insipidus, so restriction of water is the primary form of therapy for SIADH (*6, 69*). As described above, the pathology of SIADH is caused by the retention of an excess amount of water, which causes dilution of serum sodium, expansion of extracellular fluid volume, and consequent natriuresis. As was shown initially by Leaf (*33*), the entire pathogenetic sequence may be avoided if water is restricted to a sufficient degree to avoid any overhydration (*68*). However, the amount of water to be restricted and the success of this therapy will vary tremendously from patient to patient. Interestingly, even though the pattern of release of vasopressin described in the preceding section may not be helpful in making an etiologic diagnosis, it may be quite helpful in deciding on the appropriate therapeutic regimen. For example, in patients with reset osmostat, when a water load is administered the patients will dilute their urine, excrete the water, and not develop further hyponatremia; in patients with a fixed but low level of secretion of vasopressin, modest restriction of fluid intake may be appropriate to treat the disorder. In cases of sustained high levels of vasopressin in plasma, urinary dilution is so compromised that intolerable levels of fluid restriction may be required to prevent hyponatremia. In these cases additional therapy with agents that inhibit the action of vasopressin on the kidney—e.g., demeclocycline (*15*) or lithium (not recommended because of side effects) (*91*)—or orally administered urea (*19*) may be necessary.

ACKNOWLEDGMENTS

Work described here was supported by National Institutes of Health grants AM 16166, NS 17138, AM 31302, and MO1-RR00056. Joseph G. Verbalis was supported by the Research Career Development Program of the Veterans Administration. Secretarial assistance was by Jane Pennebaker.

REFERENCES

1. Amico, J. A., Seif, S. M., and Robinson, A. G. (1981). Elevation of oxytocin and the oxytocin associated neurophysin in the plasma of normal women during mid-cycle. *J. Clin. Endocrinol. Metab.* **53,** 1229–1232.
2. Amico, J. A., Seitchik, J., and Robinson, A. G. (1984). Studies of oxytocin in plasma of women during hypocontractile labor. *J. Clin. Endocrinol. Metab.* **58,** 274–279.
3. Antunes, J. L., Carmel, P. W., and Zimmerman, E. A. (1977). Projections from the paraventricular nucleus to the zona externa of the median eminence of the rhesus monkey: An immunohistochemical study. *Brain Res.* **131,** 1.
4. Armstrong, W. E., Sladek, C. D., and Sladek, J. R., Jr. (1982). Characterization of noradrenergic control of vasopressin release by the organ-cultured rat hypothalamo-neurohypophyseal system. *Endocrinology* **111,** 273–279.
5. Barker, J. L., Crayton, J. W., and Nicoll, R. (1971). Noradrenaline and acetylcholine responses of supra-optic neurosecretory cells. *J. Physiol.* (*London*) **218,** 19–32.
6. Bartter, F. C., and Schwartz, W. B. (1967). The syndrome of inappropriate secretion of antidiuretic hormone. *Am. J. Med.* **42,** 790–806.
7. Baylis, P. H., Zerbe, R. L., and Robertson, G. L. (1981). Arginine vasopressin response to insulin-induced hypoglycemia in man. *J. Clin. Endocrin. Metab.* **53,** 935–940.
8. Boylan, J. W., and Antkowiak, D. E. (1959). Mechanism of diuresis during negative pressure breathing. *J. Appl. Physiol.* **14**(1), 116–120.
9. Brownstein, M. J., and Gainer, H. (1977). Neurophysin biosynthesis in normal rats and in rats with hereditary diabetes insipidus. *Proc. Natl. Acad. Sci. U.S.A.* **74,** 4046–4049.
10. Burlet, A., Tonon, M. C., Tankosic, P., Coy, D., and Vaudry, H. (1983). Comparative immunocytochemical localization of corticotropin releasing factor (CRF-41) and neurohypophyseal peptides in the brain of Brattleboro and Long-Evans rats. *Neuroendocrinology* **37,** 64–72.
11. Buggy, J., and Johnson, A. K. (1977). Preoptic-hypothalamic periventricular lesions: Thirst deficits and hypernatremia. *Am. J. Physiol.* **233,** R44-R52.

12. Caldeyro-Barcia, R., and Poseiro, J. J. (1958). Oxytocin and contractility of the pregnant human uterus. *Ann. N.Y. Acad. Sci.* **75,** 813.
13. Chard, T. (1972). The posterior pituitary in human and animal parturition. *J. Reprod. Fertil.* **16,** 121.
14. Chard, T., Hudson, C. N., Edwards, C. R. W., and Boyd, N. R. H. (1971). Release of oxytocin and vasopressin by the human foetus during labor. *Nature (London)* **234,** 352–354.
15. Cherrell, D. A., Stote, R. M., Birge, J. R., and Senger, I. (1975). Democlocycline treatment in the syndrome of inappropriate antidiuretic hormone secretion. *Ann. Int. Med.* **85,** 654–656.
16. Czernichow, P., Pomarede, R., Brauner, R., and Rappaport, R. (1985). Diabetes insipidus (DI) in children—Etiology and associated anterior pituitary anomalies. *In* "Diabetes Insipidus" editors, (P. Czernichow and A. G. Robinson, eds.), pp. 190–209. Karger, Basel, Switzerland.
17. Czernichow, P., and Robinson, A. G. (1980). Neurohypophysis. *In* "Comprehensive Endocrinology Series (Pediatric Endocrinology)" (R. Collu, J. R. Ducharme, and H. Guyda, eds.), pp. 571–596. Raven Press, New York.
18. Daniel, P. M., and Prichard, M. M. L. (1972). The human hypothalamus and pituitary stalk after hypophysectomy or pituitary stalk section. *Brain* **95,** 813–824.
19. Decaux, G., Brimioulle, S., Genette, F., and Mockel, I. (1980). Treatment of the syndrome of inappropriate secretion of antidiuretic hormone by urea. *Am. J. Med.* **69,** 99–106.
20. Defendini, R., and Zimmerman, E. A. (1978). The magnocellular neurosecretory system of the mammalian hypothalamus. *In* "The Hypothalamus" (S. Reichlin, R. J. Baldessarini, and J. B. Martin, eds.), pp. 137–154. Raven Press, New York.
21. DeRubertis, F. R., Michelis, M., and Davis, B. B. (1974). "Essential" hypernatremia. *Arch. Intern. Med.* **134,** 889–894.
22. Dubovsky, S. L., Grabon, S., Berl, T., and Schrier, R. W. (1973). Syndrome of inappropriate secretion of antidiuretic hormone with exacerbated psychosis. *Ann. Int. Med.* **79,** 551–554.
23. Dunn, F. L., Brennan, T. J., Nelson, A. E., and Robertson, G. L. (1973). The role of blood osmolality and volume in regulating vasopressin secretion in the rat. *J. Clin. Invest.* **52,** 3212–3219.
24. Ferguson, J. K. W. (1941). A study of the motility of the intact uterus at term. *Surg. Gynecol. Obstet.* **73,** 359–366.
25. Fuchs, A. R., and Saito, S. (1971). Pituitary oxytocin and vasopressin content of pregnant rats before, during and after parturition. *Endocrinology* **106,** 1571.
26. Gardiner, T. W., Verbalis, J. G., and Stricker, E. M. (1983). Impaired vasopressin secretion in rats with electrolytic lesions of the "AV3V area" localized to nucleus medianus. *Soc. Neurosci. Abstr.* **9,** 191.
27. Gibbens, D., Boyd, N. R. H., and Chard, T. (1972). Spurt release of oxytocin during human labour. *J. Endocrinol.* **53,** liv–lv.
28. Halter, J. B., Goldberg, A. P., Robertson, G. L., and Porte, D. (1977). Selective osmoreceptor dysfunction in the syndrome of chronic hypernatremia. *J. Clin. Endocrin. Metab.* **44,** 609–616.
29. Heinbecker, P., and White, H. L. (1944). Hypothalamico-hypophyseal system and its relation to water balance in the dog. *Am. J. Physiol.* **133,** 582–593.
30. Hollinshead, W. H. (1964). The interphase of diabetes insipidus. *Mayo Clin. Proc.* **39,** 92–100.
31. Ikkos, D., Luft, R., and Olivecrona, H. (1955). Hypophysectomy in man: Effect on water excretion during the first two postoperative months. *J. Clin. Endocrinol. Metab.* **15,** 553–567.
32. Kovacs, K., Laszlo, F. A., and David, M. A. (1962). The antidiuretic phase of water metabolism in rats after lesions of the pituitary stalk. II. The role of antidiuretic hormone. *J. Endocrinol.* **25,** 397–401.

33. Leaf, A., Bartter, F. C., Santos, R. F., and Wrong, O. (1953). Evidence in man that urinary electrolyte loss induced by Pitressin is function of water retention. *J. Clin. Invest.* **32,** 868–878.
34. Lindvall, O., and Bjorklund, A. (1974). The organization of ascending catecholamine neuron systems in the rat brain. *Acta Physiol. Scand.* **412,** 1–48.
35. Lipsett, M. B., MacLean, I. P., West, C. D., Li, M. C., and Pearson, O. H. (1956). An analysis of the polyuria induced by hypophysectomy in man. *J. Clin. Endocrinol. Metab.,* **16,** 183–195.
36 Magoun, H. W., Fisher, C., and Ranson, S. W. (1939). The neurohypophysis and water exchange in the monkey. *Endocrinology* **25,** 161–174.
37 Mangiopane, M. L., Thrasher, T. N., Keil, L. C., Simpson, J. B., and Ganong, W. F. (1983). Deficits in drinking and vasopressin secretion after lesions of the nucleus medianus. *Neuroendocrinology* **37,** 73–77.
38. McEwen, B. S., Gerlach, J. L., and Micco, D. J. (1975). Putative glucocorticoid receptors in hippocampus and other brain regions of the rat brain. *In* "The Hippocampus", (S. Isaacson and J. Pribram, eds.), pp. 285–322. Plenum, New York.
39. McNeill, T. H., and Sladek, J. R., Jr. (1980). Simultaneous monoamine histofluorescence and neuropeptide immunocytochemistry II. Correlative distribution of catecholamine varicosities and magnocellular neurosecretory neurons in the rat supraoptic and paraventricular nuclei. *J. Comp. Neurol.* **193,** 1023–1033.
40. Miller, M., Dalakos, T., Moses, A. M., Fellerman, H., and Streeten, D. H. P. (1970). Recognition of partial defects in antidiuretic hormone secretion. *Ann. Int. Med.* **73,** 721–729.
41. Miller, M., and Moses, A. M. (1977). Clinical states due to alteration of ADH release and action. *In* "Neurohypophysis" (A. M. Moses and L. Share, eds.), pp. 153–166. Karger, New York.
42. Miselis, R. R., Shapiro, R. E., and Hand, P. J. (1979). Subfornical efferents to neural systems for control of body water. *Science* **207,** 1022–1025.
43. Moses, A. M. (1983). Long-standing posttraumatic diabetes insipidus. *Medical Grand Rounds* **2,** 117–128.
44. Moses, A. M. (1985). Clinical and laboratory observations in the adult with diabetes insipidus and related syndromes. "Diabetes Insipidus" (P. Czernichow and A. G. Robinson, eds.), pp. 156–175. Karger, Basel, Switzerland.
45. Moses, A. M., Howanitz, J., van Gemert, M., and Miller, M. (1973). Clofibrate-induced antidiuresis. *J. Clin. Invest.* **52,** 536–542.
46. Moses, A. M., Moses, L. K., Notman, D. D., and Springer, J. (1981). Antidiuretic responses to injected desmopressin, alone and with indomethacin. *J. Clin. Endocrinol. Metab.* **52,** 910–913.
47. Moses, A. M., Numann, P., and Miller, M. (1973). Mechanism of chlorpromide-induced antidiuretic in man: Evidence for release of ADH and enhancement of peripheral action. *Metabolism* **22,** 59.
48. Nelson, P. B., Goodman, M., Maroon, J. C., Martinez, J., and Robinson, A. G. (1983). Factors in predicting outcome from surgery in patients with prolactin secreting pituitary adenomas. *Neurosurgery* **13**(6), 634–641.
49. O'Connor, W. J. (1952). The normal interphase in the polyuria which follows section of the supraoptico–hypophyseal tracts in the dog. *Q. J. Exp. Physiol.* **37,** 1–10.
50. Parry, H. B., and Livett, B. G. (1973). A new hypothalamic pathway to the median eminence containing neurophysin and its hypertrophy in sheep with natural scrapie. *Nature (London)* **242,** 63–65.
51. Pickford, M., and Ritchie, A. E. (1945). Experiments on the hypothalamic–pituitary control of water excretion in dogs. *J. Physiol.* **104,** 105–128.

52. Pokracki, F. J., Robinson, A. G., and Seif, S. M. (1981). Chlorpropamide effect: Measurement of neurophysin and vasopressin in humans and rats. *Metabolism* **30,** 72.
53. Ramsay, D. J. (1982). Effects of circulating angiotensin II on the brain. *In* "Frontiers in Endocrinology" (W. F. Ganong and L. Martini, eds.), Vol. 7 pp. 263–285. Raven Press, New York.
54. Ramsay, D. J., Thrasher, T. N., and Keil, L. C. (1983). Osmotic release of vasopressin in dogs depends on the integrity of the OVLT. *In* "The Neurohypophysis: Structure, Function and Control, Progress in Brain Research," *vol. 60* (B. A. Cross and G. Leng, eds.), pp. 91–98. Elsevier Biomedical Press, Amsterdam.
55. Rhodes, C. H., Morrell, J. I., and Pfaff, D. W. (1981). Immunohistochemical analysis of magnocellular elements in rat hypothalamus: Distribution and numbers of neurophysin, oxytocin and vasopressin containing cells. *J. Comp. Neurol.* **198,** 45–64.
56. Richter, D., and Schmale, H. (1983). The structure of the bovine hypothalamic precursors AVP-Np II and OXT-Np I. *In* "The Neurohypophysis: Structure, Function, and Control, Progress in Brain Research" (B. A. Cross and G. Leng, eds.), *Vol. 60,* pp. 227–233. Elsevier Biomedical Press, Amsterdam.
57. Rivier, C., and Vale, W. (1983). Interaction of corticotropin-releasing factor and arginine vasopressin on adrenocorticotropin secretion *in vivo. Endocrinology* **113,** 939–942.
58. Robertson, G. L. (1978). Cancer and inappropriate antidiuresis. *In* "Biological Markers of Neoplasia: Basic and Applied Aspects" (R. W. Ruddon, ed.), pp. 277–293. Elsevier Biomedical Press, Amsterdam.
59. Robertson, G. L. (1981). Diseases of the posterior pituitary. *In* "Endrocrinology and Metabolism" (P. Felig, J. D. Baxter, A. E. Broadus, and L. A. Froliman, eds.), p. 251–277. McGraw-Hill, New York.
60. Robertson, G. L., and Athar, S. (1976). The interaction of blood osmolality and blood volume in regulating plasma vasopressin in man. *J. Clin. Endocrinol. Metab.* **42,** 613–620.
61. Robertson, G. L., and Baylis, P. H. (1980). Plasma vasopressin response to hypertonic saline infusion to assess posterior pituitary function. *J. R. Soc. Med.* **73,** 255–260.
62. Robertson, G. L., Mahr, E. A., Athar, S., and Sinha, T. (1973). Development and clinical application of a new method for the radioimmunoassay of arginine vasopressin in human plasma. *J. Clin. Invest.* **52,** 2340–2352.
63. Robertson, G. L., Shelton, R. L., and Athar, S. (1976). The osmoregulation of vasopressin. *Kidney Int.* **10,** 25–37.
64. Robertson, G. L., Aycinena, P., and Zerbe, R. L. (1982). Neurogenic disorders of osmoregulation. *Am. J. Med.* **72,** 339–353.
65. Robinson, A. G., and Verbalis, J. G. (1985). Treatment of diabetes insipidus. *In* "Diabetes Insipidus" (P. Czernichow and A. G. Robinson, eds.), pp. 292–303. Karger, Basel, Switzerland.
66. Rowe, J. W., Shelton, R. L., Helderman, J. H., Vestal, R. E., and Robertson, G. L. (1979). Influence of the emetic reflexon vasopressin release in man. *Kidney Int.* **16,** 729–735.
67. Scherbaum, W. A., and Bottazzo, G. F. (1983). Autoantibodies to vasopressin cells in idiopathic diabetes insipidus: Evidence for an autoimmune variant. *Lancet* **1,** 897–901.
68. Schrier, R. W., and Leaf, A. (1981). Effect of hormones on water, sodium, chloride, and potassium metabolism. *In* "Textbook of Endocrinology" (R. H. Williams, ed.), pp. 1032–1046. W. B. Saunders, Philadelphia.
69. Schwartz, W. B., Bennett, W., Curelop, S., and Bartter, F. C. (1957). A syndrome of renal sodium loss and hyponatremia probably resulting from inappropriate secretion of antidiuretic hormone. *Am. J. Med.* **23,** 529–542.
70. Seitchik, J., and Castillo, M. (1982). Oxytocin augmentation of dysfunctional labor. I. Clinical data. *Am. J. Obstet. Gynecol.* **144,** 899.

71. Share, L. (1967). Role of peripheral receptors in the increased release of vasopressin in response to hemorrhage. *Endocrinology* **81,** 1140–1146.
72. Sharkey, P. C., Perry, J. H., and Ehni, G. (1961). Diabetes insipidus following section of hypophyseal stalk. *J. Neurosurg.* **18,** 445–460.
73. Silverman, A. J., Hoffman, D. L., and Zimmerman, E. A. (1981). The descending afferent connections of the paraventricular nucleus of the hypothalamus. *Brain Res. Bull.* **6,** 47–61.
74. Sladek, C. D., and Joynt, R. J. (1979). Cholinergic involvement in osmotic control of vasopressin release by the organ-cultured rat hypothalamus-neurohypophyseal system. *Endocrinology* **105,** 367–371.
75. Sladek, C. D., and Joynt, R. J. (1979). Angiotensin stimulation of vasopressin release from the rat hypothalmo-neurohypophyseal system in organ culture. *Endocrinology* **104,** 148–153.
76. Sofroniew, M. V. (1983). Vasopressin and oxytocin in the mammalian brain and spinal cord. *Trends Neurosci.* **11,** 467–72.
77. Sridhas, C. B., Calvert, G. D., and Ibberson, H. K. (1974). Syndrome of hypernatremia, hypodipsia, and partial diabetes insipidus: A new interpretation. *J. Clin. Endocrinol. Metab.* **38,** 890–901.
78. Stillman, M. A., Recht, L. D. Rosario, S. L., Seif, S. M., Robinson, A. G., and Zimmerman, E. A. (1977). The effects of adrenalectomy and glucocorticoid replacement on vasopressin and vasopressin-neurophysin in the zona externa of the median eminence of the rat. *Endocrinology* **101,** 42–49.
79. Stricker, E. M. (1969). Osmoregulation and volume regulation in rats: Inhibition of hypovolemic thirst by water. *Am. J. Physiol.* **217,** 98–105.
80. Stricker, E. M. (1973). Thirst, sodium appetite, and complementary physiological contributions to the regulation of intravascular fluid volume. *In* "The Neuropsychology of Thirst" (A. Epstein, H. R. Kissileff, and E. Stellar, eds.), pp. 73–98. V. H. Winston, Washington, D.C.
81. Stricker, E. M. (1978). The renin-angiotensin system and thirst: Some unanswered questions. *Fed. Proc.* **37,** 2704–2710.
82. Swanson, L. W., and Sawchenko, P. E. (1980). Paraventricular nucleus: A site for the integration of neuroendocrine and autonomic mechanisms. *Neuroendocrinology* **31,** 410–417.
83. Swanson, L. W., and Sawchenko, P. E. (1983). Hypothalamic integration: Organization of the paraventricular and supraoptic nuclei. *In* "Annual Review of Neuroscience" (W. M. Cowan, E. M. Shotter, C. F. Stevens, and R. F. Thompson, eds.), pp. 269–324. Annual Reviews, Inc., Palo Alto, Calif.
84. Swanson, L. W., Sawchenko, P. E., Rivier, J., and Vale, W. W. (1983). Organization of ovine corticotropin-releasing factor immunoreactive cells and fibers in the rat brain: An immunohistochemical study. *Neuroendocrinology* **36,** 165–186.
85. Thompson, D. A., Campbell, R. G., Lilavanthana, V., and Robertson, G. L. (1981). Increased thirst and plasma arginine vasopressin levels during 2-deoxy-D-glucose-induced glucoprivation in humans. *J. Clin. Invest.* **60,** 1083–1093.
86. Tribollet, E., and Dreifuss, J. J. (1981). Localization of neurons projecting to the hypothalamic paraventricular nucleus area of the rat: A horseradish peroxidase study. *Neuroscience* **6,** 1315–1328.
87. Vale, W., Speiss, J., Rivier, C., and Rivier, J. (1981). Characterization of a 41-residue ovine hypothalamic peptide that stimulates secretion of corticotropin and β-endorphin. *Science* **213,** 1394–1397.
88. Vale, W., Vaughan, J., Smith, M., Yamamoto, G., Rivier, J., and Rivier, C. (1983). Effects of synthetic ovine corticotropin-releasing factor, glucocorticoids, catecholamines, neurohypophyseal peptides, and other substances on cultured corticotrophic cells. *Endocrinology* **113,** 1121–1131.
89. Valtin, H., North, W. G., Edwards, B. R., and Gellai, M. (1984). Animal models of diabetes

insipidus. *In* "Diabetes Insipidus" (P. Czernichow and A. G. Robinson, eds.), pp. 105–126. Karger, Basel, Switzerland.

90. van Gemert, M., Miller, M., Carey, R. J., and Moses, A. M. (1975). Polyuria and impaired ADH release following medical preoptic lesioning in the rat. *Am. J. Physiol.* **228,** 1293–1297.
91. Verbalis, J. G., and Robinson, A. G. (1983). Treatment of hypothalamic diabetes insipidus. *In* "Current Therapy in Endocrinology 1983–1984" (D. T. Kreiger and C. W. Bardon, eds.), pp. 1–7. B. C. Decker, Trenton, N.J.
92. Verbalis, J. G., and Robinson, A. G. (1984). Hypernatremia and hyponatremia. *In* "Topics in Emergency Medicine" (G. S. Levey and P. Paris, eds.), Vol. 5, pp. 40–44. Aspen Systems Corporation, Rockville, M.D.
93. Verbalis, J. G., and Robinson, A. G. (1984). Neurophysin and vasopressin: Newer concepts of secretion and regulation. *In* "The Pituitary, Comprehensive Endocrinology Series" (H. Imura, ed.), pp. 307–339. Raven Press, New York.
94. Verbalis, J. G., Robinson, A. G. and Moses, A. M. (1985). Postoperative and posttraumatic diabetes insipidus. *In* "Diabetes Insipidus" (P. Czernichow and A. G. Robinson, eds.), pp. 247–265. Karger, Basel, Switzerland.
95. Weitzman, R., Kleeman, C., and Vorherr, H. (1980). Water metabolism and the neurohypophysial hormones. *In* "Metabolic Control and Disease" (P. K. Bondy and L. E. Rosenberg, eds.), pp. 1241–1323. W. B. Saunders, Philadelphia.
96. White, W. A., and Bergland, R. M. (1972). Experimental inappropriate ADH secretion caused by positive-pressure respirators. *J. Neurosurg.* **36,** 608–613.
97. Wiegand, S. J., and Price, J. L. (1980). Cells of origin of the afferent fibers to the median emimence in the rat. *J. Comp. Neurol.* **192,** 1–19.
98. Zerbe, R. L., Baylis, P. H., and Robertson, G. L. (1981). Vasopressin function in clinical disorders of water balance. *In* "The Pituitary" (C. Beardwell and G. L. Robertson, eds.), pp. 297–329. Butterworth, London.
99. Zerbe, R. L., and Robertson, G. L. (1981). A comparison of plasma vasopressin measurements with a standard indirect test in the differential diagnosis of polyuria. *New Engl. J. Med.* **305,** 1539–1546.
100. Zimmerman, E. A., Carmel, P. W., Husain, M. K., Ferin, M., Tannenbaum, M., Frantz, A. G., and Robinson, A. G. (1973). Vasopressin and neurophysin: High concentrations in monkey hypophyseal portal blood. *Science* **182,** 925–927.
101. Zimmerman, E. A., Defendini, R., Sokol, H. W., and Robinson, A. G. (1975). The distribution of neurophysin-secreting pathways in the mammalian brain. Light microscopic studies using the immunoperoxidase technique. *Ann. N.Y. Acad. Sci.* **248,** 92.
102. Zimmerman, E. A., Hsu, K. C., Robinson, A. G., Carmel, P. W., Frantz, A. G., and Tannenbaum, M. (1973). Studies of neurophysin secreting neurons with immunoperoxidase techniques employing antibody to bovine neurophysin. I. Light microscopic findings in monkey and bovine tissues. *Endocrinology* **92,** 931–940.
103. Zimmerman, E. A., and Robinson, A. G. (1976). Hypothalamic neurons secreting vasopressin and neurophysin. *Kidney Int.* **10,** 12–24.
104. Zimmerman, E. A., Robinson, A. G., Husain, M. K., Acosta, M., Frantz, A. G., and Sawyer, W. H. (1974). Neurohypophyseal peptides in the bovine hypothalamus: The relationship of neurophysin I to oxytocin, and neurophysin II to vasopressin in supraoptic and paraventricular regions. *Endocrinology* **95,** 931–936.

Chapter 2

The Other Neurohypophyseal Peptides

THOMAS L. O'DONOHUE

Experimental Therapeutics Branch
National Institute of Neurological and Communicative Disorders and Stroke
National Institutes of Health
Bethesda, Maryland 20205

JOSZEF Z. KISS

Laboratory of Cell Biology
National Institute of Mental Health
Bethesda, Maryland 20205

I. INTRODUCTION

The first neuropeptides, oxytocin and vasopressin, were identified from extracts of the posterior pituitary gland almost a century ago (Oliver and Schaefer, 1895; Howell, 1898). Until very recently, these two hormones were essentially the only biologically active peptides known to exist in the mammalian brain and neurohypophysis. In the past two decades, however, over 50 other peptides have been identified and characterized in the central nervous system. These recently identified neuropeptide systems have extensive distributions throughout the nervous system and have numerous diverse functions.

It was initially thought that oxytocin- and vasopressin-containing neurons were located solely in the hypothalamo–neurohypophyseal tract (see Bergmann and Scharrer, 1951). More recent immunohistochemical investigations, however, have demonstrated that, like other central neuropeptides, vasopressin- and oxytocin-containing neurons project throughout the brain. The contents of vasopressin and oxytocin found in extrahypothalamic areas, however, is far lower than that contained in the hypothalamo–neurohypophyseal system (cf. Caffe and Van Leeuwen, 1983; Sofroniew and Weindl, 1981; Dogterom *et al.*, 1978).

THE PEPTIDES, VOLUME 8

Recent results show that the converse is also true—other neuropeptides that have been identified in relatively high concentrations in the central nervous system often exist in the neurohypophysis. In this chapter the identification of peptides other than oxytocin and vasopressin in the posterior lobe is reviewed and, where the information is known, the anatomical source and possible functions of these peptides is discussed.

As shown in Table I, there are 17 peptides in addition to oxytocin and vasopressin that have been identified in the posterior pituitary gland. Two of these peptides, vasotocin and mesotocin, are nonmammalian forms of the posterior pituitary hormones. The other 15 peptides have all been identified in the mammalian neurohypophysis. It should be noted, however, that while the number of peptides in the posterior lobe is dramatically increasing, the total peptide content in the mammalian neurohypophysis is not significantly changing. This is due to the relatively low concentrations of peptides other than vasopressin and oxytocin in the neurohypophysis. For example, the concentration of cholecystokinin in the

Table I. Concentration of Neuropeptides in the Posterior Lobe

Neuropeptide	Concentration (fmol/lobe)	Species	Reference
Vasopressin	660,000	Rat	Crowley *et al.* (1978)
Oxytocin	750,000	Rat	Crowley *et al.* (1978)
Vasotocin	18,000	Chick	Nouwen *et al.* (1983)
Mesotocin	12,000	Chick	Nouwen *et al.* (1983)
ACTH	Present	Rat	See text
α-MSH	Present	Rat	See text
Endorphin	Present	Rat	See text
α-Neoendorphin	1,000	Rat	Zamir *et al.* (1984); Zamir (1984)
β-Neoendorphin	950	Rat	Zamir *et al.* (1984); Zamir (1984)
Dynorphin A	600	Rat	Zamir *et al.* (1984); Zamir (1984)
Dynorphin B	1200	Rat	Zamir *et al.* (1984); Zamir (1984)
[Leu]Enkephalin	1200	Rat	Zamir *et al.* (1984); Zamir (1984)
[Met]Enkephalin	600	Rat	Zamir *et al.* (1984); Zamir (1984)
FMRFamide	20	Rat	O'Donohue *et al.* (1984)
Corticotropin-releasing Factor	20	Rat	Suda *et al.* (1983)
Somatostatin	180	Rat	Saavedra *et al.* (1983)
Cholecystokinin	200	Rat	Beinfeld *et al.* (1980)
Motilin	0.4	Rat	O'Donohue *et al.* (1981)
Glucagon	Present	Rat	See text
Secretin	30	Rat	Charlton *et al.* (1982)
Substance P	30	Rat	DePalatis *et al.* (1984)
Thymosin	5	Rat	Palaszynski *et al.* (1983)
Insulin-like growth factor	Present	Porcine Human	Marcinkiewicz *et al.* (1985)
Angiotensin	Present	Rat	See text

posterior lobe is only approximately 0.03% that of the oxytocin and vasopressin content. The extremely low concentrations of these peptides must be considered when making predictions of their functional significance.

In general, there are four classes of peptides other than oxytocin and vasopressin in the neurohypophysis. These include the opioid peptides, certain hypothalamic releasing or release-inhibiting factors, gastrointestinal peptides, and the fourth class—others.

II. OPIOID PEPTIDES

All of the known opioid peptides have been reported to be contained in the neurohypophysis. These include peptides derived from proenkephalin, prodynorphin, and proopiomelanocortin (POMC).

Moderate concentrations of [Met]- and [Leu]enkephalin have been identified in the rat neurointermediate lobe by radioimmunoassay (RIA) and by RIA combined with gel-filtration chromatography (Rossier *et al.*, 1979, 1980b). The enkephalin immunoreactivity has been localized in fine varicose fibers in the neurohypophysis (Rossier *et al.*, 1980b). Immunoreactive fibers could be traced through the pituitary stalk and the internal zone of the median eminence, as well as in the paraventricular and supraoptic nuclei. Pretreatment of rats with colchicine allowed visualization of enkephalin-positive perikarya in the magnocellular nuclei of both the paraventricular and supraoptic nuclei (Jacobwitz *et al.*, 1978). Lesion studies confirmed that the source of most of the enkephalin in the posterior pituitary gland is the paraventricular and supraoptic nuclei (Rossier *et al.*, 1980b).

Dynorphin and related peptides have more recently been identified in terminals of the neurohypophysis (Goldstein and Ghazzarossian, 1980). The two most likely sources for the dynorphin-containing fibers in the neurohypophysis were the paraventricular and supraoptic nuclei, as dynorphin-positive magnocellular perikarya were observed in both regions (Goldstein and Ghazzarossian, 1980; Watson *et al.*, 1982; Weber *et al.*, 1981). The main source of dynorphin for the posterior pituitary seems to be the supraoptic nucleus, as a lesion of the paraventricular nucleus failed to decrease neurohypophyseal or median-eminence dynorphin or alpha-neoendorphin concentrations (Palkovits *et al.*, 1984).

Until recently, it was quite difficult to evaluate the relationship of enkephalins and dynorphins to each other and to the prohormone from which the enkephalin peptides were derived. The first problem was that most of the enkephalin antisera that were used cross-reacted with both the Met and Leu variants of the peptide. The second and greater difficulty was due to the fact that [Leu]enkephalin can be biosynthetically derived from two different precursors—proenkephalin, which contains four copies of [Met]- and one copy of [Leu]enkephalin, and prodynorphin, which gives rise to alpha- and beta-neoendorphin and dynorphin A and B, which each contain one copy of [Leu]enkephalin. Thus, [Met]enkephalin can be derived only from proenkephalin, while [Leu]enkephalin can be generated from both proenkephalin and prodynorphin. Recent studies of prodynorphin-containing systems of the striato–nigral pathway have, in fact, demonstrated that [Leu]enkephalin is probably cleaved from prodynorphin (Zamir *et al.*, 1986). It is therefore clear that immunohistochemical staining using antibodies directed against [Leu]enkephalin would likely stain both proenkephalin- and prodynorphin-containing systems. The difficulties in distinguishing these two systems has to a large extent been overcome by using antisera generated against peptides specific for the proenkephalin or prodynorphin systems. The results of these studies demonstrate that proenkephalin- and prodynorphin-derived peptides are located in two distinct neuronal populations in the hypothalamo–neurohypophyseal system. Proenkephalin was localized in oxytocin-containing magnocellular neurons, in addition to a large number of parvocellular neurons in the paraventricular nucleus, while prodynorphin was found to be contained exclusively in neurons that produce vasopressin (Vanderhaegen *et al.*, 1983; Watson *et al.*, 1982; Weber *et al.*, 1981). In fact, dynorphin has actually been shown to be stored in the same secretory vesicles as vasopressin in the neurohypophysis (Whitnall *et al.*, 1983). This distinction between the two systems confirms previous studies that demonstrated the coexistence of oxytocin with [Met]enkephalin and vasopressin with [Leu]enkephalin in axon terminals of the neural lobe (Martin and Voight, 1981; Martin *et al.*, 1983).

Further studies have demonstrated that all of the opioid peptides in the neurohypophysis are not derived from the paraventricular and supraoptic nuclei. While transection of the pituitary stalk depleted essentially all of the opioid peptides from the neural lobe, combined lesions of the paraventricular and supraoptic nuclei produced only a 45% decrease in the [Met]enkephalin content of the neurohypophysis (Zamir *et al.*, 1986). This data is consistent with morphological data that show that enkephalin immunoreactivity was found in terminals of the posterior lobe which do not contain oxytocin or vasopressin (Van Leeuwen *et al.*, 1983). A possible source of this [Met]enkephalin may be the arcuate nucleus, which contains enkephalin-positive perikarya (Jacobowitz *et al.*, 1978). Monosodium glutamate treatment, which destroys the arcuate but spares the rest of the hypothalamus, markedly depletes the [Met]enkephalin content of the posterior pituitary (N. Zamir *et al.*, personal communication).

The sequence of another enkephalin-like peptide, Phe-Met-Arg-Phe-amide or FMRFamide, exists in the proenkephalin molecule. This peptide, also known as molluscan cardioactive tetrapeptide, was originally identified in the heart ganglia of the clam. Although it is clear that FMRFamide is not derived from proenkephalin (see O'Donohue *et al.*, 1984), there are relatively high concentrations of FMRFamide-like peptide in the posterior lobe (O'Donohue *et al.*, 1984). Although the concentrations of the FMRFamide-like peptide in the posterior lobe, as shown in Table I, are extremely low, these values may be misleading, as the endogenous FMRFamide-like peptide is clearly different in structure from molluscan cardioactive peptide (O'Donohue *et al.*, 1984; Dockray and Williams, 1983) and probably only cross-reacts weakly in the FMRFamide RIA (Dockray *et al.*, 1983). Immunohistochemical studies of the pituitary gland with FMRFamide antisera demonstrated many immunoreactive fibers in the neural lobe but no fibers or cells in the intermediate or anterior lobes (Williams and Dockray, 1983). Likely sources for the FMRFamide in the neural lobe are the immunoreactive cells in the paraventricular or arcuate nuclei (see Fig. 1) (Chronwall *et al.*, 1984). Whether the FMRFamide-like peptide is derived from yet another opioid prohormone is unknown.

The third characterized prohormone that can give rise to opioid peptides, proopiomelanocortin (POMC), is different from the other opioid prohormones in that it also gives rise to corticotropic and melanotropic peptides (see O'Donohue and Dorsa, 1982). There are two potential sources for POMC-derived peptides in the neural lobe. The first source is the intermediate lobe, which contains over 99% of the POMC in the body. The close anatomical juxtaposition of the intermediate and posterior lobes makes measurement of POMC-derived peptides in the posterior lobe quite difficult. In fact, a number of POMC-containing "intermediate-lobe" cells called invasion cells actually reside in the neural lobe. The second potential source is the POMC-containing cells of the arcuate nucleus (O'Donohue *et al.*, 1979). Although it has been known for some time that POMC-containing cells in this region have fibers that project throughout the brain (O'Donohue *et al.*, 1979), it has only recently been found that POMC neurons may project to the pituitary stalk and perhaps to the gland itself (Knigge and Joseph, 1981; Kiss *et al.*, 1984). It must be noted that this ventral projection appears to give rise primarily to an innervation of structures in the internal zone of the median eminence (Kiss *et al.*, 1984). Ultrastructural investigations have demonstrated that these POMC immunoreactive boutons are in close contact with the capillary loops of the subependymal plexus (Hisano *et al.*, 1982), which is consistent with the demonstration of α-melanotropin (α-melanocyte stimulating hormone, alpha-MSH) in portal blood (Oliver *et al.*, 1977). Relatively few POMC fibers descend further into the infundibulum, and those that do appear to form terminal fields as opposed to bundles *en passage* (Knigge and Joseph, 1981; Kiss *et al.*, 1984). From this region, a few fibers may descend further to innervate the intermediate and posterior lobes, but only a low density of POMC-positive fibers has been reported in the neural lobe (Knigge and Joseph, 1981). The contribution of opioid peptides to the neural lobe from POMC appears to be minor when compared to the innervation provided by the proenkephalin and prodynorphin systems.

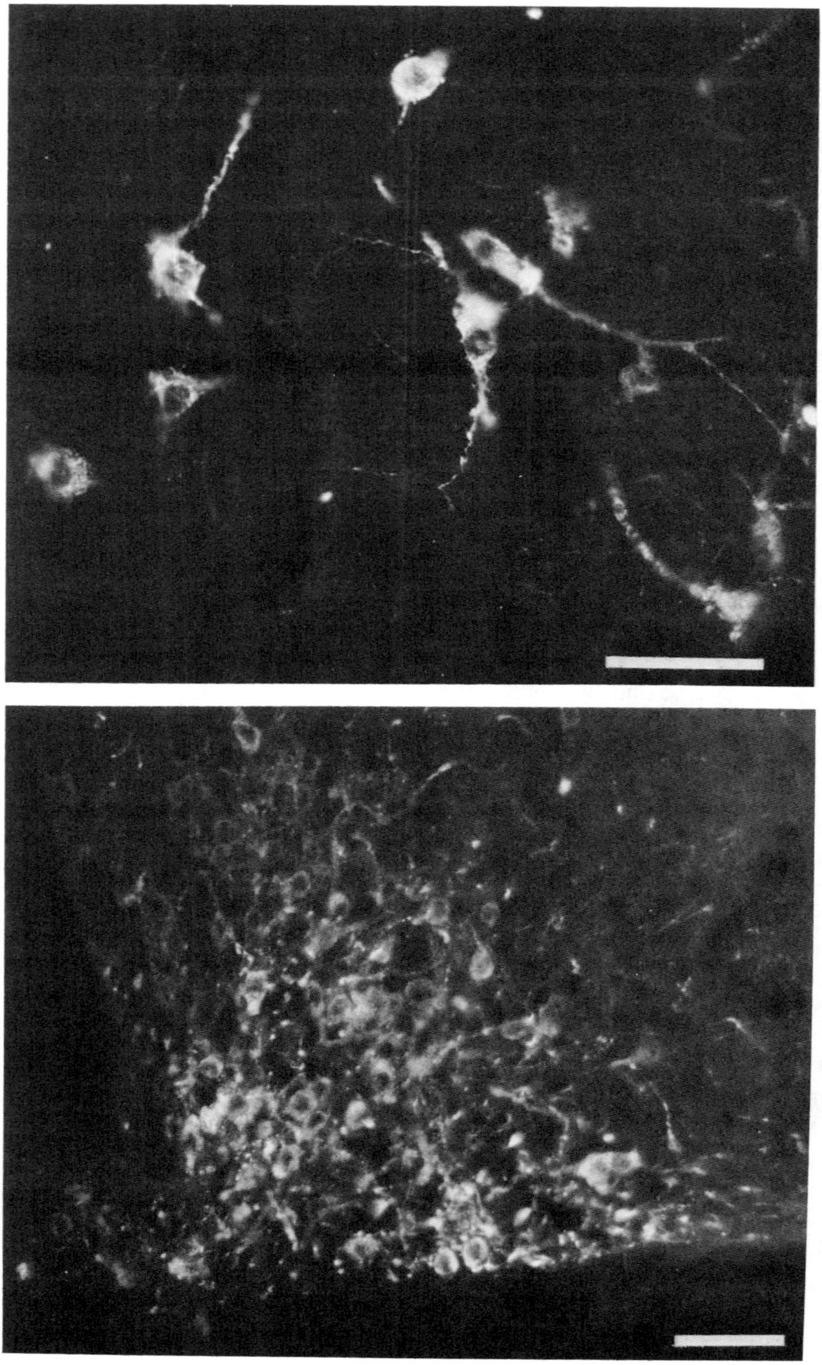

III. HYPOTHALAMIC RELEASING AND RELEASE-INHIBITING FACTORS

Two hypothalamic trophic factors have been identified in the neural lobe: corticotropin-releasing factor (CRF) and somatostatin. The possibility of the existence of CRF in the posterior lobe had been raised by a number of early studies using a bioassay to quantitate CRF. The results of the bioassay were controversial, as vasopressin also had CRF activity. The identification of a specific CRF (Vale *et al.*, 1981) and the development of antisera to this peptide have allowed definitive quantitation of CRF in the posterior lobe, as shown in Table I. Initial immunocytochemical studies identified CRF-containing perikarya only in the parvocellular part of the paraventricular nucleus and only in scattered cells in the supraoptic nucleus (Olschowka *et al.*, 1982). After adrenalectomy, however, far greater numbers of CRF cells can be observed in the parvocellular part of the paraventricular and supraoptic nuclei (Roth *et al.*, 1983; Swanson *et al.*, 1983). Using immunocytochemistry after colchicine treatment (Fig. 2), CRF immunoreactivity has been found to be colocalized with oxytocin in the paraventricular and supraoptic nuclei as well as in the internal zone of the median eminence and the pituitary stalk (Burlet *et al.*, 1983).

As shown in Table I, a moderate concentration of somatostatin can be detected in the posterior lobe. Transection of the hypothalamo–hypophyseal pathway markedly depletes somatostatin from the posterior lobe, demonstrating that the somatostatin is of central origin. The major somatostatin-containing cell group is located in the periventricular nucleus beginning rostrally at the level of the median eminence and extending caudally to the level of the rostral arcuate nucleus (Elde and Parsons, 1975; Hokfelt *et al.*, 1975; Krisch, 1977). Dubois and Kolodziejczyk (1975) have reported that somatostatin was contained within magnocellular perikarya of the supraoptic nucleus, although Dierickx and Vandesande (1977) have concluded that this staining was due to cross-reactivity with neurophysin. It is clear that further work is required to define the origin of the somatostatin in the posterior pituitary.

IV. GASTROINTESTINAL PEPTIDES

Many gastrointestinal peptides seem to serve hormonal roles in the gut and neurotransmitter roles in the central nervous system. A number of these neu-

---

Figure 1. FMRFamide immunoreactive perikarya in the paraventricular (top) and arcuate (bottom) nuclei after colchicine pretreatment of rats. Bar = 50 μm.

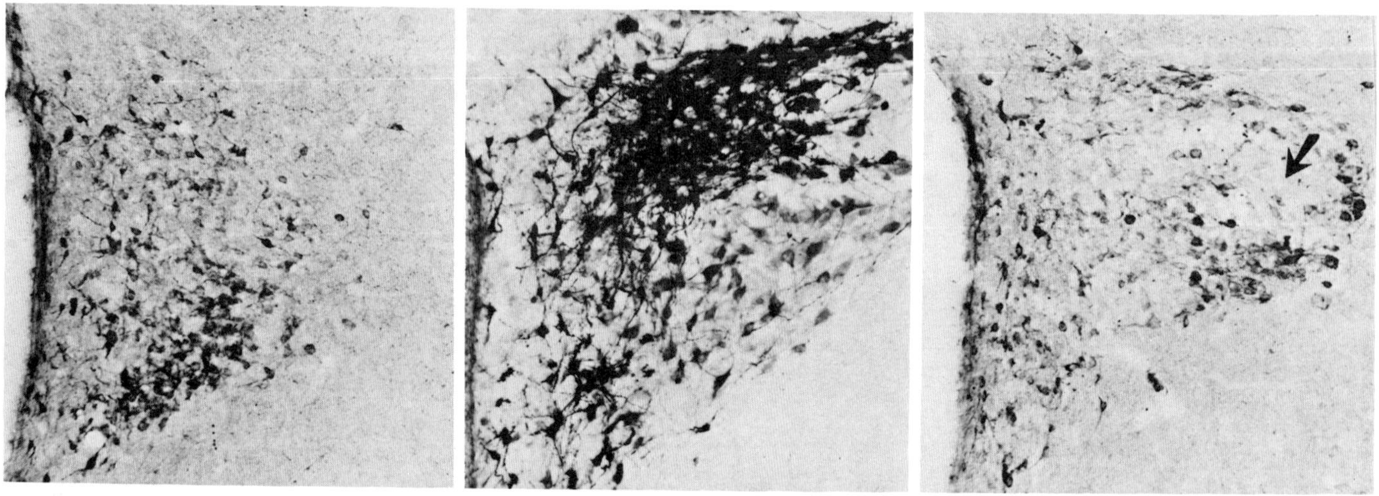

Figure 2. CCK- and CRF-containing neurons in the paraventricular nucleus of colchicine-treated nucleus. From left to right: CCK, CRF, and CCK immunostaining in the rostral to caudal paraventricular. Note that two separate populations of neurons contain CCK and CRF. Note that the CCK staining (in left-hand figure) is in magnocellular neurons in the lateral aspect of nucleus, which also contains oxytocin and forms a ring around the region of vasopressin neurons (marked with arrow). (×130).

ropeptides have also been identified in the posterior lobe. As shown in Table I, these include cholecystokinin (CCK), secretin, glucagon, and motilin.

A number of investigations have presented data that demonstrate that peptides of the gastrin–cholecystokinin family of peptides are present in the posterior pituitary gland of several vertebrate species (Beinfeld *et al.*, 1980, 1981; Rehfeld, 1978; VanderHaegen et al., 1975). Chromatographic studies have demonstrated that the rat pituitary does not contain gastrin but primarily contains the sulfated C-terminal octapeptide of CCK-33 (Beinfeld *et al.*, 1980, 1981). Immunocytochemical studies demonstrated that CCK fibers pass through the internal zone of the median eminence, via the infundibulum, and terminate in the neural lobe (Rehfeld, 1978; Vanderhaegen *et al.*, 1980). Two likely sources of neural-lobe CCK were identified when rats were treated with colchicine, as CCK-positive magnocellular perikarya were found in both the paraventricular and supraoptic nuclei (see Fig. 2) (Vanderhaegen *et al.*, 1980). Lesion experiments indicate that 60–80% of the CCK in the pituitary is derived from the paraventricular nucleus (Beinfeld *et al.*, 1980; Palkovits *et al.*, 1984). The fact that similar depletions of CCK were obtained after lesioning the lateral retrochiasmatic area, which destroys both paraventricular and supraoptic afferents to the neural lobe, indicates that most of the CCK in the neural lobe is derived from the paraventricular nucleus (Palkovits *et al.*, 1984). Subsequent colocalization studies suggest that CCK is present in some of the oxytocin-containing neurons of the paraventricular and supraoptic nuclei (Van Leeuwen, 1981). In fact, the precise topographical map of CCK magnocellular neurons in the paraventricular nucleus corresponds very closely to the distribution of oxytocin cells (Kiss *et al.*, 1984). This coexistence was further confirmed by Martin and Voight (1981), who showed the coexistence of oxytocin and CCK in posterior-lobe terminals. These latter ultrastructural studies also demonstrated that CCK immunoreactivity is contained in neurosecretory granules, which indicates that these two peptides are costored, as are vasopressin and dynorphin.

A number of peptides from the secretin–glucagon family of peptides have been identified in the central nervous system. The family of structurally related peptides currently consists of five members: secretin, glucagon, vasoactive intestinal peptide, gastric inhibitory polypeptide, and PHI. Of this family, there are four peptides that have been reported to exist in the brain: secretin (O'Donohue *et al.*, 1981b), glucagon (Tager *et al.*, 1980), vasoactive intestinal peptide (Said and Rosenberg, 1976), and PHI (Beinfeld *et al.*, 1984). Of this family, secretin and glucagon have been reported to exist in the posterior pituitary gland. The secretin is clearly of central origin, as transection of the pituitary stalk markedly decreases the secretin content of the neural lobe (Charlton *et al.*, 1982). Some of the secretin may come from neurons in the paraventricular nucleus, as lesions of this region result in a 50% decrease in the secretin content of the neural lobe (C. G. Charlton *et al.*, unpublished findings). Staining for pancreatic glucagon in

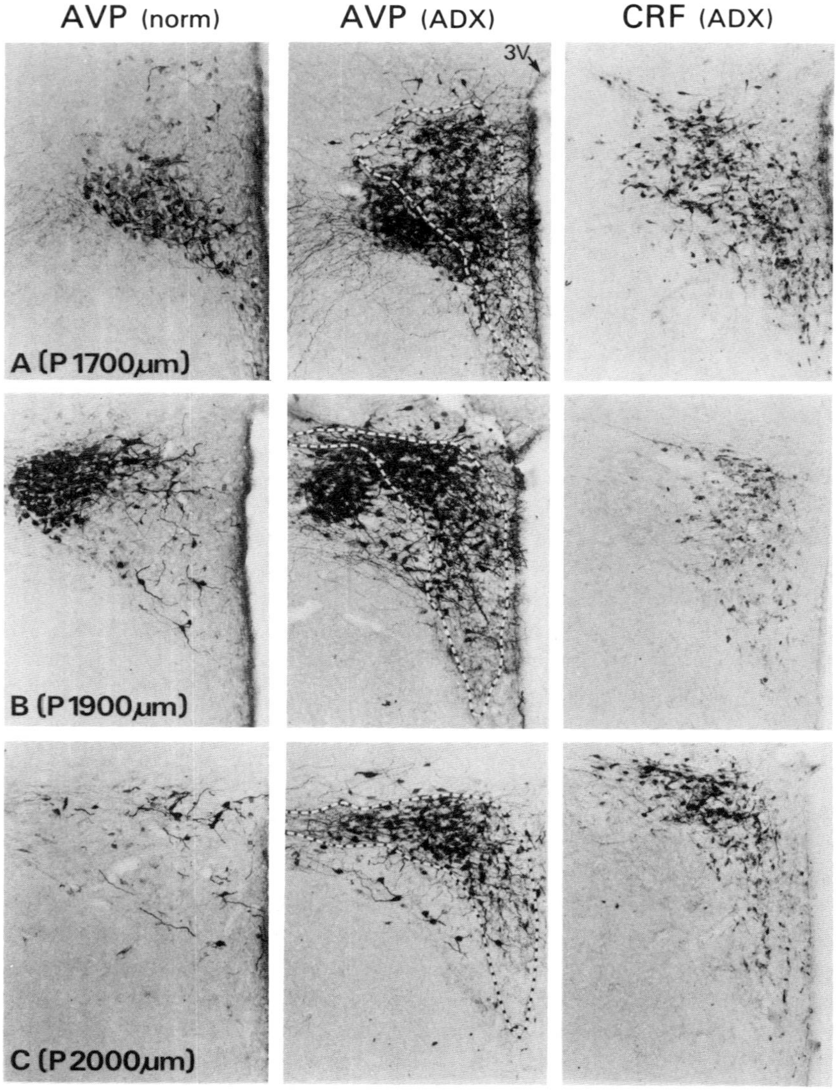

Figure 3. Topography of (vasopressin-) VP- and CRF-immunoreactive neurons in the PVN of normal and adrenalectomized (ADX) rats. Photomicrographs show coronal sections of the PVN at various levels: row A, 1700 μm; row B, 1900 μm; and row C, 2000 μm behind the bregma. The white dashed line outlines the parvocellular subdivision of the PVN. Note the large increase in number of VP parvocellular neurons in ADX rats (second column, white, dashed line) over the relatively few, immunostained parvocellular neurons of normal animals (first column), and that there is a considerable overlap of the distribution of CRF- (third column) and VP-containing parvocellular neurons.

magnocellular neurons in vasopressinergic parts of the paraventricular parts of the paraventricular and supraoptic nuclei has been demonstrated (Joseph and Sternberger, 1979) (Fig. 3); although there is no evidence as to whether glucagon actually exists in the posterior lobe.

As is also shown in Table I, a motilin-immunoreactive peptide exists in the posterior lobe of the pituitary gland. It is clear that this immunoreactive peptide is different in structure from motilin, although the precise structure of the peptide has not been determined (O'Donohue *et al.*, 1981a).

V. OTHER PEPTIDES

A number of other peptides have been identified in the posterior pituitary, although the sources of these peptides are unknown. As shown in Table I, these include substance P, thymosin, and an insulin-like growth factor. Renin immunoreactive cells have also been localized in the oxytocinergic region of the supraoptic and paraventricular nuclei (Fuxe *et al.*, 1980), and angiotensin II immunoreactivity seems to be present in vasopressin-containing neurons (Phillips *et al.*, 1979; Zimmerman *et al.*, 1980). It is of interest that angiotensin II immunoreactivity is absent in homozygous Brattleboro rats (Zimmerman *et al.*, 1980).

VI. PHYSIOLOGICAL SIGNIFICANCE OF OTHER NEUROHYPOPHYSEAL PEPTIDES

As discussed in the introduction to this chapter, the quantities of peptides other than vasopressin and oxytocin in the posterior pituitary are relatively low. There are three possible functions for these peptides in the neural lobe. The first is that they, like oxytocin and vasopressin, are secreted into the blood stream to act on distant targets. Consistent with such a thought is the recent demonstration of secretin receptors on the kidney and an antidiuretic action of this peptide (Charlton *et al.*, 1984). Opiate receptors have also been demonstrated to exist in the kidney (Quirion *et al.*, 1983). The extremely low concentrations of secretin, known opioids, and the other peptides discussed in the neural lobe, however, make their roles as peripherally acting hormones doubtful. Perhaps at some time in development or evolution these peptides may have played roles as hormones secreted from the neurohypophysis. Another possibility is that these peptides may have actions in certain pathological states. For instance, the reason why homozygous Brattleboro rats that lack vasopressin are able to concentrate urine better than posterior-lobectomized rats remains unknown. Perhaps other neurohypophyseal peptides in addition to vasopressin and oxytocin may have functional actions on urine output. A second possible action of the other neural-lobe peptides is that they may have local actions in the neurohypophysis. The recent demonstration of opiate receptors on neural-lobe pituicytes (Lightman *et al.*, 1983), the pituitary form of glial cells, supports such a possibility. The work of Hamprecht and others has demonstrated actions of many peptides, including alpha-MSH and secretin, on brain glia (Van Calker *et al.*, 1980, 1983). Perhaps the other peptides act locally and have a trophic or regulatory function on the glial cells, which then help maintain a suitable environment for neurohypophyseal function. The third possible function for these trace peptides, and the least attractive, is that they have no function at all and are in actuality "genetic noise" in the vasopressin- and oxytocin-containing neurons. It is clear that a considerable effort will be required to determine the role of the "other peptides" in the physiology of the hypothalamo–neurohypophyseal system.

REFERENCES

Armstrong, W. E., Warach, S., Hatton, G. I., and McNeill, T. H. (1980). *Neuroscience* **5,** 1931–1958.

Beinfeld, M. C. (1980). *Neuropeptides* **1,** 203–209.

Beinfeld, M. C., Meyer, D. K., and Brownstein, M. J. (1980). *Nature (London)* **288,** 376–378.

Beinfeld, M. C., Meyer, D. K., Eskay, R. L., Jensen, R. T., and Brownstein, M. J. (1981). *Brain Res.* **212,** 51–57.

Beinfeld, M. C., Korchak, D. M., Roth, B. L., and O'Donohue, T. L. (1984). *J. Neurosci.* **4,** 2681–2688.

Bergmann, W., and Scharrer, E. (1951). *Am. Sci.* **3,** 255–259.

Bjorklund, A., Falck, B., Hromek, F., Owman, C., and West, A. (1970). *Brain Res.* **17,** 1–23.

Bugnon, C., Block, B., Lergs, D., Gouget, A., and Fellman, D. (1979). *Neurosci. Lett.* **14,** 43–48.

Bujis, R. M. (1978). *Cell Tissue Res.* **192,** 423–435.

Burlet, A., Tonon, M. C., Tankosic, P., Coy, D., and Vaudry, M. (1983). *Neuroendocrinology* **37,** 73–77.

Caffe, A. R., and van Leeuwen, F. W. (1983). *Cell Tissue Res.* **233,** 23–33.

Charlton, C. G., O'Donohue, T. L., Miller, R. L., Jacobowitz, D. M. (1982). *Peptides* **3,** 565–567.

Charlton, C. G., Quirion, R., Handelmann, G. C., Miller, R. L., Jensen, R. T., Finkel, M. S., and O'Donohue, T. L. (1986). *Peptides* (in press).

Chronwall, B. M., Olschowka, J. A. and O'Donohue, T. L. (1984). *Peptides* 5: 569–584.

Crowley, W. R., O'Donohue, T. L., George, J. M., and Jacobowitz, D. M. (1978). *Life Sci.* **23,** 2579–2586.

DePalatis, L. R., Khorram, O., Ho, R. H., Negro-Vilar, A., and McCann, S. M. (1984). *Life Sci.* **34,** 225–238.

Dierickx, K., and Vandesande, F. (1977). *Cell Tissue Res.* **201,** 349–359.

Dockray, G. J., and Williams, R. G. (1983). *Brain Res.* **266,** 295–303.

Dockray, G. J., Reeve, J. R. Jr., Shively, J., Gayton, R. J., and Barnard, C. S. (1983). *Nature (London)* **305,** 328–330.

Dogterom, J., Snijdewint, F. G. M., and Buijis, R. M. (1978). *Neurosci. Lett.* **9,** 341–346.

Dubois, M. P., and Kolodziejczyk, E. (1975). *C. R. Acad. Sci. (D) (Paris)* **281,** 1737–1740.

Elde, R. P., and Parsons, J. A. (1975). *Am. J. Anat.* **144,** 541–548.

Fuxe, K., Ganten, D., Hokfelt, T., Locatelli, V., Paulsen, K., Stock, G., Rix, E., and Taugner, R. (1980). *Neurosci. Lett.* **18,** 245–250.

Goldstein, A., and Ghazarossian, V. E. (1980). *Proc. Natl. Acad. Sci. U.S.A.* **77,** 6207–6210.

Hisano, S., Kawano, H., Nishiyama, T., and Daikoku, S. (1982). *Cell Tissue Res.* **124,** 303–314.

Hokfelt, T., Efendic, S., Hellerstrom, C., Johanson, O., Luft, R., and Arimura, A. (1975). *Acta Endocrinol. (Copenh.)***80,** 5–41.

Hollt, V., Haarmann, I., Bovermann, K., Jerlicz, M., and Herz, A. (1980). *Neurosci. Lett.* **18,** 149–153.

Hollt, V., Haarmann, I., Seizinger, B. R., and Herz, A. (1981). *Neuroendocrinology* **33,** 333–339.

Howell, W. H. (1898). *J. Exp. Med.* **3,** 245–258.

Jacobowitz, D. M., Silver, M. A., and Soden, W. G. (1978). *In* "Endorphins in Mental Health Research" (E. Udsin and W. Bunney, eds.), pp. 62–73. Oxford Univ. Press, London and New York.

Joseph, S. A., and Sternberger, L. A. (1979). *J. Histochem. Cytochem.* **27,** 1430–1437.

Kiss, J. Z., Williams, T. M., and Palkovits, M. (1984). *J. Comp. Neurol.* **277,** 173–181.

Knigge, K. M., and Joseph, S. A. (1981). *Cell Tissue Res.* **215,** 333–340.

Krisch, B. (1977). *Cell Tissue Res.* **179,** 211–224.

Larson, L. I. (1977). *Lancet* **2,** 1321–1323.

Lauber, M., Nicolas, P., Bousetta, M., Faby, C., Beguin, P., Camier, M., Vaudry, M., and Cohen, P. (1981). *Proc. Natl. Acad. Sci. U.S.A.* **76,** 6086–6090.

Lightman, S. L., Ninkovic, M., Hunt, S. P., and Iverson, L. L. (1983). *Nature (London)* **305,** 235–237.

Marcinkiewicz, M., Benjannet, S., Seidah, N. G., Cantin, M., Chretien, M. (1985). *J. Histochem. Cytochem.* **33,** 1219–1226.

Martin, R., and Voigt, K. H. (1981). *Nature (London.)* **289,** 502–504.

Martin, R., Geis, R., Holl, R., Schafer, M., and Voigt, K. H. (1983). *Neuroscience* **8,** 213–227.

Micevych, P., and Elde, R. (1980). *J. Comp. Neurol.* **190,** 135–146.

Nakao, K., Yoshimasa, T., Oki, S., Tanaka, I., Nakai, Y., Wakimasu, M., Fujino, M., and Imura, H. (1981). *Regul. Pept.* **2,** 201–208.

Nouwen, E. J., Decuypere, E., Michels, H., and Kuhn, E. R. (1983). *Gen. Comp. Endocrinol.* **3,** 445–451.

O'Donohue, T. L., and Dorsa, D. M. (1982). *Peptides* **3,** 353–395.

O'Donohue, T. L., Miller. T. L., and Jacobowitz, D. M. (1979). *Brain Res.* **176,** 101–123.

O'Donohue, T. L., Beinfeld, M. C., Chey, W. Y., Chang, T. M., Nilaver, G., Zimmerman, E. A., Yajima, H., Adachi, H., Poth, M., McDevitt, R. P., Jacobowitz, D. M. (1981a). *Peptides* **2,** 467–477.

O'Donohue, T. L., Charlton, C. G., Miller, R. L., Boden, G., and Jacobowitz, D. M. (1981b). *Proc. Natl. Acad. Sci. USA* **78,** 5221–5224.

O'Donohue, T. L., Bishop, J. F., Chronwall, B. M., Watson, W. H., and Groome, J. R. (1984). *Peptides* **5,** 563–568.

Oliver, C., Mical, R. S., and Porter, J. C. (1977). *Endocrinology* **101,** 598–604.

Oliver, G., and Schafer, E. A. (1895). *J. Physiol.* **18,** 277–279.

Olschowka, J. A., O'Donohue, T. L., Mueller, G. P., and Jacobowitz, D. M. (1982). *Peptides* **3,** 995–1015.

Palaszynski, E. W., Moody, T. W., O'Donohue, T. L., and Goldstein, A. L. (1983). *Peptides* **4,** 463–467.

Palkovits, M., Kiss, J. Z., Beinfeld, M. C., and Brownstein, M. J. (1984). *Brain Res.* **299,** 186–189.

Phillips, M. L., Weyhenmeyer, J., Felix, J., Ganten, D., and Coffman, W. E. (1979). *Fed. Proc.* **38,** 2260–2266.

Quirion, R., Finkel, M. S., Mendelsohn, F. A. O., and Zamir, N. (1983). *Life Sci.* **33,** 299–302.

Rehfeld, J. F. (1978). *Nature (London)* **271,** 771–773.

Rossier, J., Battenberg, E., Pittman, Q., Bayon, A., Koda, L., Miller, R., Guillemin, R., and Bloom, F. (1979). *Nature (London)* **277,** 653–655.

Rossier, J., Audigier, Y., Ling, N., Cros, J., and Udenfriend, S. (1980a). *Nature (London)* **288,** 88–90.

Rossier, J., Pittman, Q., Bloom, F., and Guillemin, R. (1980b). *Proc. Fed. Am. Soc. Exp. Biol.* **39,** 2555–2560.

Roth, K. A., Weber, E., Barchas, J. D., Chang, D., Chang, J. K. (1983). *Science* **219,** 189–191.

Saavedra, J. M., Rouseot, C., and Dray, F. (1983). *Neuroendocrinology* **37,** 164–165.

Said, S. I., and Rosenberg, R. (1976). *Science* **192,** 907–908.

Seidah, N. G., Hsi, K. L., De Serres, G., Rochemont, J., Hamelin, J., Antakly, T., Cantin, M., and Chretien, M. (1983). *Arch. Biochem. Biophys.* **225,** 525–534.

Sofroniew, M. V., and Weindl, A. (1981). *In* "Endogenous Peptides and Learning and Memory Processes" (J. L. Martinez, R. A. Jensen, R. B. Messing, H. Rigten, and J. L. McGaugh, eds.), pp. 327–369. Academic Press, New York.

Suda, T., Tomori, N., Tozawa, F., Mouri, T., Demura, H., and Shizume, K. (1983). *Endocrinology* **113,** 1182–1184.

Tager, H., Hohenboken, M., Markese, L., and Dinerstein, R. L. (1980). *Proc. Natl. Acad. Sci. USA* **77,** 6229–6233.

Vale, W., Spiess, J., Rivier, C., and Rivier, J. (1981). *Science* **213,** 1394–1397.

van Calker, D., Muller, M., and Hamprecht, B. (1980). *Proc. Natl. Acad. Sci. USA* **77,** 6907–6911.

van Calker, D., Loffler, F., and Hamprecht, B. (1983). *J. Neurochem.* **40,** 418–427.

Vanderhaeghen, J. J., Signeau, J. C., and Gept, W. (1975). *Nature* (London) **257,** 604–605.

Vanderhaeghen, J. J., Lotstra, F., DeMey, J., and Gilles, C. (1980). *Proc. Natl. Acad. Sci. USA* **77,** 1190–1194.

Vanderhaeghen, J. J., Lotstra, F., Vandesande, F., and Dierickx, K. (1980). *Cell Tissue Res.* **221,** 227–231.

Vanderhaeghen, J. J., Lostra, F., Liston, D. R., and Rossier, J. (1983). *Proc. Natl. Acad. Sci. USA* **80,** 5139–5143.

van Leeuwen, F. W. (1981). *Neurosci. Lett. (Suppl.)* **7,** 182.

van Leeuwen, F., and Caffe, R. (1983). *Cell Tissue Res.* **228,** 525–534.

van Leeuwen, F. W., and Swaab, D. F. (1977). *Cell Tissue Res.* **177,** 493–501.

van Leeuwen, F. W., DeRaay, C., Swaab, D. F., and Fisser, B. (1979). *Cell Tissue Res.* **202,** 189–201.

van Leeuwen, F. W., Pool, C., and Sluiter, A. (1983). *Neuroscience* **8,** 229–241.

Watkins, W. B. (1980). *Cell Tissue Res.* **207,** 65–80.

Watson, S. J., Akil, H., Fischli, W., Goldstein, A., Zimmerman, E., Nilaver, G., and van Wimersma Greidanus, T. B. (1982). *Science* **216,** 85–87.

Weber, E., Roth, K. A., and Barchas, J. D. (1981). *Biochem. Biophys. Res. Commun.* **103,** 951–958.

Whitnall, M. H. Gainer, H., Cox, B. M., and Molineaux, C. J. (1983). *Science* **222,** 1137–1139.

Williams, R. G., and Dockray, G. J. (1983). *Brain Res.* **276,** 213–229.

Zamir, N. (1985). *Endocrinology* **117,** 1687–1692.

Zamir, N., Palkovits, M., Brownstein, M. J. (1984). *J. Neurosci.* **4,** 1240–1247.

Zamir, N., Zamir, D., Eiden, L., Palkvits, M., Brownstein, M. J., Eskay, R. L., Weber, E., Fadon, A. I., and Feuerstein, G. (1986). (submitted).

Zimmerman, E. A., Krupp, L., Hoffman, D. L., Matthew, E., and Nilaver, G. (1980). *Peptides* **1,** 3–10.

Chapter 3

Biochemistry and Biology of Vasopressin, Oxytocin, and Their Corresponding Neurophysins

DIETMAR RICHTER

Institut für Zellbiochemie
und klinische Neurobiologie
Universität Hamburg
2 Hamburg 20, Federal Republic of Germany

I. INTRODUCTION

Various studies have focused on the functional role of the hypothalamus and on the question of how this organ controls, integrates, and coordinates complex responses and expression in the central nervous system as well as in peripheral target organs. Neuropeptides play a central part in many of these processes by modulating neural signals or converting them into specific physiological responses (Krieger, 1983).

In order to understand how these responses are controlled and regulated and how these peptides are involved in these processes, it is essential to analyze their biosynthetic pathways as well as their effects on the respective target systems. This understanding is invaluable in situations where organs are malfunctioning. For instance, by recent studies of Alzheimer's disease it was shown that tissues of the hippocampus contained a highly active ribonuclease that is normally under regulatory control. It is thought that in patients with this disease, the out-of-control ribonuclease rapidly degrades information essential in these cells. Based on recombinant-DNA technology, methods are now available to investigate various neurohypophyseal disorders at a molecular level, initially in animals and later in clinical situations.

This chapter focuses on two characteristic hypothalamic neuropeptides, the

THE PEPTIDES, VOLUME 8

hormones vasopressin and oxytocin. Their expression as high-molecular-weight precursors—polyproteins—is discussed, with emphasis on recombinant-DNA techniques as a new approach in molecular neuroendocrinology. The chapter also deals with the structural organization and a comparison of the genes encoding vasopressin and oxytocin. It includes the analysis of a genetic defect known as diabetes insipidus, which is due to the lack of vasopressin. First, some general remarks are made on the structural aspects of polyproteins.

II. STRUCTURAL ASPECTS—POLYPROTEINS

Although many of the neuropeptides studied have been analyzed by classical Edman amino acid sequencing, the elucidation of their biosynthetic pathways has only been achieved with the help of recombinant-DNA methods. Based on these approaches, it has been shown that many—if not all—neuropeptides are initially synthesized as much larger precursors. These precursors may comprise other related or unrelated biologically active peptides and thus in analogy to the similarly constructed viral polyproteins have been termed cellular polyproteins (Koch and Richter, 1980; Herbert, 1981).

The conversion of the polyprotein into the mature peptide(s) is a crucial step in the biosynthetic pathway of neuropeptides. It requires precise signals within the sequence of the polyprotein, as well as specific enzyme(s) recognizing these signals. In general these signals are pairs of basic amino acids that flank the respective peptide and that are recognized by proteolytic enzymes sometimes tissue-specifically expressed (Herbert, 1981; Douglass *et al.*, 1984).

A few examples of polyproteins are illustrated in Fig. 1 and discussed below. They reflect the range of complexity achieved in the composition of the precursors and indicate how diverse the mechanisms must be in order to express these peptides.

Simple forms of polyproteins are represented by the precursors to the releasing factors such as corticotropin-releasing factor (CRF) (Furutani *et al.*, 1983), luteinizing hormone–releasing hormone (LHRH) (Seeburg and Adelman, 1984),

Figure 1. Polyproteins as precursors to neuropeptides. Dotted bars, signal sequence; open bars, "spacer" sequences (bars are not drawn to scale); black bars, the physiological active hormones or neuropeptides; hatched bars, auxiliary sequences known to modify the physiology of the associated peptide hormone. ACTH, adrenocorticotropin; CGRP, calcitonin gene-related peptide; CLIP, corticotropn-like intermediate-lobe peptide; GLP-1, GLP-2, glucagon-like peptide 1 and 2; β-LPH, β-lipotropin; PHM-27, a 27-amino-acid peptide with an N-terminal histidine and a C-terminal methionine; Met*, C-terminally extended methionine enkephalin; MSH, melanotropin-stimulating hormone; TRH, thyrotropin-releasing factor; VIP, vasoactive intestinal peptide. Amino acids occuring as processing signals or as extension of peptides are listed as follows: A, alanine; G, glycine; H, histidine; K, lysine; I, isoleucine; R, arginine; S, serine. For further details, see text.

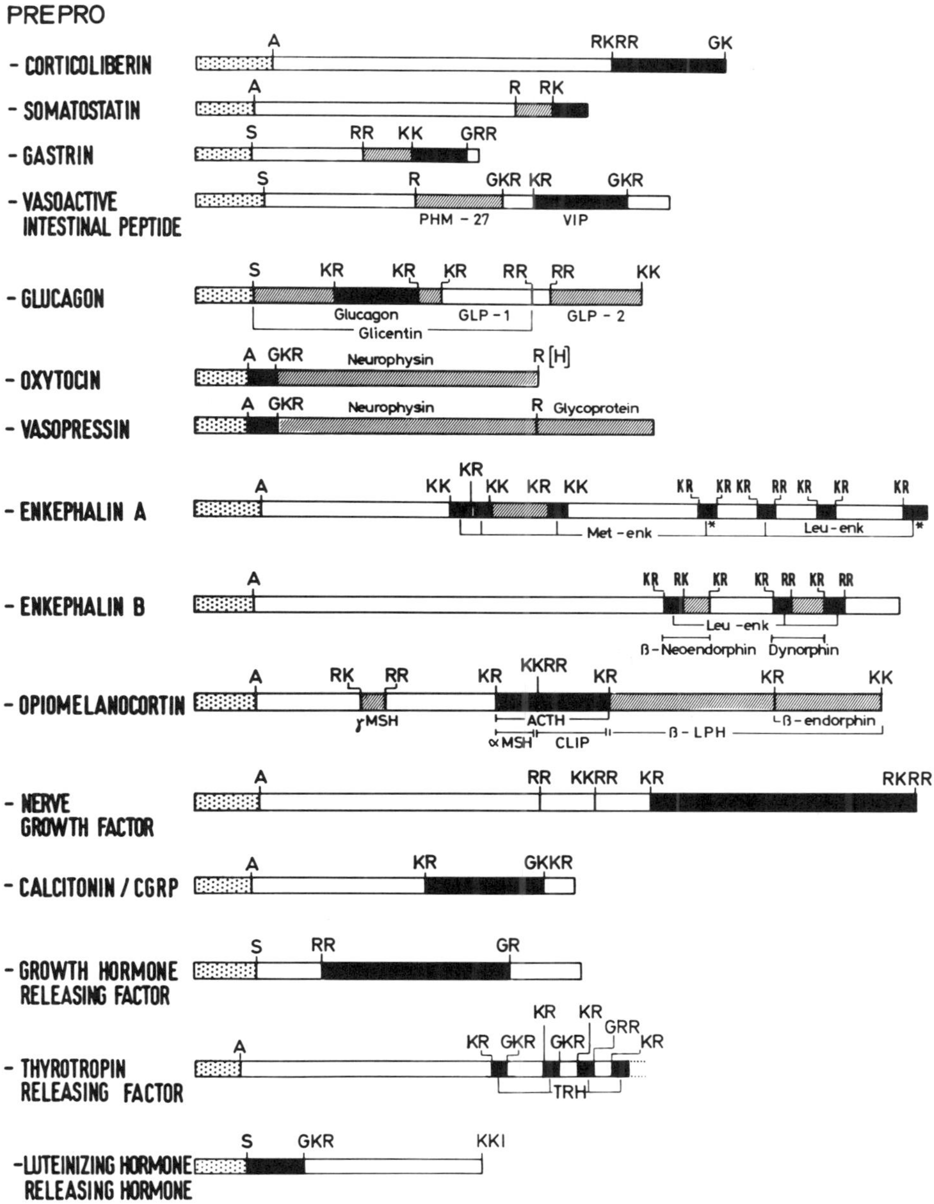

PREPRO
- CORTICOLIBERIN
A
RKRR
GK
- SOMATOSTATIN
A
R
RK
- GASTRIN
S
RR
KK
GRR
- VASOACTIVE INTESTINAL PEPTIDE
S
R
GKR
KR
GKR
PHM - 27
VIP
- GLUCAGON
S
KR
KR
KR
RR
RR
KK
Glucagon
GLP - 1
GLP - 2
Glicentin
- OXYTOCIN
A
GKR
Neurophysin
R [H]
- VASOPRESSIN
A
GKR
Neurophysin
R
Glycoprotein
- ENKEPHALIN A
A
KK
KR
KK
KR
KK
KR
KR
KR
RR
KR
KR
KR
Met - enk
Leu - enk
- ENKEPHALIN B
A
KR
RK
KR
KR
RR
KR
RR
Leu - enk
β - Neoendorphin
Dynorphin
- OPIOMELANOCORTIN
A
RK
RR
KR
KKRR
KR
KR
KK
γ MSH
ACTH
β - endorphin
α MSH
CLIP
β - LPH
- NERVE GROWTH FACTOR
A
RR
KKRR
KR
RKRR
- CALCITONIN / CGRP
A
KR
GKKR
- GROWTH HORMONE RELEASING FACTOR
S
RR
GR
- THYROTROPIN RELEASING FACTOR
A
KR
GKR
KR
GKR
KR
GRR
KR
TRH
- LUTEINIZING HORMONE RELEASING HORMONE
S
GKR
KKI

or thyrotropin-releasing factor (TRH) (Richter *et al.*, 1984). There a single peptide—CRF, LHRH, or TRH—is present within a precursor (Fig. 1). The remaining sequence may have no other function than to serve as "spacer" in the synthesizing and packaging process. Alternatively, the "spacer" may be of physiological relevance, as suggested by the somatostatin "spacer" peptide, which is not quickly hydrolyzed but remains *in situ,* possibly with a specific function (Lechan *et al.*, 1983).

The TRH precursor in addition belongs to a category of polyproteins that contain several copies of identical or closely related units. Repetitive units have been found in the precursors to the two enkephalins (Douglass *et al.*, 1984) or to caerulein (Hoffman *et al.*, 1983), while closely related peptides are present in preproglucagon (Bell *et al.*, 1983). The multiplicity of a peptide hormone within its precursor may have the effect of amplifying the transcriptional control.

More complex polyproteins are exemplified by somatostatin (Taylor *et al.*, 1981; Rutter *et al.*, 1983), gastrin (Yoo *et al.*, 1982; Boel *et al.*, 1983), or vasoactive intestinal peptide (Itoh *et al.*, 1983). In these cases, the hormone can be excised from its precursor in more than one way to give a shorter or longer peptide with different functions.

The multifunctional polyproteins, such as proopiomelanocortin (Douglass *et al.*, 1984), provasopressin (Richter, 1983), or prooxytocin (Richter, 1983), contain different peptides with somewhat related functions that are placed under the same transcriptional and translational control element. The proopiomelanocortin precursor provides an added sophistication, in that different hormones overlap within the precursor and are differentially expressed in different tissues (Douglass *et al.*, 1984).

It is interesting to note that in most cases where the functional peptide is located at the C-terminus of the precursor it is extended by one or more basic amino acids that are absent in the biologically active peptide [e.g., CRF, nerve growth factor (NGF)]. An extra basic amino acid is also found at the C-terminus of the neurophysin that is part of the oxytocin precursor. These basic amino acids apparently have to be removed during posttranslational processing and most likely require specific cleavage enzyme(s). One could speculate that these amino acids are remnants of former C-terminal extensions to these precursors, which had been eliminated during evolution because of lack of function.

The class of precursors represented by the calcitonin (Amara *et al.*, 1982) or substance P (Nawa *et al.*, 1984) polyproteins are particularly remarkable since their genes give rise to pre-RNAs from which alternatively spliced mRNAs are generated in a tissue-specific way. For instance, translation of the mRNA in the thyroid yields the precursor to calcitonin; in the brain, however, it yields calcitonin gene-related protein (CGRP). It has been suggested that the two mRNAs of the calcitonin/CGRP gene result from a tissue-specific utilization of alternative polyadenylation sites regulated by a selective endonuclease activity, although the underlying mechanism remains to be resolved (Evans *et al.*, 1982).

In the case of substance P, a single preprotachykinin gene gives rise to two distinct mRNAs resulting from alternative splicing: one mRNA encodes substance P alone (α-PPT), and the other encodes substance P plus another neuropeptide, the tachykinin substance K (β-PPT). The relative amounts of the two mRNAs vary; α-PPT mRNA is predominantly made in the nervous system, whereas β-PPT mRNA is found preferentially in the thyroid, intestines, and also in the nervous system (Nawa *et al.*, 1984). The different mRNAs of the preprotachykinin gene may be due to tissue-specific regulation of the splicing mechanism—e.g., by alteration in the specificity of the splicing enzymes (Nawa *et al.*, 1984). Clearly, this tissue-specific expression of neuropeptide genes provides a further way to increase the diversity of the neuroendocrine system.

What is the advantage of a cell or organ to express peptides as polyproteins? Doubtless this mode of gene expression adds an extra level of regulation between translation and secretion of the functional peptide. Tissue-specific conversion of the polyprotein offers a subtle control mechanism; without needing to generate new genes, different or related functions can be carried out by a family of peptides using the same gene product. Slightly modified members could then differentially activate different receptor systems.

In the case of the precursors to oxytocin and vasopressin, functionally linked polypeptides (hormone and neurophysin carrier) are coordinately synthesized on the same precursor. The products of proopiomelanocortin are linked for similar functional reasons. An additional benefit that might be gained from such a system is the amplification by using repetitive identical units (TRH, enkephalins).

For the near future, studies of the events leading from the polyprotein to the biologically active peptide will be an intriguing problem. The folding and the three-dimensional structure of the nascent polyproteins that—together with specific amino acids as processing signals—most likely are integral parts in the proteolytic processing events.

III. VASOPRESSIN AND OXYTOCIN—FUNCTIONAL ASPECTS

The biosynthesis of vasopressin and oxytocin has been traced from the level of the genome in the magnocellular neurons of the hypothalamus through to release of the physiological peptides by neurosecretory nerve endings in the posterior pituitary. The biosynthetic pathway for both hormones involves the prior synthesis of composite precursors, which—besides the hormones—also includes the neurophysin proteins (Acher, 1983). The latter are thought to play a carrier or protective role in helping the hormones to reach intact their site of release in the posterior pituitary (Breslow, 1979).

The vasopressin, but not the oxytocin precursor, contains additionally a glycopeptide moiety of 39 amino acids of unknown function at its C-terminus. The two neurophysins—one associated with vasopressin, the other with oxytocin—consist of 93–95 amino acids with a central part more or less identical for all neurophysins sequenced so far; sequence variations occur at the N- and C-terminus of the carrier proteins.

As hormones, vasopressin controls water resorption in the distal kidney tubuli and oxytocin the contraction of the uterus during birth and milk ejection. Vasopressin and oxytocin—besides their regulatory role as hormones—appear to be involved in many other processes such as learning and memory; responses to tolerance development and physical dependence on alcohol, opiate, or heroin; addiction; rewarded behavior; cardiovascular regulation; control of body temparature; and brain development (De Wied, 1983). Since vasopressin and oxytocin are expressed in magnocellular neurons of various hypothalamic areas, it will be interesting to find out how the numerous physiological conditions listed above affect the expression of the two genes: in particular, whether they are expressed differently depending on the anatomical location of the neuronal cells and their projection to other parts of the brain (Burbach *et al.*, 1984; Zerbe and Palkovits, 1984).

IV. GENE STRUCTURE

To date, the oxytocin and vasopressin genes have been determined from the rat (Schmale *et al.*, 1983; Ivell and Richter, 1984b) and cow (Ruppert *et al.*, 1984). Between as well as within the two species, the vasopressin and oxytocin genes are highly conserved, confirming their derivation from a common ancestral gene, probably early in vertebrate evolution. Each gene comprises three exons separated by two intervening sequences (Fig. 2). Each exon encodes one of the principal functional domains of the polyprotein—hormone, carrier protein, glycopeptide. In the case of the oxytocin gene, the third exon comprises only the C-terminal variable part of the neurophysin plus an extra basic amino acid.

The first exon encodes the signal peptide, the hormone, either vasopressin or oxytocin, and the first few variable amino acids of the corresponding neurophysin molecule.

The second exon encodes the central 67 amino acids of the neurophysin carrier. The DNA for this highly conserved sequence is so closely homologous

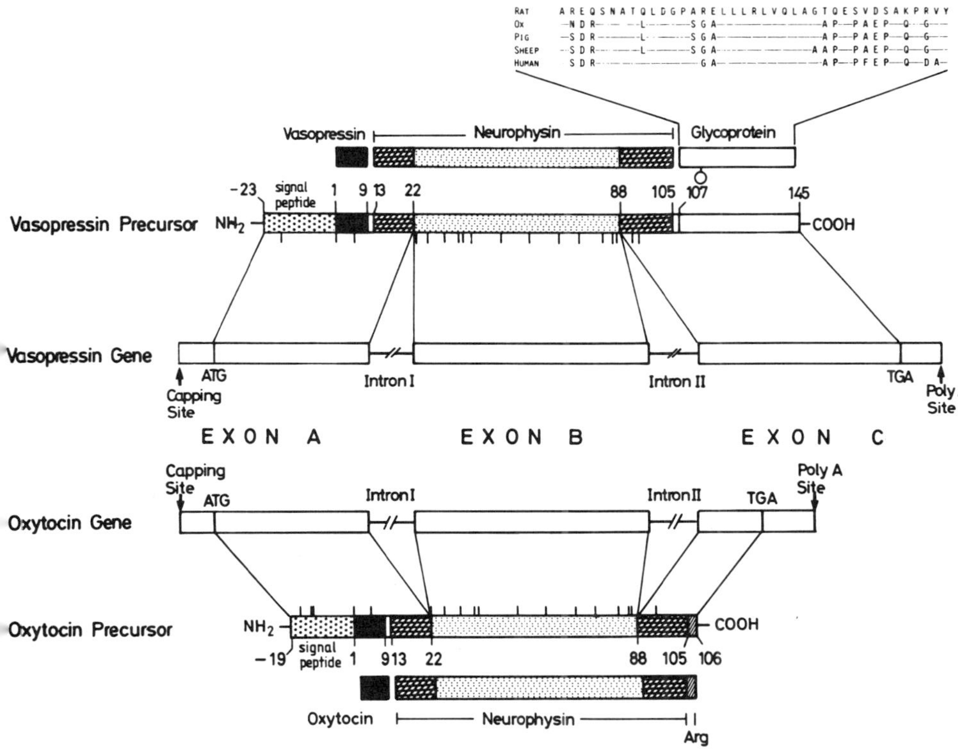

Figure 2. Comparison of the oxytocin and vasopressin genes from the rat and their precursor products. Protein structure: dotted bars, putative signal sequences; black bars, peptide hormones; hatched and dark hatched bars, neurophysins (light hatching shows the conserved regions); open bar, glycopeptide of the vasopressin precursor; ○, carbohydrate chain; significant amino acids are numbered, such that negative numbers refer to signal sequences and positive ones to the prohormone sequences; Arg, arginine residue present at the C-terminus of the oxytocin precursor; cysteine residues are indicated by vertical lines. The insert shows the amino acid sequences of glycopeptides from various species. Gene structure: the exons are shown as open bars; capping and polyadenylation sites are indicated by arrows.

between the oxytocin and vasopressin genes that a "gene conversion" (Baltimore, 1981) must be invoked as explanation. Although still poorly understood, gene conversion appears to act via mutual recombination or correction between closely related gene sequences on the same or different chromosomes, so as to ensure conservation of an important DNA sequence. When counting the silent substitutions occuring in the rat and calf genes as the result of the mutation rate, gene conversion took place in the calf 1 Myr (10^6 years) ago and in the rat 10 Myr ago (Ruppert *et al.*, 1984).

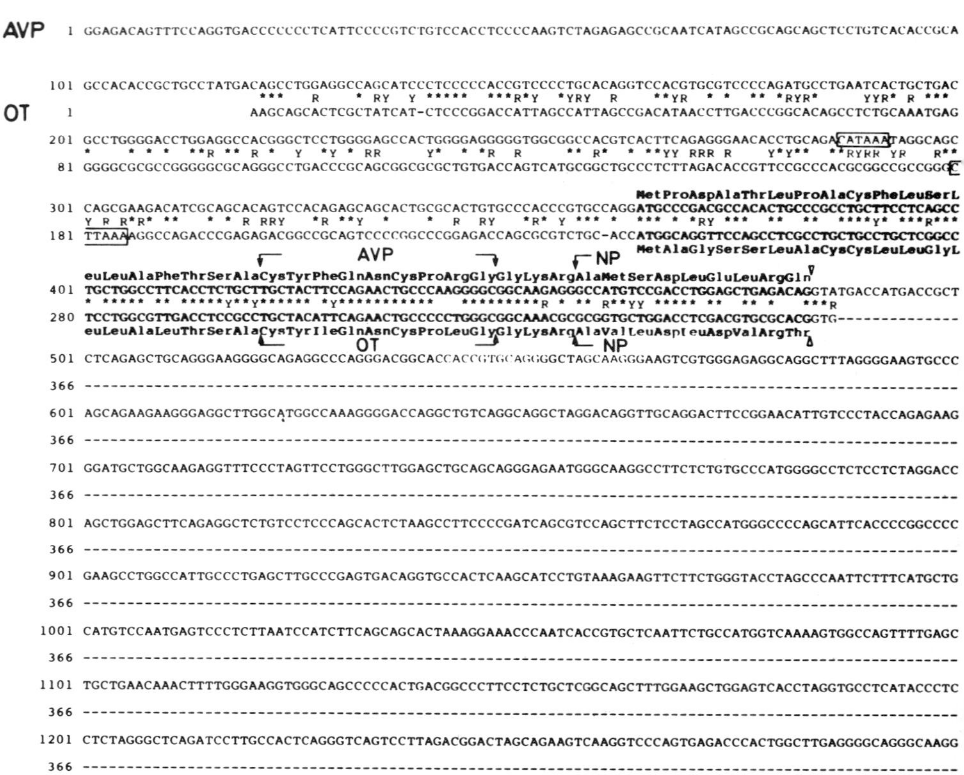

Figure 3. Comparison of the bovine genes for arginine vasopressin (AVP) and oxytocin (OT). Boxed areas, modified TATA box; open triangles, exon–intron junctions; underlined sequences, polyadenylation signal sequence: AVP, arginine vasopressin; OT, oxytocin; NP, neurophysin; GP, glycoprotein. Identical nucleotides in both sequences are shown by asterisk, purine changes by R, pyrimidine changes by Y; dashed lines indicate computer-induced spaces to get better nucleotide fitting for other parts of the two genes.

The third exon includes the variable C-terminus of the neurophysins, and in the case of the vasopressin gene a glycopeptide. The oxytocin gene predicts an extra basic amino acid (His or Arg) following immediately after the known neurophysin sequence. This basic amino acid corresponds precisely with the cleavage signal separating the neurophysin and glycopeptide in the vasopressin precursor. Whether the ancestral oxytocin gene also possessed a C-terminal peptide and whether this part was discarded following gene duplication, leaving the terminal basic amino acid in this precursor as a rudiment of the ancestral cleavage signal, remains speculation.

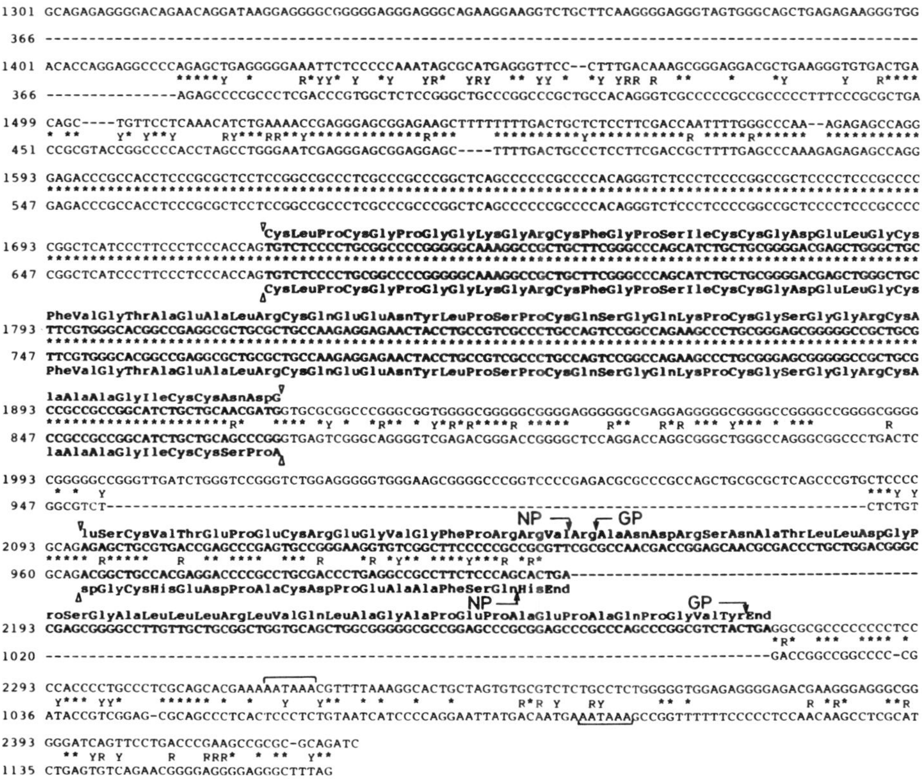

Figure 3 *(Continued)*

V. COMPARISON OF THE BOVINE AND RAT GENES

The structural organization of the vasopressin and oxytocin genes are in line with other eukaryotic split genes and include the consensus sequences at exon–intron junctions, a modified ''Goldberg–Hogness'' sequence, CATAAA located 29 nucleotides upstream of the presumptive transcription start site, an A residue bounded by a pyrimidine-rich region at the transcription start site, and a poly-adenylation site AATAAA at the 3′ end of the sequence (Figs. 3 and 4). Noteworthy is the strongly modified TATA box of the bovine oxytocin gene, which reads CTTAAA.

The 5′ untranslated region that precedes the protein-coding region shows little homology between the vasopressin and oxytocin genes in each species (Figs. 5

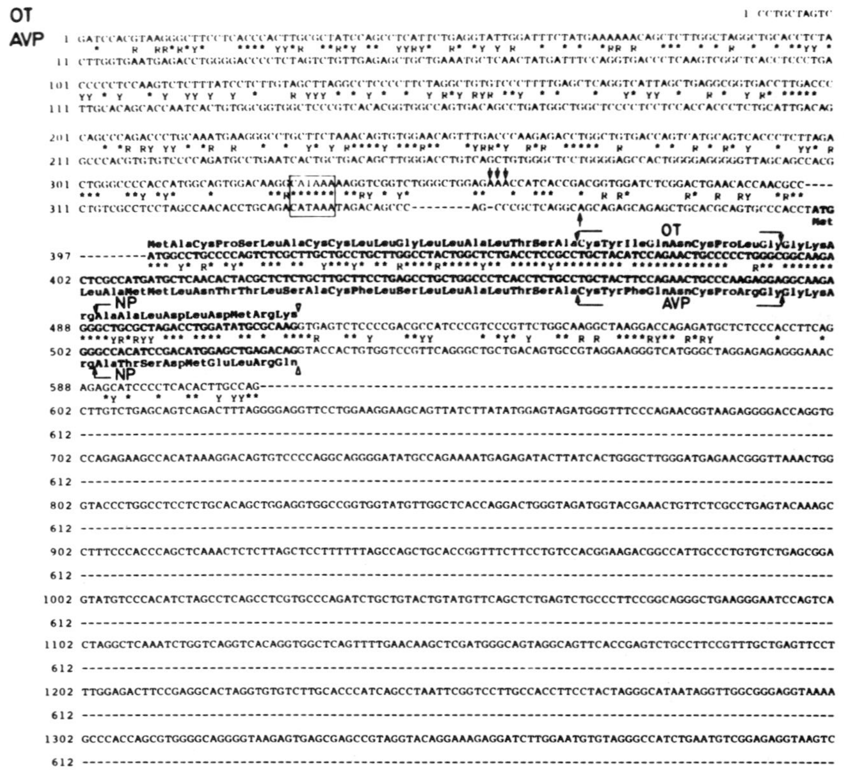

Figure 4. Comparison of the rat genes for arginine vasopressin (AVP) and oxytocin (OT) (see also legend to Fig. 3).

and 6); since this region is thought to be important for the regulation of transcription, it would emphasize the independent regulatory control of the vasopressin and oxytocin genes.

However, the sequence homology in the 5′ noncoding region is remarkably high when comparing the vasopressin gene from the cow and rat. For instance, within a stretch of 128 bases preceding directly the modified TATA box (CATAAA) of the bovine vasopressin gene, 118 bp (base pairs) are identical with the corresponding rat gene sequence (except for eight pyrimidine and seven purine exchanges). This extraordinary conservation may point to an important promotor region for the control of the vasopressin gene in both species (Fig. 5). The conservation is not as quite as high when comparing the respective promotor region of the oxytocin gene from both species (Fig. 6).

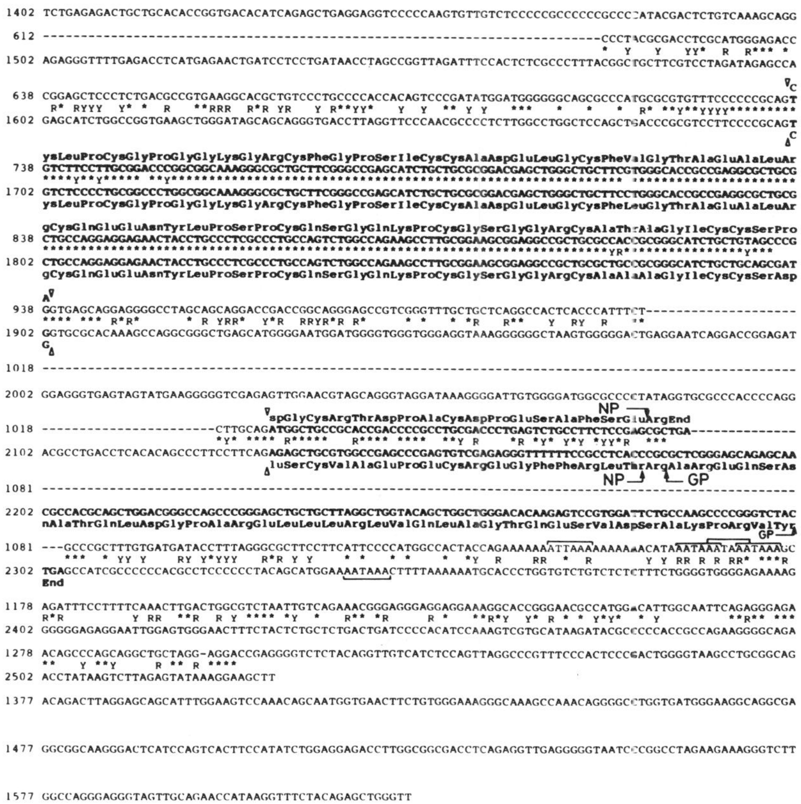

Figure 4 *(Continued)*

When comparing the protein-coding region of the two genes from the rat and cow, the homologies in the sequence for the second exon are particularly striking. While the vasopressin and oxytocin gene from the rat are entirely homologous within exon B for a stretch of 143 nucleotides with one base difference (Fig. 3), the homology is even more striking for the respective bovine sequences (Fig. 4). There the homology extends into the preceding intron, covering 332 bp without any interruption. As already discussed, a recent gene-conversion event may explain this rather unusually high degree of sequence homology.

No significant sequence homology has been observed in the intervening sequences between the two genes of either species. Again, some homologies can be found when comparing the 3′ untranslated regions where the sequences around the polyadenylation site(s) are highly conserved for the bovine and rat

Figure 5. Comparison of the bovine (B) and rat (R) vasopressin gene (see also legend to Fig. 3).

vasopressin gene. Remarkable again is that the rat oxytocin gene contains more than one potential polyadenylation signal, possibly giving rise to mRNAs with differently extended 3′ ends. Whether alternative polyadenylation signals (Heidecker and Messing, 1983) play a role in the regulation of the expression of the two genes has not yet been resolved.

VI. TRANSCRIPTIONAL CONTROL

With DNA probes specific for either hormone precursor, it is now possible to study their regulation at the transcriptional level. Two methods have been

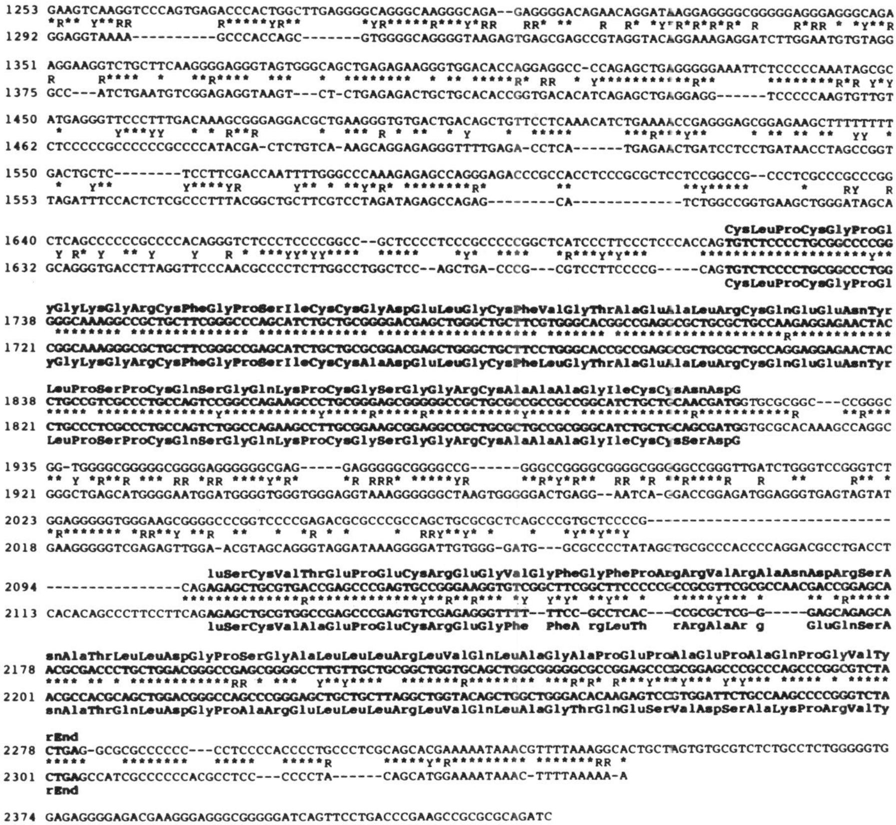

Figure 5 *(Continued)*

adopted to quantify the mRNA encoding the vasopressin precursor; one is based on liquid or blot hybridization assays, and the other makes use of thin-layer sections of the hypothalamus that are then *in situ* hybridized to specific labeled DNA probes.

In the following some examples are given for elevated mRNA levels under different physiological conditions. One should be cautious, however, with the interpretation of the data. Although the elevated mRNA levels found in one hypothalamic area compared to others may reflect increased transcription rate, it is too premature to conclude that a gene is specifically switched on. Differences in mRNA levels under given physiological conditions, for instance, may be due just to different RNase activities.

```
R    1 GATCCACGTAAGGGCTTCCTCACCCACTTGCGC

R   34 TATCCAGCCTCATTCTGAGGTATTGGATTTCTATGAAAAAACAGCTCTTGGCTAGGCTGCACCTCTACCCCCTCCAAGTCTCTTTATCCTCTTGTAGCTT
B    1 AAGCAGCA----CTC-GCTATCATCTCCCGGACCATTAGCCATTAGCCGAC---ATAACCTTGACCCGGCACAGC-CTCTGCAAATGAGGGG------GC
       *R**   *      **  ***R*  Y*YYYY ** * Y R*Y*******Y**    R*R********** R  * R  *Y*********R***      *
R  134 AGGCCTCCCCTTCTAGGCTGTGTCCCTTTTGAGCTCAGGTCATTAGCTGAGGCGGTGACCTTGACCCCAGCCCAGACCCTGCAAATGAAGGGCCTGCTTC

    86 GCGCCGGGGGCGCAGGGCCTGACCCGCAGCGGCGCGCTGTGACCAGTCATGCGGCTGCCCTCTTAGACACCGTTCCGCCCACGCGGCCGCCGGGCTT--A
         R *R* *   * * R*YY******R RR R *  ****************R*YYR***********   *YY**R**   **R*Y * *RR* Y   *
   234 TAAACAGTGTGGAACAGTTTGACCCAAGAGACCTGGCTGTGACCAGTCATGCAGTCACCCTCTTAGACTGGGCCCCACCATGGCAGTGGACAAGGCATAA

                                                                             MetAlaGlySerSerLeuAlaCysCysLeuLeuGly
   184 AAAGGCCAG-----ACCCGAGAGACGGCCGCAGTCCCCGGCCCGGAGACCAGCGCGTCTGCACCATGGCAGGTTCCAGCCTCGCCTGCTGCCTGCTCGGC
       *****Y*R*     R*Y ****R** RY*R* * *  Y** Y*  R***YRR    Y  R*R*******  *YY****Y*****Y***********Y***
   334 AAAGGTCGGTCTGGGCTGGAGAAACCATCACCGACGGTGGATCTCGGACTGAACACCA-ACGCCATGGCCTGCCCCAGTCTCGCTTGCTGCCTGCTTGGC
                                                                             MetAlaCysProSerLeuAlaCysCysLeuLeuGly

       LeuLeuAlaLeuThrSerAlaCysTyrIleGlnAsnCysProLeuGlyGlyLysArgAlaValLeuAspLeuAspValArgThr
   279 CTCCTGGCGTTGACCTCCGCCTGCTACATTCAGAACTGCCCCCTGGGCGGCAAACGCGCGGTGCTGGACCTCGACGTGCGCACGGTGAG-AGCCCCGCCC
       ** ***** Y******************Y************************R * ** *Y***R***** **YR****** ******   ***** *
   433 CTACTGGCTCTGACCTCCGCCTGCTACATCCAGAACTGCCCCCTGGGCGGCAAGAGGGCTGCGCTAGACCTGGATATGCGCAAGGTGAGTCTCCCCGACG
       LeuLeuAlaLeuThrSerAlaCysTyrIleGlnAsnCysProLeuGlyGlyLysArgAlaAlaLeuAspLeuAspMetArgLys

   378 TCGACCCGTGGCTCTCCGGGCTGCCCGGCCCGCTGCCACAGGGTCGCCCCCGCCGCCCCCTTTCCCGCGCTGACCGCGTACCGGCCCCACCTAGCCTGGG
       Y*R *****  * Y**Y** * R   ***Y R  R*** **R  * *Y***R** Y*      * Y * ** **  Y *R**R****Y**   *
   533 CCATCCCGTCCCGTTCTGG-CAA---GGCTAAGGACCAGAGATGCTCTCCCACCTTCAGAGAGCATCCCCTCACACT-TGCCAGCCCTAC---G------

   478 AATCGAGGGAGCGGAGGAGCTTTTGACTGCCCTCCTTCCACCGCTTTTGAGCCCAAAGAGAGAGCCAGGGAGACCCGCCACCTCCCGCGCTCCTCCGGCC
          ***                      ****    **Y                          **********  *       ** YY****Y*R*
   619 ---CGA-----------------------CCTC---GCAT-------------------------GGGAGACCCG-GA------GC-TCCCTCTGAC-
                                                                                                  CysLeu
   578 GCCCTCGCCCGCCCGGCTCAGCCCCCCGCCCCACAGGGTCTCCCTCCCCGGCCGCTCCCCTCCCGCCCCCGGCTCATCCCTTCCCTCCCACCAGTGTCTC
       *** * *    ** * ***  * Y***Y*******    R    ****   Y Y** Y*      *   *****   ** *R*   **Y**Y***R ********Y
   654 GCCGT-GAAGGCAC-GCT--G-TCCCTGCCCCAC-CACAGTCCCGATATGGATGGGGGGCAG-CGCCC-ATGCGCGTG--TTTCCCCCCG-CAGTGTCTT
                                                                                                  CysLeu

       ProCysGlyProGlyGlyLysGlyArgCysPheGlyProSerIleCysCysGlyAspGluLeuGlyCysPheTrpGlyThrAlaGluAlaLeuArgCysG
   678 CCCTGCGGCCCCGGGGGCAAAGGCCGCTGCTTCGGGCCCAGCATCTGCTGCGGGGACGAGCTGGGCTGCTTCTGGGCACGGCCGAGGCGCTGCGCTGCC
       **Y***** ***** ******** ************** ************ ****************** ****** ******************
   743 CCTTGCGGACCCGGCGGCAAAGGGCGCTGCTTCGGGCCGAGCATCTGCTGCGCGGACGAGCTGGGCTGCTTCGTGGGCACCGCCGAGGCGCTGCGCTGCC
       ProCysGlyProGlyGlyLysGlyArgCysPheGlyProSerIleCysCysAlaAspGluLeuGlyCysPheValGlyThrAlaGluAlaLeuArgCysG

       lnGluGluAsnTyrLeuProSerProCysGlnSerGlyGlnLysProCysGlySerGlyGlyArgCysAlaAlaAlaGlyIleCysCysSerProA
   778 AAGAGGAGAACTACCTGCCGTCGCCCTGCCAGTCCGGCCAGAAGCCCTGCGGGAGCGGGGGCCGCTGCGCCGCCGCCGGCATCTGCTGCAGCCCGGGTGA
       *R***************** *************Y***********Y*****R*****R************R**** *********Y***********
   843 AGGAGGAGAACTACCTGCCCTCGCCCTGCCAGTCTGGCCAGAAGCCTTGCGGAAGCGGAGGCCGCTGCGCCACCGCGGCATCTGCTGTAGCCCGGGTGA
       lnGluGluAsnTyrLeuProSerProCysGlnSerGlyGlnLysProCysGlySerGlyGlyArgCysAlaThrAlaGlyIleCysCysSerProA

                                                                                             spGlyCysHisG
   878 GTCGGGCAGGGGTCGA--GACGGGACCGGGGCTCCAGGACCAGGCGGGCTGGGCCAGGGCGGCCCTGACTCGGCGTCTCTCTGTGCAGACGGCTGCCACG
       *Y **  *****Y* *  RR*R******R    * R*** * * ****Y*  **YR     ****     * *  *R*Y***** ******Y*******R*R
   943 GCAGG--AGGGGCCTAGCAGCAGGACCGACCGGCAGGGAGCCGTCGGGTT-TGCTGCTCAGGCC-ACTCACC-CATTTCTCT-TGCAGATGGCTGCCGCA
                                                                                             spGlyCysArgT

       luAspProAlaCysAspProGluAlaAlaPheSerGlnHisEnd
   976 AGGACCCCGCCTGCGACCCTGAGGCCGCCTTCTCCCAGCACTGAGACCGGCCGGCCCCCGATACCGTCGGAGCGCAGCCCTCATCCCCT--CTGTAATCA
         ********************** *Y********* ***R***** *** YY *Y   Y****** *YR*R****  **Y****Y***Y     Y* *
  1037 CCGACCCCGCCTGCGACCCTGAGTCTGCCTTCTCCGAGCGCTGAGCCCGCTTTGTGA-TGATACCTTTAGGGCGCTTCCTTCATTCCCCATGGCCACTAC
       hrAspProAlaCysAspProGluSerAlaPheSerGluArgEnd

  1074 TCCCCAGGAATTATGACAATGAAATAAA--------GCCGGTTTTTCCCCCTCCAA-CAAGCCTCGCATCTGAGTGTCA-GAACGGGGAGGGGAGGGCT
       Y    *RR***** R* ** R* *****             R*Y    **Y**YY*Y*** *  * ** **R***R* ***** R******RR**R*RR**
  1136 CAGAAAAAAATTAAAAAAAAAACATAAATAAATAAATAAAGCAGATTTCCTTTTCAAACTTGACTGGCGTCTAATTGTCAGAAACGGAGGGAGGAGG--

  1164 TTAG
         **
  1234 AAAGGCACCGGGAACGCCATGGACATTGGCAATTCAGAGGGAGAACAGCCCAGCAGGCTGCTAGGAGGACCGAGGGGTCTCTACAGGTTGTCATCTCCAG

  1334 TTAGGCCCGTTTCCCACTCCCGACTGGGGTAAGCCTGCGGCAGACAGACTTAGGAGCAGCATTTGGAAGTCCAAACAGCAATGGTGAACTTCTGTGGGAA

  1434 AGGGCAAAGCCAAACAGGGGCCTGGTGATGGGAAGGCAGGCGAGGCGGCAAGGGACTCATCCAGTCACTTCCATATCTGGAGGAGACCTTGGCGGCGACC

  1534 TCAGAGGTTGAGGGGGTAATCCCGGCCTAGAAGAAAGGGTCTTGGCCAGGGAGGGTAGTTGCAGAACCATAAGGTTTCTACAGAGCTGGGTT
```

Figure 6. Comparison of the bovine (B) and rat (R) oxytocine gene (see also legend to Fig. 3).

A. Osmotic Stress and Vasopressin-Encoding mRNA Levels

Rats placed under osmotic stress by drinking 2% saline respond with an elevated plasma osmolality. To maintain the water balance, the stores of vasopressin in the posterior pituitary are depleted while the hormone level in the plasma rises (Robertson, 1977). When studying the vasopressin levels in different hypothalamic areas, increased hormone levels are found in the supraoptic nucleus (SON) and paraventricular nucleus (PVN), not in the suprachiasmatic nucleus (SCN) (Zerbe and Palkovits, 1984).

These data coincide with the finding that rats exposed to saline show a significant increase of mRNA encoding vasopressin in the SON and PVN but not in the SCN (Burbach et al., 1984). The increase is fivefold in the SON and twofold in the PVN (Table I). It is known that vasopressin-producing neurons of the SON and PVN project toward the posterior pituitary, and those of the SCN toward other brain areas. Hence, it has been speculated that these data point to a different regulation in the expression of the vasopressin gene in SON and PVN versus SCN (Burbach *et al.*, 1984), relating to peripheral and central functions of vasopressin, respectively.

Whether these data reflect indeed differential regulation of the vasopressin gene must be examined more rigorously—for instance, by transcription run-on experiments measuring the levels of nascent vasopressin mRNA in the cell nucleus. Also other physiological parameters are needed that may affect specifically the vasopressin mRNA level in the SCN but not in the other areas.

A preliminary estimate of the vasopressin mRNA level under osmotic stress conditions has been made by using total rat hypothalami and Northern blot analysis (Majzoub *et al.*, 1983). In these experiments, a 20-fold increase of the

Table I. Levels of Rat Hypothalamic mRNA Specific for the Vasopressin Precursor in Response to Osmotic Stress[a]

Region	Control	Osmotic stress
Supraoptic nucleus	21.4 ±5.6	97.9 ±44.7
Paraventricular nucleus	16.4 ±2.5	35.1 ±2.1
Suprachiasmatic nucleus	19.8 ±5.6	17.0 ±4.5

[a]Rats were induced to osmotic stress by drinking of 2% saline for 2 weeks. mRNA levels are expressed as total pg vasopressin mRNA per microdissected region + SEM. [From Burbach *et al.* (1984), with permission.]

vasopressin encoding mRNA is observed. A number of reasons may account for this rather high increase, e.g., the rat strains used may respond more dramatically to osmotic stress than others.

B. Luteal Cycle and Oxytocin-Encoding mRNA Levels

A quantitative determination has been done for the oxytocin-encoding mRNA of the bovine corpus luteum (Ivell and Richter, 1984a; Ivell *et al.*, 1985). Here it would appear that transcription of the oxytocin gene is maximal accompanying ovulation and decreases thereafter. Compared to the hypothalamus, the corpus luteum can produce up to 250 times more oxytocin mRNA per organ (Fig. 7). Again, it is unknown whether the higher levels of oxytocin mRNA are due to higher transcription rate and/or different degradation of the mRNA in the respective tissues.

C. *In Situ* Hybridization

This method combines a number of advantages. It allows not only the cells expressing the respective gene to be identified but also the mRNA in these cells

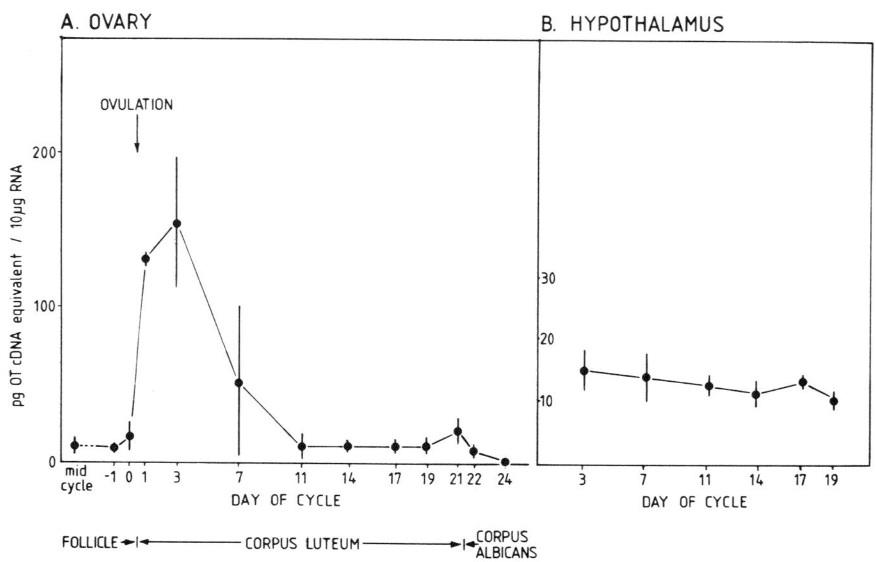

Figure 7. Follicular luteal (A) and hypothalamic (B) levels of oxytocin mRNA through the oestrous cycle of the nonpregnant cow (Ivell *et al,* 1985).

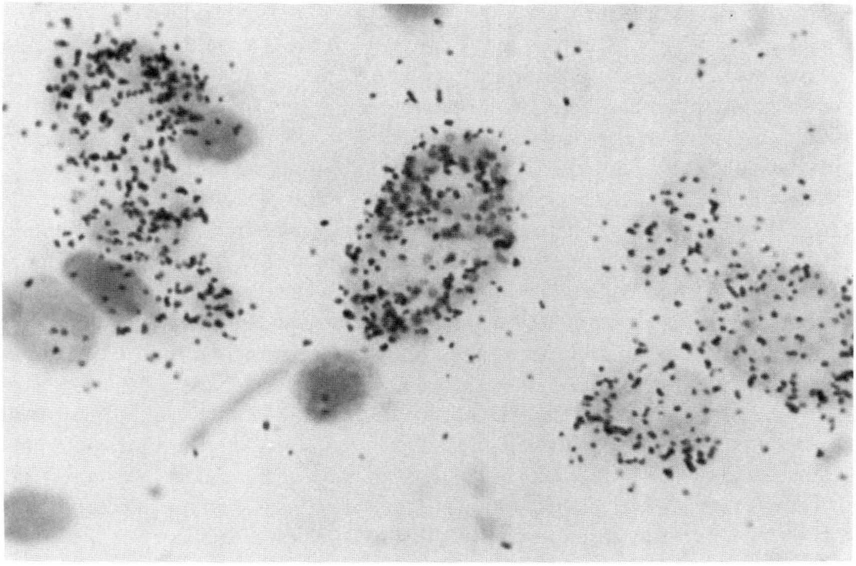

Figure 8. *In situ* hybridization of a ^{3}H-labeled cDNA probe specific for the rat oxytocin precursor to a thin section of the supraoptic nucleus from a rat hypothalamus. Sections were exposed for 3 weeks.

to be quantified, as well as a study of the effect of endogenously induced factors on the anatomy and number of cells expressing the respective gene.

The expression of the rat oxytocin gene in the SON is demonstrated in Fig. 8 (McCabe *et al.*, 1986). It shows that some neurons are specifically decorated with silver grains, whereas others are not, suggesting that the former are oxytocin- the latter vasopressin-producing cells. Thin sections obtained from rats where osmotic stress is induced by drinking 2% saline show a significant increase in the grain numbers specific for vasopressin-encoding mRNA; under these conditions, the oxytocin-producing cells appear to show no change.

VII. PRECURSOR STRUCTURE

According to the nucleotide sequence analysis, the bovine preprovasopressin precursor consists of 166 amino acid residues with a molecular weight of 17,310; the rat counterpart that could be predicted from its gene sequence was almost similar in size, with 168 amino acid residues and a molecular weight of 17,826. The deduced bovine and rat preprooxytocin precursors both contain 125 amino acid residues with a molecular weight of 12,826 (bovine).

In general, the sequence of the vasopressin- and oxytocin-associated neurophysins predicted from the cDNA agrees with the one obtained by convential Edman sequence determination, with one exception (Acher, 1983). The DNA sequence predicts that the rat neurophysin of the oxytocin precursor contains a glutamic acid instead of a glutamine residue in position 92—a difference that may be due to a microheterogeneity between various rat strains.

As with other hormone precursors, those for vasopressin and oxytocin include a signal peptide that is separated from the pro-hormone by a small neutral amino acid, alanine. Except for the rat vasopressin precursor, the putative signal sequences consist of 19 amino acids each. Only the gene sequence of the rat vasopressin precursor predicts a longer signal sequence, with 23 amino acid residues and three possible translation start sites (Schmale *et al.*, 1983).

The rule that in most cases an A is found three bases upstream and a G residue four bases downstream of the AUG translation start codon (Kozak, 1981) would predict that the rat vasopressin preprohormone starts at the third methionine residue with an 19-amino-acid-long signal peptide (Table II). However, as indicated in Table II, there are exceptions to this rule. The rat oxytocin precursor has a G and a U residue in the respective positions. Microsequencing of the protein precursors should solve this problem.

Inspection of the amino acid sequence of the two precursors shows that the peptide units within one polyprotein are separated either by pairs (between hormone and carrier neurophysin) or by single basic amino acid residues (between carrier and glycopeptide). This may suggest the existence of different sets of processing enzymes discriminating between the two cleavage sites (see also below). While the paired basic residues represent the frequent form of cleavage signals, the one comprising a single residue is also found in a few cases [e.g., precursors to growth-hormone release factor (Gubler *et al.*, 1983) or to somatostatin (Rutter *et al.*, 1983)].

The nucleotide sequence encoding the calf and rat oxytocin precursors predicts that there is a supernumerary basic amino acid at the carboxy-terminus that is not

Table II. Nucleotide Sequence of the Translation Start Sites of the Vasopressin and Oxytocin Precursors from Calf and Rat[a]

			−23 Met	Leu	Ala	Met	−19 Met		
AVP	RAT	ACCT	ATG	CTC	GCC	**A**TG	ATG	CTC	**A**AC
AVP	CALF					**A**GG	ATG	CCC	**G**AC
OT	RAT					**G**CC	ATG	GCC	**T**GC
OT	CALF					**A**CC	ATG	GCA	**G**GT

[a]For the rat AVP precursor, the first five putative amino acids are shown (−23 to −19). AVP, arginine vasopressin; OT, oxytocin.

found in any of the isolated neurophysins and that is either an arginine [rat: Ivell and Richter (1984b)] or histidine [calf: Land *et al.* (1983)] residue. As outlined above in the rat, this extra arginine residue is found at the same position as the single arginine separating neurophysin and the glycopeptide in the vasopressin precursor. Possibly the extra amino acid is removed by posttranslational processing, predicting an exopeptidase that accepts arginine or histidine as signal.

The rat and bovine vasopressin precursors (Figs. 2–4) include a C-terminal glycopeptide with a glycosylation site Asn-Ala-Thr (Schmale *et al.*, 1983; Land *et al.*, 1982, 1983). Comparing the glycopeptide sequence from calf with those from rat, pig, and human showed remarkable homologies. The glycosylation site and the leucine-rich center part are well conserved in all species so far known. The consecutive leucine residues may represent alternative processing signals for converting the glycopeptide into subfractions (Smyth and Massey, 1979). The glycopeptide, initially isolated from posterior pituitary extracts (Holwerda, 1972), has been localized in vasopressin-producing magnocellular neurons (Watson *et al.*, 1982), yet its biological function remains to be solved.

VIII. PROCESSING AND MODIFICATIONS

A. Kinetic Studies: Transport and Release

A still-unresolved problem is the processing of the precursors and the order of the processing steps that lead to the mature peptides. In general, when precursor synthesis has been accomplished in the rough endoplasmic reticulum, the protein is transferred to the Golgi apparatus. There it becomes concentrated, and formation of neurosecretory granules begins with the precursor undergoing a number of maturation steps. In the case of the vasopressin and oxytocin precursors, there is ample time for these steps during their transport along the axons of the neuronal cells. Early kinetic studies (Sachs and Takabatake, 1964) indicate that in the dog this transport requires about 1.5 hr from the time of synthesis to release of the physiological nonapeptide.

Pulse chase experiments *in vivo* in the rat (Gainer *et al.*, 1977) also show that cleavage of the polyproteins occurs continously during axonal transport, with both cleaved and uncleaved products appearing in the neurohypophysis after a few hours. This is in line with immunohistochemical studies using vasopressin antisera or oxytocin antisera, which specifically require the C-terminal amide for recognition and have located substantial quantities of the free hormones in hypothalamic cell bodies. High-performance liquid chromatography (HPLC) analysis of hypothalamic nuclear extracts also shows both hormones and their respective neurophysins in the proximal regions of the neurones (Swann *et al.*, 1982).

On the other hand, the uncleaved prohormones can be located in neurosecretory granules of the posterior pituitary (Russell *et al.*, 1980). Apparently, processing of the two precursors occurs throughout axonal transport between the cell body and the nerve terminals in the neurosecretory granules, probably at neutral or slightly acidic pH.

The two hormones are normally released upon electrical stimulation of the respective neurons. This has been well documented for oxytocin production during lactation when the pulsatile release of the hormone can be correlated with electrical activity in the supraoptic nucleus (Moos *et al.*, 1982).

It is in the nature of the neurosecretory response that only a percentage of the synthesized peptides are ever released. What occurs to the unreleased peptides is not clearly understood. It has been reported that vasopressin-containing granules undergo an aging process, changing their physical properties, moving away from the nerve endings and accumulating in axonal swellings [Hering bodies (Nordmann and Labonesse, 1981)]. The contents of these aged granules are no longer available for release, possibly undergo proteolytic degradation, and may become targets for the lysosomal system.

B. Intermediate Forms

Little information is yet available on intermediate forms of the vasopressin or oxytocin precursors. A glycopeptide has been isolated by concanavalin A Sepharose chromatography from neurosecretory granules, which contains a neurophysin but not a vasopressin antigenic moiety (North *et al.*, 1983). A similar intermediate has been detected in the guinea pig posterior pituitary (Robinson and Jones, 1984), implying that in the neurohypophysis the first processing step occurs between vasopressin and the neurophysin.

In a heterologous system, such as the oocytes of *Xenopus laevis* that have been programmed with hypothalamic mRNA by microinjection, the cleavage of the vasopressin precursor occurs between neurophysin and the glycopeptide at the single arginine residue (Richter, 1983). The second cleavage step between hormone and neurophysin cannot be observed. This would imply that the processing enzymes of the oocyte recognize a single arginine but not a pair of basic amino acids, suggesting that there are at least two different processing systems in the cell, one acting on pairs and the other on single basic amino acids.

C. Signals for Processing and Modifications

A number of signals can be recognized within the vasopressin and oxytocin precursors that serve either for the proteolytic cleavage process (single or pairs of basic amino acids), for glycoslyation (Asn-X-Thr), or for amidation (Gly). The significance of these signals for the processing event is stressed by the finding that the nucleotide sequence encoding glycine, lysine, and arginine—the three amino acids that separate the hormones from their respective carrier neurophysins—are highly conserved between the two genes.

Conversion of the two precursors into the active hormones must include their amidation. Glycine has been found to serve as a signal in the amidation reaction of many oligopeptides (Richter, 1983). The finding of an extra glycine residue adjacent to the carboxy-terminus of either vasopressin or oxytocin is therefore consistent with other precursors to amidated oligopeptides (Table III). In this reaction, glycine serves as nitrogen donor, leaving glyoxylate to be released as the other product. An enzyme catalyzing the amidation reaction has been isolated from soluble fractions of secretory granules of the rat pituitary (Bradbury *et al.*, 1982; Eipper *et al.*, 1983; Mains *et al.*, 1984). It requires a glycine residue with a free carboxyl group, suggesting that removal of the basic amino acids is a precondition before the amidation reaction can take place. Ascorbic acid stimulates, whereas divalent metal chelators such as EDTA inhibit, the amidation of model peptides (Mains *et al.*, 1984).

A significant difference between the vasopressin and oxytocin precursors is that the former becomes N-glycosylated, presumably at the Asn-X-Thr site with-

Table III. Partial Amino Acid Sequences of Precursors to Oligopeptides Amidated at the Carboxy-Terminus[a]

Melittin	-Lys -Arg -Gln -Gln- **Gly**-COOH
α-MSH	-Gly -Lys -Pro -Val- **Gly**-Lys-Lys-
γ-MSH	-Trp -Asp -Arg -Phe- **Gly**-Arg-Arg-
Vasopressin	-Cys -Pro -Arg -Gly- **Gly**-Lys-Arg-
Oxytocin	-Cys -Pro -Leu -Gly- **Gly**-Lys-Arg-
Caerulein	-Trp -Met -Asp -Phe- **Gly**-Arg-Arg-
Gastrin	-Trp -Met -Asp -Phe- **Gly**-Arg-Arg-
Calcitonin	-Val -Glu -Ala -Pro- **Gly**-Lys-Lys-
Human joining peptide	-Gly -Pro -Arg -Glu- **Gly**-Lys-Arg-
Egg-laying hormone	-Leu -Glu -Lys -Lys- **Gly**-Lys-Arg-
CRF	-Leu -Asp -Ile -Ala- **Gly**-Lys-
TRH	-Arg -Gln -His -Pro- **Gly**-Lys-Arg-
LHRH	-Arg -Pro -Gly -Gly- **Gly**-Lys-Arg-
Substance P	-Phe -Gly -Leu -Met- **Gly**-Lys-Arg-
Substance K	-Val -Gly -Leu -Met- **Gly**-Lys-Arg-

$$\text{-X-Gly}\overset{\downarrow}{\ }\text{- Lys}\overset{\downarrow}{\ }\text{- Arg}\overset{\downarrow}{\ }\text{-} \longrightarrow \text{-X-Gly} \xrightarrow[\text{O}_2\text{; ascorbate}]{\text{amidating enzyme}} \text{-X} \cdot \text{NH}_2 + \text{glyoxylate}$$

[a]For abbreviations see legend to Figure 1. Further details are given in text. The vertical arrows indicate the positions where proteolytic cleavage takes place.

in the C-terminal glycopeptide moiety. Glycosylation of the vasopressin precursor has been demonstrated *in vivo* and *in vitro* (Gainer, 1982). The purpose of the glycosylation is unclear, especially since the oxytocin precursor does not depend on this modifying step for synthesis or secretion.

Experiments with tunicamycin—an antibiotic known to block glycosylation—have suggested that glycosylation may be involved with transport mechanisms. Following intraventricular tunicamycin injection in the rat, the vasopressin components do not appear to be properly axonally transported (Gonzalez *et al.*, 1981). The Golgi cisternae of the magnocellular neurons become grossly distended, pointing to an excessive accumulation in this organelle.

D. Processing Enzymes

In recent years a number of processing enzymes with activity like trypsin or like carboxypeptidase B have been described (Steiner *et al.*, 1983). So far, either these studies lack a detailed analysis of the processed product(s) or, quite often, the polyproteins to be studied are replaced by model peptides. Basically, two groups of processing enzymes have been proposed as candidates for catalyzing the conversion of polyproteins into mature peptides: serine proteases, including kallikreins and plasmin or plasminogen activator, and thiol proteases with properties like cathepsin B. In the following, some of the properties of these two types of enzymes are discussed, although they have not yet been used for studying the conversion of the authentic vasopressin or oxytocin precursors *in vitro*. It is generally assumed that these processing enzymes contain a broad substrate specificity—e.g., when cDNA encoding human proinsulin is expressed in a tumor cell line normally secreting ACTH, processing of the proinsulin precursor into insulin occurs correctly (Moore *et al.*, 1983).

1. Serine Proteases

These appear to have a neutral or slightly basic pH optimum; they may be involved in cleavage at single as well as at pairs of basic amino acids, and they have one or more sensitive thiol groups. Some of these enzyme preparations have been obtained from secretory granules from pituitaries (Smyth *et al.*, 1977) or adrenal chromaffin granules (Lindberg *et al.*, 1982). One serine protease has been isolated from yeast. It cleaves peptide precursors such as proenkephalin between two basic amino acids (Mizuno and Matsuo, 1984).

2. Thiol Proteases

These enzymes have a slightly acid pH with properties similar to cathepsin B. Again, most studies have been done with preparations obtained from secretory granules of pituitaries (Gainer, 1982). The enzyme(s) cleave between pairs of basic amino acids. The model peptide used in many studies is proopiomelanocortin (Douglass *et al.*, 1984). It is claimed that after the first cleavage between the two basic residues a second enzyme—one like carboxypeptidase B—removes the C-terminal and an aminopeptidase-like enzyme the N-terminal arginine or lysine residue (Loh and Parish, 1984). Several thiol proteases have been isolated from granules of the pancreatic islets with cathepsin B-related activities that may take part in the processing of proinsulin (Steiner *et al.*, 1983). Another thiol protease has been found in secretory granules from anglerfish pancreas; this enzyme cleaves proinsulin, prosomatostatin, and proglucagon and has a preference for arginine residues (Noe, 1981).

Though the few experimental data on processing enzymes are far from complete, they clearly indicate the complex situation of this posttranslational step. There may well be a protease gene family expressing different but related gene products that become specialized as tissue-specific proteases. Analysis of the genes encoding the various proteases no doubt will shed more light in this field.

IX. GENETIC DEFECT—DIABETES INSIPIDUS

Deficiency of vasopressin is characterized by the excretion of large amounts of a very dilute urine accompanied by excessive thirst. This disorder is known as vasopressin-sensitive diabetes insipidus, in contrast to the nephrogenic form, which is due to vasopressin resistance of the kidney (Sokol and Valtin, 1982).

Patients deficient in vasopressin drink considerable quantities of water and excrete proportionately large amounts of urine. The defect can be reversed by applying arginine vasopressin. In humans, X-linked and autosomally dominant forms of diabetes insipidus exist (Green *et al.*, 1967). In some patients, neurons of the supraoptic and paraventricular nucleus of the hypothalamus are strikingly decreased (Braverman *et al.*, 1965).

An appropriate model for studying the genetic defect is the so-called Brattleboro rat, with an autosomally recessive form of diabetes insipidus (Sokol and Valtin, 1982). These animals lack vasopressin as well as its neurophysin, while oxytocin and its associated carrier are not affected. The observed lower levels of oxytocin in the neurohypophysis are probably due to the osmotic stress, since normal levels can be restored when the animals are treated with vasopressin.

A. *In Vivo* Studies

Studies of Brattleboro rats revealed that neurons of the supraoptic and paraventricular nuclei are hypertrophied (Valtin *et al.*, 1974). Cells supposed to produce vasopressin are devoid of normal neurosecretory granules; instead, they have smaller, dense-cored granules, which, however, have also been observed in normal rats, and thus may not contain the mutated vasopressin precursor. A possible precursor candidate, however, could be the "peptide X," an unglycosylated protein isolated in small quantities from Brattleboro rats and with neurophysin- and vasopressin-like immunoreactivity (Russell *et al.,* 1980).

B. Gene Structure and Consequences of the Mutation

The gene for the vasopressin precursor from the mutant rats has been isolated and its sequence determined (Schmale and Richter, 1984; Schmale *et al.*, 1984). It contains a deletion of a single G residue in exon B that encodes the conserved part of neurophysin. This mutation gives rise to an open reading frame predicting a hormone precursor with a different C-terminus. Because of the open reading frame, there is no stop codon for terminating the translation process. Hence theoretically, translation of the mRNA could lead to a product with a polylysine tail at the C-terminus.

Starting from amino acid residue 76 of the precursor, almost all of the remaining amino acid residues have been replaced (Fig. 9). Five out of 14 cysteine residues present originally in the neurophysin molecule are substituted by other amino acids—a change that should have severe impacts on the folding of the protein.

The new reading frame of the mutated gene would also predict the replacement of a basic amino acid (arginine) normally separating the neurophysin from the glycopeptide, as well as the loss of the glycosylation site. The diagram in Fig. 10 shows the prediction of the secondary structure of the vasopressin precursor from normal rats and diabetes insipidus rats deduced from a computer program developed by Rawlings *et al.* (1983). Basically, from amino acid residue 99 of the preprohormone sequence the secondary structure changes significantly with more alpha-helical and beta-turn structures in the mutated precursor.

C. Transcription, mRNA Sequence, and Translation

A number of experiments demonstrate that the mutated gene is correctly transcribed and that the resulting mRNA contains a single nucleotide deletion in the neurophysin-encoding part as predicted from the gene sequence (Schmale and Richter, 1984; Schmale *et al.*, 1984, Ivell *et al.*, 1986; Krisch *et al.*, 1986):

1. Northern blot analysis of hypothalamic poly(A)+ RNA from normal and Brattleboro rats yields a vasopressin-specific mRNA, a result that could be

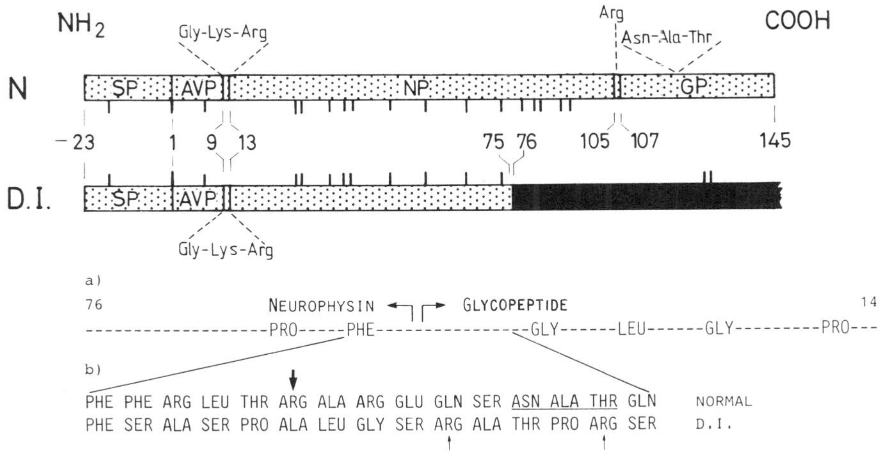

Figure 9. Comparison of the vasopressin precursors from normal (N) and diabetes insipidus (D.I.) rats, deduced from their gene structures. The negative number indicates the putative translation start site, and the other numbers show significant positions within the prohormone: 1, first cysteine residue of vasopressin; 9–13, the amino acids Gly, Lys, Arg separating vasopressin from neurophysin; 75–76, the site where in the predicted D.I. precursor the new reading frame starts; 105, last amino acid residue of the neurophysin; 107, first amino acid residue of the glycopeptide; 145, last amino acid of the prohormone. The basic amino acid arginine separating neurophysin and the glycopeptide as well as the glycosylation site Asn-Ala-Thr are indicated. The vertical lines beneath (N) or on top (D.I.) of the precursor schemes show the positions of the cysteine residues. SP, signal peptide; AVP, arginine vasopressin; NP, neurophysin; GP, glycopeptide. Part (a) shows a partial amino acid sequence of the vasopressin precursor from normal rats starting from residue 76 up to 145 of the prohormone. Only those amino acids are listed that occur at identical positions in both precursors; dashes indicate residues that are different in the predicted precursor sequence of the D.I. rat. Part (b) shows a sequence comparison of the region where—in the normal precursor—the neurophysin binds to the glycopeptide; it includes the glycosylation site (underlined), which is absent in the D.I. precursor. The dark arrowhead points to the arginine residue known to serve as processing signal separating neurophysin and the glycopeptide. This processing signal is absent in the D.I. precursor, and possible alternative processing sites are indicated by the two light arrowheads. [From Schmale and Richter (1984), with permission.]

confirmed recently by using an oligonucleotide DNA probe for screening the respective poly(A)+ RNAs (Majzoub *et al.*, 1984).

2. That the two mRNAs are identically spliced is supported by microinjection of the vasopressin gene from normal and Brattleboro rats into the nucleus of oocytes (*Xenopus laevis*), as well as by transfection experiments in 3T3 cells (Schmale *et al.*, 1984).

3. Sl mapping of mRNAs from normal and Brattleboro rats gives identical results, excluding the possibility of alternative transcription start sites, which

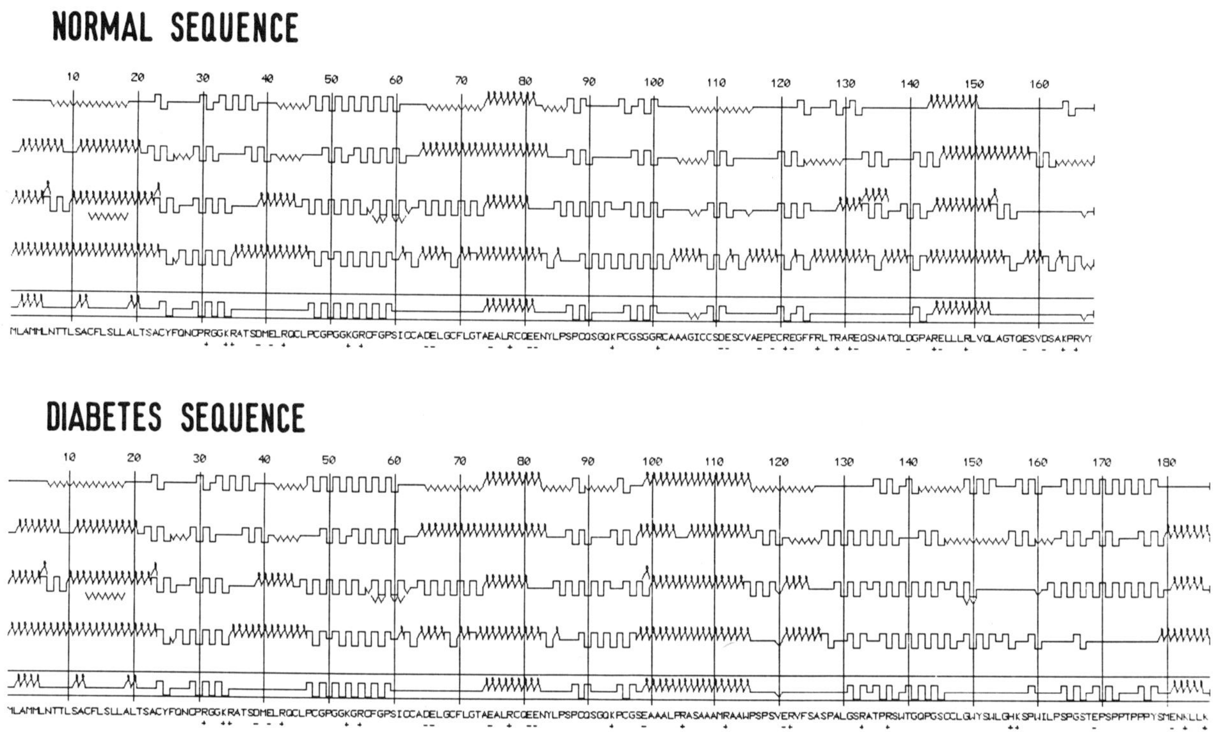

Figure 10. Diagram plot of the prediction of secondary structure of the preprovasopressin precursor from normal and diabetes insipidus (Brattleboro) rats. For calculations and other details, see Rawlings *et al.* (1983). Symbols: ⊓⊔⊓, beta turn; ʌʌʌʌ, alpha helix; ⏝⏝⏝, beta sheet; ———, random coil. Each line represents different calculations; in the bottom line, three of the four calculations are in agreement in their prediction.

could affect the secondary structure of the Brattleboro mRNA and thus its translation efficiency.

4. Finally, cloning and sequencing of hypothalamic mRNA from Brattelboro rats shows no difference to the coding sequence of the corresponding gene (Schmale *et al.*, 1984). The mRNA lacks a single G nucleotide in the neurophysin encoding part (Schmale and Richter, 1984) (Fig. 11).

Inspection of Northern blots of hypothalamic poly(A)+ RNA from wild-type (Wistar) and Brattleboro rats suggests comparable amounts of vasopressin and oxytocin precursor-specific mRNA (Schmale and Richter, 1984). Quantitative studies by dot-blot analysis of hypothalamic poly(A)+ RNA from Long-Evans, Wistar, and homozygous Brattleboro rats indicate the following relative amounts of vasopressin precursor-specific mRNA: Long-Evans 100%, Wistar 66%, and Brattleboro 27%. Based on the cloning experiments for hypothalamic poly(A)+ RNA from Brattleboro rats described above, out of 5000 transformants, two contain the vasopressin precursor sequence (Ivell *et al.*, 1986). Taking this information into account, the vasopressin precursor-specific mRNA amounts to 0.04% of total poly(A)+ RNA in the Brattleboro, to 0.1% in the Wistar, and 0.15% in the Long-Evans rat.

Recent reports claim very low numbers of vasopressin-specific mRNA in normal (0.01% for Long-Evans rats) and Brattleboro rats [0.001% (Majzoub *et al.*, 1984)]. From these studies, it appears that saline-treated anaimals have been used for calculation; in that case the numbers for vasopressin-specific mRNA in untreated Long-Evans rats will be even lower (Majzoub *et al.*, 1983), since the same group reports that saline treatment increases vasopressin-specific mRNA production to about 20-fold. Taking this into consideration, only 0.0005% of the total poly(A)+ RNA should be vasopressin-specific in the untreated animals—a number that is in conflict with the numbers of positive transformants outlined above. One possible explanation of these low numbers may result from the short oligonucleotides used in evaluating the mRNA content (Majzoub *et al.*, 1984); this probe may be less reliable than longer and more specific cDNA probes.

Although vasopressin-encoding mRNA can be isolated from hypothalami of Brattleboro rats, synthesis of a normal-sized vasopressin precursor (MW 19,000) in a cell-free protein synthesizing system has not been observed (Schmale *et al.*, 1984).

The lack in synthesizing authentic vasopressin precursor is due to the deletion of a nucleotide that shifts the reading frame in a way that a stop codon signal is no longer read; consequently ribosomes may be able to proceed further on even translating the poly(A) sequence. If one calculates the maximal size of such a potential precursor, then roughly 90 amino acids have to be added to the normal-sized sequence giving rise to a molecular weight between 26,000 and 28,000. The C-terminus of this precursor should consist of ~70 lysine residues corresponding to the ~200 adenosines of the poly(A) tail (Ivell *et al.*, 1986).

```
D.I.                           MetLeuAlaMetMetLeuAsnThrThrLeuSerAlaCysPheLeuSerLeuLeuAlaLeuThrSerAlaCysTyrPheGl   ←AVP
      1 GCTGCACGCAGTGCCCACCTATGCTCGCCATGATGCTCAACACTACGCTCTCTGCTTGCTTCCTGAGCCTGCTGGCCCTCACCTCTGCCTGCTACTTCCA
        ****************************************************************************************************
N.    1 GCTGCACGCAGTGCCCACCTATGCTCGCCATGATGCTCAACACTACGCTCTCTGCTTGCTTCCTGAGCCTGCTGGCCCTCACCTCTGCCTGCTACTTCCA
                            MetLeuAlaMetMetLeuAsnThrThrLeuSerAlaCysPheLeuSerLeuLeuAlaLeuThrSerAlaCysTyrPheGl
                                                                                                        AVP
          AVP→        ←NpII
        nAsnCysProArgGlyGlyLysArgAlaThrSerAspMetGluLeuArgGlnCysLeuProCysGlyProGlyGlyLysGlyArgCysPheGlyProSer
    101 GAACTGCCCAAGAGGAGGCAAGAGGGCCACATCCGACATGGAGCTGAGACAGTGTCTCCCCTGCGGCCCTGGCGGCAAAGGGCGCTGCTTCGGGCCGAGC
        ****************************************************************************************************
    101 GAACTGCCCAAGAGGAGGCAAGAGGGCCACATCCGACATGGAGCTGAGACAGTGTCTCCCCTGCGGCCCTGGCGGCAAAGGGCGCTGCTTCGGGCCGAGC
        nAsnCysProArgGlyGlyLysArgAlaThrSerAspMetGluLeuArgGlnCysLeuProCysGlyProGlyGlyLysGlyArgCysPheGlyProSer
          AVP→        ←NpII

        IleCysCysAlaAspGluLeuGlyCysPheLeuGlyThrAlaGluAlaLeuArgCysGlnGluGluAsnTyrLeuProSerProCysGlnSerGlyGlnL
    201 ATCTGCTGCGCGGACGAGCTGGGCTGCTTCCTGGGCACCGCCGAGGCGCTGCGCTGCCAGGAGGAGAACTACCTGCCCTCGCCCTGCCAGTCTGGCCAGA
        ****************************************************************************************************
    201 ATCTGCTGCGCGGACGAGCTGGGCTGCTTCCTGGGCACCGCCGAGGCGCTGCGCTGCCAGGAGGAGAACTACCTGCCCTCGCCCTGCCAGTCTGGCCAGA
        IleCysCysAlaAspGluLeuGlyCysPheLeuGlyThrAlaGluAlaLeuArgCysGlnGluGluAsnTyrLeuProSerProCysGlnSerGlyGlnL

                       ▼
        ysProCysGlySerGluAlaAlaAlaLeuProArgAlaSerAlaAlaAlaMetArgAlaAlaTrpProSerProSerValGluArgValPheSerAlaS
    301 AGCCTTGCGGAAGCG-AGGCCGCTGCGCTGCCGCGGGCATCTGCTGCAGCGATGAGAGCTGCGTGGCCGAGCCCGAGTGTCGAGAGGGTTTTTCCGCCT
        ************** *************************************************************************************
    301 AGCCTTGCGGAAGCGGAGGCCGCTGCGCTGCCGCGGGCATCTGCTGCAGCGATGAGAGCTGCGTGGCCGAGCCCGAGTGTCGAGAGGGTTTTTCCGCCT
        ysProCysGlySerGlyGlyArgCysAlaAlaAlaGlyIleCysCysSerAspGluSerCysValAlaGluProGluCysArgGluGlyPhePheArgLe

        erProAlaLeuGlySerArgAlaThrProArgSerTrpThrGlyGlnProGlySerCysCysLeuGlyTrpTyrSerTrpLeuGlyHisLysSerProTr
    400 CACCCGCGCTCGGGAGCAGAGCAACGCCACGCAGCTGGACGGGCCAGCCCGGGAGCTGCTGCTTAGGCTGGTACAGCTGGCTGGGACACAAGAGTCCGTG
        ****************************************************************************************************
    401 CACCCGCGCTCGGGAGCAGAGCAACGCCACGCAGCTGGACGGGCCAGCCCGGGAGCTGCTGCTTAGGCTGGTACAGCTGGCTGGGACACAAGAGTCCGTG
        uThrArgAlaArgGluGlnSerAsnAlaThrGlnLeuAspGlyProAlaArgGluLeuLeuLeuArgLeuValGlnLeuAlaGlyThrGlnGluSerVal
     NpII→  ←GP

        pIleLeuProSerProGlySerThrGluProSerProProThrProProProTyrSerMetGluAsnLysLeuLeuLysAsnAlaLysLysLysLys
    500 GATTCTGCCAAGCCCCGGGTCTACTGAGCCATCGCCCCCCACGCCTCCCCCCTACAGCATGGAAAATAAACTTTTAAAAAATGCAAAAAAAAAAAAA
        *************************************************************************************************
    501 GATTCTGCCAAGCCCCGGGTCTACTGAGCCATCGCCCCCCACGCCTCCCCCCTACAGCATGGAAAATAAACTTTTAAAAAATGCAAAAAAAAAAAAA
        AspSerAlaLysProArgValTyr
                   GP→
```

Figure 11. Nucleotide sequence of the vasopressin-encoding mRNA obtained from a Brattleboro hypothalamic cDNA library (Ivell *et al.*, 1986), comparing normal (N) and Brattleboro (diabetes insipidus, D.I.) types. Identical nucleotides are indicated by asterisk. Nucleotides are numbered; the sequence shown lacks the first 11 nucleotides of the transcription start site. The single deletion found in the Brattleboro cDNA sequence is marked by an arrowhead. See also legends to Figs. 3–6. The line on top of the sequence indicates the 14-mer peptide.

Indeed, this vasopressin-like precursor has been identified using antisera raised either against vasopressin, neurophysin, or a 14-mer peptide (Krisch *et al.*, 1986; Guldenaar *et al.*, 1986); the latter can be predicted from the new reading frame (Ivell *et al.*, 1986).

Recently, it has been shown by electrophoretic analysis of vasopressin mRNA that (1) the hypothalamic mRNA from Brattleboro rats contains a longer stretch of poly(A) sequence than the wild-type strains; (2) vasopressin mRNA is also present in the adrenal, ovary, testis, and cerebellum, at very low levels; however, (3) the extra-hypothalamic mRNA is considerably shorter than that in the hypothalamus, because of a curtailed poly(A) sequence. Thus, similar vasopressin gene transcripts are subject to a tissue-specific differential polyadenylation (Ivell *et al.*, 1986).

At present it is not clear why the hypothalamic vasopressin mRNA from Brattleboro rats has a longer poly(A) tail than its nonmutant counterpart. One possibility is that, because the ribosomes read through into the noncoding 3′ region and presumably only leave the messenger slowly, this region including the poly(A) tail is more protected against cytoplasmic RNase activity. Hence, cytoplasmic clipping of the poly(A) tail would be inhibited in the case of the mutant mRNA.

Polyadenylation and the length of the poly(A) sequence may also explain the apparent paradox that vasopressin, although blocked in the hypothalamus of the mutant rat, may nevertheless be produced in peripheral tissues (Lim *et al.*, 1984; Nussey *et al.*, 1984). The vasopressin mRNA isolated from the adrenal or from gonads possesses a very short poly(A) tail, which when translated in the mutant rat would yield ~10 lysine residues at the C-terminus of the mutated precursor. This precursor may then appear sufficiently normal to avoid being diverted from the regular processing pathway, with the result that vasopressin, whose immediately flanking cleavage signals are intact, could still be packaged, excised, and released just as in nonmutant secretory systems.

D. Immunocytochemical Studies

The finding that the vasopressin mRNA has a markedly longer poly(A) tail in the hypothalamus than that in peripheral tissues (Ivell *et al.*, 1986) suggests a tissue-specific differential polyadenylation process. Given that a vasopressin-like precursor is synthesized in the hypothalamus of Brattleboro rats, albeit at a reduced level, why does no vasopressin reach the pituitary? Because of the lysine tail at the C-terminus, the mutated precursor may appear "too foreign" to be correctly handled by the posttranslational processing system of the magnocellular neurons. In fact, the highly charged material may "stick" to membranes and, eventually, may trigger the scavenging mechanisms of the lysosomal system.

Evidence for this view is provided by immunohistochemistry at the light and electron microscope levels (Krisch *et al.*, 1986).

Using antibodies against the 14-mer peptide, the cells of the supraoptic nucleus can be specifically stained in the Brattleboro but not in normal rats. No positive staining is seen in axons or nerve terminals. At the electron microscope level, immunoreactivity is observed in the endoplasmic reticulum and trans-Golgi, but not in neurosecretory granules. Instead, there is heavy staining of lysosomal bodies.

A recent immunocytochemical study has indicated, however, that in Brattleboro rats not only vasopressin but also the glycopeptide of the normal precursor can be identified in a few hypothalamic neurons (Richards *et al.*, 1986). This is in conflict with data reported by Watson *et al.* (1982), who showed that hypothalamic neurons from Brattleboro rats lack the vasopressin-associated glycopeptide. Although the absolute specificity of the antiglycopeptide antibody used in those studies is not yet proven, the data imply that the normal vasopressin precursor can be synthesized even in the Brattleboro rat. Since there is no evidence for the existence of a second vasopressin gene in rats, one would have to assume that somehow there is a reversion to the initial reading frame. Although a shift in reading frame during translation is unknown in eukaryotes, such mechanisms have been described in bacteria and, recently, in viral systems and could represent another way to regulate gene expression (Jacks and Varmus, 1985).

The results presented here offer a relatively simple hypothesis that explains many of the hitherto paradoxical phenomena associated with the expression of vasopressin in the Brattleboro rat. These observations are consistent with a model whereby the single vasopressin gene with its frame-shift mutation is expressed differentially at the posttranscriptional level, due to differences in the length of the poly(A) tail.

X. EXPRESSION IN PERIPHERAL TISSUES

Synthesis of both vasopressin and oxytocin has recently been shown to occur also in cells other than the hypothalamic magnocellular neurons. Production of the two hormones has been claimed for lung carcinomas, adrenal gland, testis, and ovarian corpus luteum (Wathes *et al.*, 1983; Flint and Sheldrick, 1982; Nicholson *et al.*, 1984). The substantiation of these claims would involve a reconsideration of the relative role of the hypothalamus in peripheral functions dependent on circulating oxytocin or vasopressin.

Studies of mRNA from bovine corpus luteum have indicated that this organ indeed expresses the oxytocin gene (Ivell and Richter, 1984a). Depending on the phase of the luteal cycle, up to 250 times more oxytocin mRNA per organ is produced than in a single hypothalamus (Ivell *et al.*, 1985). This result is clearly

```
                                                       5' G T C C C C G G C C C G G A G A C C

                             MET   ALA   GLY   SER   SER   LEU   ALA   CYS   CYS   LEU   LEU   GLY   LEU   LEU
A G C G C G T C T G C A C C A T G G C A G G T T C C A G C C T C G C C T G C T G C C T G C T C G G C C T C C T G

                                          O  x  y  t  o  c  i  n                                         Np I
ALA   LEU   THR   SER   ALA   CYS   TYR   ILE   GLN   ASN   CYS   PRO   LEU   GLY   GLY   LYS   ARG   ALA   VA
G C G T T G A C C T C C G C C T G C T A C A T T C A G A A C T G C C C C C T G G G C G G C A A A C G C G C G G T

L  LEU   ASP   LEU   ASP   VAL   ARG   THR   CYS   LEU   PRO   CYS   GLY   PRO   GLY   GLY   LYS   GLY   ARG
G C T G G A C C T C G A C G T G C G C A C G T G T C T C C C C T G C G G C C C C G G G G G C A A A G G C C G C T

CYS   PHE   GLY   PRO   SER   ILE   CYS   CYS   GLY   ASP   GLU   LEU   GLY   CYS   PHE   VAL   GLY   THR   ALA
G C T T C G G G C C C A G C A T C T G C T G C G G G G A C G A G C T G G G C T G C T T C G T G G G C A C G G C C

 GLU   ALA   LEU   ARG   CYS   GLN   GLU   GLU   ASN   TYR   LEU   PRO   SER   PRO   CYS   GLN   SER   GLY   GL
G A G G C G C T G C G C T G C C A A G A G G A G A A C T A C C T G C C G T C G C C C T G C C A G T C C G G C C A

N  LYS   PRO   CYS   GLY   SER   GLY   GLY   ARG   CYS   ALA   ALA   ALA   GLY   ILE   CYS   CYS   SER   PRO
G A A G C C C T G C G G G A G C G G G G G C C G C T G C G C C G C C G C C G G C A T C T G C T G C A G C C C G G

ASP   GLY   CYS   HIS   GLU   ASP   PRO   ALA   CYS   ASP   PRO   GLU   ALA   ALA   PHE   SER   GLN   HIS   ***
A C G G C T G C C A C G A G G A C C C C G C C T G C G A C C C T G A G G C C G C C T T C T C C C A G C A C T G A

G A C C G G C C G G C C C C C G A T A C C G T C G G A G C G C A G C C C T C A C T C C C T C T G T A A T C A T C

C C T A G G A A T T A T G A C A A T G A A A T A A A G C C G G T T T T T T C C C C C T C C (A) 3'
```

Figure 12. The cDNA and amino acid sequences corresponding to the bovine luteal mRNA for the oxytocin precursor.

of considerable significance for the oxytocin-dependent control of reproductive function.

The sequence of the cloned luteal oxytocin mRNA is identical to that already described from the bovine hypothalamus (Fig. 12). Also, when translated in a cell-free system, the luteal mRNA gives the same common precursor to oxytocin and neurophysin I as is produced from hypothalamic mRNA.

One can therefore conclude that the oxytocin-like peptide found in the corpus luteum is indeed the same as hypothalamic oxytocin and that it is synthesized there via a precursor of the same form as in the hypothalamus. It will now be interesting to develop luteal-cell systems to investigate *in vitro* the regulation of the oxytocin gene expression, which *in vivo* appears to be linked to the hormonal status of the estrous cycle, as well as to look at other potential oxytocin-containing tissues.

XI. PERSPECTIVES

The hypothalamus has until recently been generally considered as the principal source of circulating oxytocin, which is involved in the milk-ejection reflex and uterine contraction. Now that oxytocin as well as vasopressin has been found in other organs, one must reconsider the control and regulation as well as the function of these and similar hormones when expressed peripherally. This also raises the question of the transcriptional control of these genes in the different organs. Regulation may also occur at the level of posttranslational processing and release and hence will require more detailed studies of the involved mechanisms. Finally, how are the signals relayed from the cell membrane to the nucleus? Are there feedback signals from the precursor synthesis toward the transcription of a gene? Genetic defects as found in the Brattleboro rat strain may be helpful in finding some of the answers to these problems.

ACKNOWLEDGMENT

The author is grateful to Dr. R. Ivell for helpful suggestions and discussions throughout the preparation of the manuscript. This work was financially supported by grants from Deutsche Forschungsgemeinschaft and Fonds der Chemischen Industrie.

REFERENCES

Acher, R. (1983). "Brain Peptides" (D. Krieger, M. Brownstein, and J. Martin, eds.), pp. 135–163.

Amara, S. G., Jonas, V., Rosenfeld, M. G., Ong, E. S., and Evans, R. M. (1982). *Nature (London)* **298,** 240–244.

Baltimore, D. (1981). *Cell* **26,** 295–296.

Bell, G. I., Santerre, R. F., Mullenbach, G. T. (1983). *Nature (London)* **302,** 716–718.

Boel, E., Vuust, J., Norris, F., Norris, K., Wind, A., Rehfeld, J., and Marcker, K. (1983). *Proc. Natl. Acad. Sci. USA* **80,** 2866–2869.

Bradbury, A. F., Finnie, M. D. A., and Smyth, D. G. (1982). *Nature (London)* **298,** 686–688.

Braverman, L. E., Mancini, J. P., and McGoldrick, D. M. (1965). *Ann. Intern. Med.* **63,** 503–508.

Breslow, E. (1979). *Ann. Rev. Biochem.* **48,** 251–274.

Burbach, J. P. H., De Hoop, M. J., Schmale, H., Richter, D., De Kloet, E. R., Ten Haaf, J. A., and De Wied, D. J. (1984). *Neuroendocrinology* **39,** 582–584.

De Wied, D. (1983). *Prog. Brain Res.* **60,** 155–167.

Douglass, J., Civelli, O., and Herbert, E. (1984). *Annu. Rev. Biochem.* **53,** 665–715.

Eipper, B. A., Mains, R. E., and Glembotski, C. C. (1983). *Proc. Natl. Acad. Sci. USA* **80,** 5144–5148.

Evans, R. M., Amara, S. G., and Rosenfeld, M. G. (1982). *DNA* **1,** 323–328.

Flint, A. P. F., and Sheldrick, E. L. (1982). *Nature (London)* **297,** 587–588.

Furutani, Y., Morimoto, Y., Shibahara, S., Noda, M., Takahashi, H., Hirose, T., Asai, M., Inayama, S., Hayashida, H., Miyata, T., and Numa, S. (1983). *Nature (London)* **301,** 537–540.

Gainer, H. (1982). *Prog. Brain Res.* **60,** 205–215.

Gainer, H., Sarne, Y., and Brownstein, M. J. (1977). *J. Cell Biol.* **73,** 366–381.

Gonzalez, C. B., Swann, R. W., and Pickering, B. T. (1981). *Cell Tissue Res.* **217,** 199–210.

Green, J. R., Buchan, G. C., Alvord, E. C., Jr., and Swanson, A. G. (1967). *Brain* **90,** 707–714.

Gubler, U., Monahan, J. J., Lomedico, P. T., Bhatt, R. S., Collier, K. J., Hoffman, B. J., Boehlen, P., Ling, N., Zeytin, F., Brazeau, P., Poonian, M. S. and Gace, L P. (1983). *Proc. Natl. Acad. Sci. U.S.A.* **80,** 4311–4314.

Guldenaar, S. E. F., Nahke, P., and Pickering, B. T. (1986). *Cell Tissue Res.* **244,** 359–364.

Heidecker, G., and Messing, J. (1983). *Nucleic Acids Res.* **11,** 4891–4906.

Herbert, E. (1981). *Trends Biochem. Sci.* **6,** 184–188.

Hoffmann, W., Richter, K., Hutticher, A., and Kreil, G. (1983). *In* "Biochemical and Clinical Aspects of Neuropeptides" (G. Koch and D. Richter, eds.), pp. 175–183. Academic Press, New York.

Holwerda, D. A. (1972). *Eur. J. Biochem.* **28,** 340–346.

Itoh, N., Obata, K., Yanaihara, N., and Okamoto, H. (1983). *Nature (London)* **304,** 547–549.

Ivell, R., and Richter, D. (1984a). *EMBO J.* **3,** 2351–2354.

Ivell, R., and Richter, D. (1984b). *Proc. Natl. Acad. Sci. USA* **81,** 2006–2010.

Ivell, R., Brackett, K. H., Fields, M. J., and Richter, D. (1985). *FEBS Lett.* **190,** 263–26.

Ivell, R., Schmale, H., Krisch, B., Nahke, P., and Richter, D. (1986). *EMBO J.* **5,** 971–977.

Jacks, T., and Varmus, H. E. (1986). *Science* **230,** 1237–1242.

Koch. G., and Richter, D., eds. (1980). "Biosynthesis, Modification, and Processing of Cellular and Viral Polyproteins." Academic Press, New York.

Kozak, M. (1981). *Nucleic Acid Res.* **9,** 5233–5253.

Krieger, D. T. (1983). *Science* **222,** 975–985.

Krisch, B., Nahke, P., and Richter, D. (1986). *Cell Tissue Res.* **244,** 351–358.

Land, H., Schutz, G., Schmale, H., and Richter, D. (1982). *Nature (London)* **295,** 299–303.

Land, H., Grez, M., Ruppert, S., Schmale, H., Rehbein, M., Richter, D., and Schutz, G. (1983). *Nature (London).* **302,** 342–344.

Lechan, R., Goodman, R., Rosenblatt, M., Reichlin, S., and Habener, J. (1983). *Proc. Natl. Acad. Sci. USA* **80,** 2780–2784.

Lim, A., Lolait, S., Barlow, J., Autelitano, D., Toh, B., Boublik, J., Abraham, J., Johnston, C., and Funder, J. (1984). *Nature (London)* **310,** 61–64.
Lindberg, I., Yang, H. Y. I., and Costa, E. (1982). *Biochem. Biophys. Res. Commun.* **106,** 186–193.
Loh, Y. P., and Parish, D. (1984). 7th International Congress of Endocrinology, Quebec, Canada, no. S72.
McCabe, J. T., Morrell, J. I., Ivell, R., Schmale, H., Richter, D., and Pfaff, D. W. (1986). *J. Histochem. Cytochem.* **34,** 45–50.
Mains, R. E., Glembotski, C. C., and Eipper, B. A. (1984). *Endocrinology* **114,** 1522–1530.
Majzoub, J. A., Rich, A., v. Boom, J., and Habener, J. F. (1983). *J. Biol. Chem.* **258,** 14061–14064.
Majzoub, J. A., Pappey, A., Burg, R., and Habener, J. (1984). *Proc. Natl. Acad. Sci. USA* **81,** 5296–5299.
Mizuno, K., and Matsuo, H. (1984). *Nature (London)* **309,** 558–560.
Moore, H. P. H., Walker, M. D., Lee, F., and Kelly, R. B. (1983). *Cell* **35,** 531–538.
Moos, F., Freund-Mercier, M. J., and Richard, P. (1982). *Colloq INSERM* **110,** 121–144.
Nawa, H., Kotani, H., and Nakanishi, S. (1984). *Nature (London)* **312,** 729–734.
Nicholson, H. D., Swann, R. W., Burford, G. D., Wathes, D. C., Porter, D. G., and Pickering, B. T. (1984). *Regul. Pept.* **8,** 141–146.
Noe, B. D. (1981). *J. Biol. Chem.* **256,** 4940–4946.
Nordmann, J., and Labonesse, J. (1981). *Science* **211,** 595–597.
North, W. G., Mitchell, T. I., and North, G. M. (1983). *FEBS Lett.* **152,** 1, 29–34.
Nussey, S., Ang, V., Jenkins, J., Chowdrey, H., and Bisset, G. (1984). *Nature (London)* **310,** 64–66.
Rawlings, N., Ashman, K., and Wittmann-Liebold, B. (1983). *Int. J. Pept. Protein Res.* **22,** 515–524.
Richards, S.-J., Morris, R., and Raisman, G. (1985). *Neuroscience* **16,** 617–623.
Richter, D. (1983). *In* "Trends in Biochemistry," Vol. 8 pp. 278–281. Elsevier, Amsterdam.
Richter, K., Kawashima, E., Egger, R., and Kreil, G. (1984). *EMBO J.* **3,** 617–621.
Robertson, G. L. (1977). *Recent Prog. Horm. Res.* **33,** 333–385.
Robinson, I. C. A. F., and Jones, P. M. (1984). *Neurosci. Lett.* **39,** 273–278.
Ruppert, D., Ścherer, G., and Schutz, G. (1984). *Nature (London)* **308,** 554–557.
Russell, J. T., Brownstein, M. J., and Gainer, H. (1980). *Endocrinology* **107,** 1880–1891.
Rutter, W. J., Scott, J., Selby, M., Crawford, R. J., Shen, L. U., Hobart, P., Sanchez-Pescador, R., and Bell, G. I. (1983). *In* "Biochemical and Clinical Aspects of Neuropeptides (G. Koch and D. Richter, eds.), pp. 293–308. Academic Press, New York.
Sachs, H., and Takabatake, Y. (1964). *Endocrinology* **75,** 943–948.
Schmale, H., and Richter, D. (1984). *Nature (London)* **308,** 705–709.
Schmale, H., Heinsohn, S., and Richter, D. (1983). *EMBO J.* **2,** 763–767.
Schmale, H., Ivell, R., Breindl, M., Darmer, D., and Richter, D. (1984). *EMBO J.* **3,** 3289–3293.
Seeburg, P. H., and Adelman, J. P. (1984). *Nature (London)* **311,** 666–668.
Smyth, D. G., and Massey, D. E. (1979). *Biochem. Biophys. Res. Commun.* **87,** 1006–1010.
Smyth, D. G., Austen, B. M., Bradbury, A. F., Geisow, M. J. and Snell, C. R. (1977). *In* "Centrally Active Peptides" (J. Hughes, ed.), pp. 231–239. Macmillan, London.
Sokol, H. W., and Valtin, H., eds. (1982). The Brattleboro rat. *Ann. NY Acad. Sci.* **394,** 1–828.
Steiner, D. F., Docherty, K., Chan, S. J., Segundo, B. S., and Carroll, R. (1983). *In* "Biochemical and Clinical Aspects of Neuropeptides" (G. Koch and D. Richter, eds.), pp. 3–13. Academic Press, New York.
Swann, R. W., Gonzales, C. B., Birkett, S. D., and Pickering, B. T. (1982). *Biochem. J.* **208,** 339–349.

Valtin, H., Stewart, J., and Sokol, H. W. (1974). *In* "Handbook of Physiology," Vol. 7, (E. Knobil, and W. H. Sawyer, eds.), pp. 131–171. Williams & Wilkins, Baltimore, Maryland.

Wathes, D. C., Swann, R. W., Birkett, S. D., Porter, D. G., and Pickering, B. T. (1983). *Endocrinology* **113,** 693–698.

Watson, S. J., Seidah, N. G., and Chretien, M. (1982). *Science* **217,** 853–855.

Yoo, O. J., Powell, C. T., and Agarwal, K. L. (1982). *Proc. Natl. Acad. Sci. USA* **79,** 1049–1053.

Zerbe, R. L., and Palkovits, M. (1984). *Neuroendocrinology* **38,** 285–289.

Chapter **4**

Structure–Activity Relationships of Neurohypophyseal Peptides

VICTOR J. HRUBY
Department of Chemistry
University of Arizona
Tucson, Arizona 85721

CLARK W. SMITH
Biopolymer Chemistry
The Upjohn Company
Kalamazoo, Michigan 49001

I. INTRODUCTION

The neurohypophyseal hormones oxytocin (**I**), arginine vasopressin (**II**), and their related naturally occurring analogs (Table I) have been at the forefront of developments of modern molecular endocrinology and the efforts to develop a chemical–physical basis for understanding peptide hormone and neurotransmitter function. On the one hand, we have made remarkable progress in that for many peptide hormones and neurotransmitters we have acquired some insights into which of the functional groups in these peptides are primarily responsible for interactions with cellular membrane receptors. On the other hand, we have made less progress toward *a priori* designing analogs of the native hormones to have specific properties at specific receptors. In large measure this is because, although the principle of chemical complimentarily must certainly apply to hormone–receptor systems, only recently have we begun to develop approaches to study the conformational and topological features of the peptide hormone that are crucial for their interaction with receptors. The necessity for determining these features cannot be overestimated. Nonetheless, a critical first step, before the detailed insights into structure–function relationships that allow considerations of conformation and topology critical for biological activity can be developed, is

THE PEPTIDES, VOLUME 8

Table I. Structure and Biological Activities of Naturally Occurring Neurohypophyseal Hormones

Name	Structure	Biological Activities					Ref.
		Rat Uterus	Avian Vasodepressor	Milk Ejecting	Anti-diuretic	Pressor	
Oxytocin	Cys-Tyr-Ile-Gln-Asn-Cys-Pro-Leu-Gly-NH_2	546	507	410	2.7	3.1	a,b
Arginine vasotocin	Cys-Tyr-Ile-Gln-Asn-Cys-Pro-Arg-Gly-NH_2	127	493	210	231	160	c,d
Mesotocin	Cys-Tyr-Ile-Gln-Asn-Cys-Pro-Ile-Gly-NH_2	289	498	328	1.1	6.3	e
Valitocin	Cys-Tyr-Ile-Gln-Asn-Cys-Pro-Val-Gly-NH_2	199	278	308	0.8	9	e
Aspartocin	Cys-Tyr-Ile-Asn-Asn-Cys-Pro-Val-Gly-NH_2	158	201	382	0.17	0.12	f,g
Glumitocin	Cys-Tyr-Ile-Ser-Asn-Cys-Pro-Gln-Gly-NH_2	10		53	0.41	0.35	h
Isotocin	Cys-Tyr-Ile-Ser-Asn-Cys-Pro-Ile-Gly-NH_2	145	382	375	0.75	0.05	h,i
Arginine vasopressin	Cys-Tyr-Phe-Gln-Asn-Cys-Pro-Arg-Gly-NH_2	12.0	100	30-120	503	487	j
Lysine vasopressin	Cys-Phe-Phe-Gln-Asn-Cys-Pro-Lys-Gly-NH_2	4.8	48	31	203	243	k
Phenylpressin	Cys-Phe-Phe-Gln-Asn-Cys-Pro-Arg-Gly-NH_2	0.2	<1	3	350	122	l

[a]Chan and Kelley (1967); [b]Chan and duVigneaud (1962); [c]Manning *et al.* (1973c); [d]Berde *et al.* (1962); [e]Jaquenoud and Boissonnas (1961); [f]Turan *et al.* (1977); [g]Berde and Boissonnas (1968); [h]Sawyer *et al.* (1969); [i]Rudinger *et al.* (1969); [j]Meienhofer *et al.* (1970); [k]Kimbrough *et al.* (1963); [l]Huguenin and Boissonnas (1962).

obtaining information about the relationships of primary structure to biological activity.

Actually, the situation is even more complex, because it is now clear that hormone–receptor interaction generally involves at least three different states for the hormone-receptor complex. The first is an initial hormone–receptor binding interaction (recognition, the binding message) taking place. This state is a necessary but not sufficient condition for a biological response initiated by the peptide. This must be followed by the transduction state; this involves a hormone–receptor interaction (often interacting with a regulatory protein), which leads to activation of the biological response via a second messenger such as cyclic AMP, Ca^{+2}, etc. (the biological activity message). Finally, reversal of the process—that is dissociation, internalization, or translocation of the hormone–receptor complex—takes place so that the cell eventually returns to its basal state or a new basal state.

The first two of these processes related to hormone action are well recognized (Rudinger, 1971; Schwyzer, 1977; Walter *et al.*, 1977; Hruby, 1981a,b) to involve different structural (and conformational) properties of the peptide hormone. The nature of these different interactions can lead to different kinds of biological activity, including full agonist activity on the one hand or full competitive antagonist activity on the other. Of course, intermediate states also can occur that lead to various levels of partial agonist and/or partial antagonist biological activities. There is much evidence with several peptide hormone– (or neurotransmitter–) receptor systems that suggests these two steps utilize different structural properties of the peptide, and we shall see this in neurohypophyseal hormone structure–function relationships. The concept that reversal of peptide hormone– (neurotransmitter–) receptor interactions may involve still different structural (and conformational) properties of the hormone is of more recent origin. However, Hruby *et al.* have suggested (Hruby *et al.*, 1983a; Hruby, 1984) that prolongation of biological activity (or reversal) may be related to structural (and conformational) properties of the hormone in the hormone–receptor complex after the transduction step (for antagonists this would involve a state after the inhibitory complex if formed) that are different than those structural features important for binding or transduction. Thus, prolongation of the biological response can result from hormone–receptor interactions that are distinct from those that determine potency or efficiency. This concept has been applied to both oxytocin and α-melanotropin (α-MSH) (Hruby, 1984) with success.

While it is clear that conformational and dynamic properties of oxytocin are critical to its biological activities, whether agonist (Walter *et al.*, 1971; Walter, 1977; Hruby, 1981b) or antagonist (Meraldi *et al.*, 1977; Hruby, 1981a,b), the major focus of this review is on more classical structure–function analysis of neurohypophyseal peptides. Such an analysis is a necessary prelude to considerations of conformation that can provide a more rational basis for design of peptide hormone and neurotransmitter analogues (Vida and Gorden, 1984). In our discussion of structure and function, we will primarily emphasize the five "classical" biological activities attributed to the neurohypophyseal hormones:* (1) rat uterus contraction (oxytocic response); (2) avian vasodepressor activity; (3) the milk-ejecting response in guinea pig, rat, or rabbit; (4) rat antidiuretic activity; and (5) the rat pressor response. It should be noted that the first three of these activities are thought to be primarily "oxytocin-like" activities, while the latter two activities are thought to be primarily "vasopressin-like" activities. As we shall see, however, each of these receptors has *different* structural requirements for hormone–receptor interaction and biological activity. To facilitate discussion, we examine structure–function relationships for each of these receptors systems separately. In addition, we consider agonist and antagonist activities separately, since these activities generally follow quite different structure–function relationships even *at the same receptor system* (Hruby and Mosberg, 1982).

Finally, since comprehensive reviews of the early structure–function studies have been published previously (Berde and Boissonnas, 1968; Rudinger and Krejci, 1968; Walter and Schwartz, 1968), many of these earlier results will not be repeated here except as needed for development of our discussion.

II. NATURALLY OCCURRING NEUROHYPOPHYSEAL HORMONES—AN OVERVIEW

Oxytocin (Table I) was the first peptide hormone whose primary structure was determined and proven by total synthesis. This classical work of du Vigneaud and co-workers (1953) opened the way to prepare structurally modified synthetic hormones by total synthesis, and since then several hundred analogs of the neurohypophyseal hormones have been prepared. At this time 10 naturally occurring neurohypophyseal hormones have been isolated and each primary structure determined (Table I). The naturally occurring analogs, some of which were prepared by total synthesis *before* they were isolated from natural sources, provide an excellent starting point for discussing structure–function relationships and the effect of primary structure change on biological activity at the five receptor systems to be examined in this review.

Nature has modified the neurohypophyseal hormones in the 2, 3, 4, and 8 positions. Nonetheless, all of these naturally occurring peptide hormones appear to have some activity in each of the assay systems. Interestingly, however, the structural changes are sufficient to modify biological potencies as much as 1000- to 10,000-fold at different receptors. The one exception is the milk-ejecting

*Agonist potencies expressed in the tables are in Units/mg.

assay system, where only a 100-fold or so change in potency occurs for all of the naturally occurring analogs.

A second general observation is that except for arginine vasotocin (AVT) and to a lesser extent arginine vasopressin (AVP), there is generally a 100- to 1000-fold difference in potencies at oxytocin-like and vasopressin-like receptor systems. Thus, for example, oxytocin (OT) has 546 units/mg of rat uterotonic activity but only 31 units/mg of rat pressor activity. This level of receptor specificity is apparently sufficient such that under most physiological conditions, OT is primarily active only at its smooth-muscle receptors and not at the vasopressin-like receptors.

A third observation is that each of the naturally occurring neurohypophyseal hormones are nonapeptides with a 20-membered disulfide-containing ring and a carboxamide C-terminal amino acid, glycinamide.

Examination of the structural modifications that nature has made and of their effects on biological activity at the different receptors reveals some interesting insights into the properties of the different neurohypophyseal hormone receptor systems. First, Tyr^2 is replaced only in phenylpressin by a Phe residue, and this leads to a significant loss in potency at all receptor systems. In the 3 position, oxytocin-like analogs all have Ile, while vasopressin analogs have a Phe residue. Comparing AVT with AVP, which otherwise have identical primary structures, it is seen that the $Ile^3 \rightarrow Phe^3$ substitution leads to a decreased potency at the oxytocin receptor systems (Table I). In general the substitution of Gln^4 for Asn or Ser has only minor effects on potency at the oxytocin receptors but has more extensive effects at vasopressin receptors. The major structural change that appears to account for the biological activity differences at oxytocin-like versus vasopressin-like receptor systems in naturally occurring neurohypophyseal hormones is the change from Leu^8 to Arg^8 (or Lys^8). Compounds that have more lipophilic amino acids at position 8 (Leu, Ile, Val, Gln) possess oxytocin-like activities, while those with basic amino acids (Arg, Lys) at this position have vasopressin-like activities (Table I). It is particularly interesting to note that the aliphatic substitutions lead to only small differences in potency at the oxytocin-like receptors [compare especially OT with mesotocin (MT), valitocin (VT) and isotocin (IT)]. On the other hand, these same changes result in considerably larger variations in potency in the antidiuretic and pressor assays. In the discussion that follows, we will see numerous other examples of the importance of the 8 position for the biological activity of neurohypophyseal hormones.

We will now systematically investigate structure–activity relationships in neurohypophyseal hormones. For the sake of a more cohesive discussion we will investigate structure–activity relationship at the various receptor systems, taking first oxytocin agonist analogs, then oxytocin antagonist analogs, then vasopressin agonist and antagonist analogs that have been investigated in these assay systems.

III. STRUCTURE–ACTIVITY RELATIONSHIPS OF OXYTOCIN AGONISTS

A. At the Uterine Receptor

Since the earliest efforts to determine the structure(s) of the uterotonic principals in the brain, the *in vitro* rat uterus preparation has been a standard assay for oxytocin and its analogs. Since it has been the most widely used assay system for oxytocin analogs, we will first examine structure–function relationships at this receptor system and then turn to the avian vasodepressor (AVD) and milk ejecting assay systems.

1. *Fragment Analogs of the Neurohypophyseal Hormones*

A classic question of peptide hormone and neurotransmitter structure–function analysis is, what is the minimal sequence that will give full agonist activity at a particular receptor? The fragments of oxytocin that have been examined are shown in Table II. The results in Table II suggest that the intact 20-membered disulfide-containing ring of oxytocin, i.e., oxytocin$_{1-6}$ (tocinoic acid), is sufficient for full biological activity at the rat uterine receptor. Interestingly, the ring of vasopressin (pressinamide, see Table XII), which is about 10-fold less potent than the corresponding oxytocin analog (tocinamide, Table II) also does not have full agonist activity at the rat uterus. N-Terminal truncated analogs of oxytocin appear to have little or no oxytocin biological activity. Perhaps most interesting is that oxytocinoic acid (deamido-oxytocin), although more potent then tocinoic acid (Table II) is actually less active than tocinamide. It is interesting to note that removal of the N-terminal amino group of tocinamide and replacing it with a hydrogen atom leads to a 10-fold increase in potency (Table II). Deamino-oxytocin also is more active than oxytocin, but only about twofold (see Table IV). We shall also see that modifications to oxytocin that lead to inhibitors at the uterine receptors, when applied to the tocinamide or tocinoic acid structures, also yield inhibitors, although very little has been done to examine this in any detail.

2. *Structurally Extended Analogs of the Neurohypophyseal Hormones*

Early work on extending the N-terminal Cys[1] residue of oxytocin with an additional amino acid or peptide led in some cases to analogues with low potency but prolonged activity [this led to the hormonogen concept (Rudinger, 1971; Rudinger *et al.*, 1972; Berde and Boissonnas, 1968)] and in other cases led to inhibitors (Rudinger and Krejci, 1968). As illustrated in Tables II and XIII, there have been very few recent efforts to examine the effect of N- or C-terminal chain

Table II. Fragments of Oxytocin

	Rat Uterus	Avian Vasodepressor	Milk Ejecting	Antidiuretic	Pressor	Ref.
Oxytocinoic acid	1.3	0.2	12	0.02	inactive	Ferrier and du Vigneaud (1966)
Oxytocinoic acid (deamino)	<0.2	8.6				Takashima *et al.* (1970)
Tocinamide	2.7	inactive	5.9	1.3	inactive	Zaoral and Flegel (1972a); Hruby *et al.* (1971)
[Ser^4]tocinamide	inactive	inactive			inactive	Smith *et al.* (1974)
Tocinamide (deamino)	34.2	inactive				Hruby *et al.* (1971)
Tocinamide (deamino)	16.5	inactive		0.21	inactive	Zaoral and Flegel (1972b)
[Ser^4]tocinamide (deamino)	inactive	inactive			inactive	Smith *et al.* (1974)
[$(C_2H_5)_2Mpr^1$]tocinamide	pA_2 6.68	pA_2 7.42			inhibitor	Nestor *et al.* (1974)
Tocinoic acid	0.2-0.3	<0.07				Hruby *et al.* (1972)
Tocinoic acid (deamino)	3.7	inactive				Hruby *et al.* (1972)
[Ser^4]tocinoic acid	inactive	inactive			inactive	Smith *et al.* (1974)
[Ser^4]tocinoic acid (deamino)	inactive	inactive			inactive	Smith *et al.* (1974)
[$(C_2H_5)_2Mpr^1$]tocinoic acid	pA_2 6.48	pA_2 6.75			inhibitor	Nestor *et al.* (1974)
[des(Leu^8,Gly-NH_2)]oxytocin	<3% OT	<3% OT		<3% OT	<3% OT	Havran *et al.* (1973)
Ile-Gln-Asn-Cys(Cys)-Pro-Leu-Gly-NH_2						van Nispen *et al.* (1984)
<Glu-Asn-Cys(Cys)-Pro-Leu-Gly-NH_2						van Nispen *et al.* (1984)
<Glu-Asn-Cys(Cys)-Pro-Leu						van Nispen *et al.* (1984)
Asn-Cys(Cys)-Pro-Leu-Gly-NH_2						van Nispen *et al.* (1984)
<Glu-Asn-Cys-Pro-Leu-Gly-NH_2 | <Glu-Asn-Cys-Pro-Leu-Gly-NH_2						van Nispen *et al.* (1984)
<Glu-Asn-Cys-Pro-Leu | <Glu-Asn-Cys-Pro-Leu						van Nispen *et al.* (1984)
<Glu-Asn-Cys(Cys)Pro						van Nispen *et al.* (1984)

Table III. Extended Analogs of Oxytocin

Analogs extended at the NH-Terminus						
	Rat Uterus	Avian Vasodepressor	Milk Ejecting	Antidiuretic	Pressor	Ref.
N-Bromoacetyl						Walter *et al.* (1973)
N-Carbamoyl	~0.5					Smyth (1970)
N-Oxytocinoyl	1	0.44				Takashima and du Vigneaud (1970)
N-Deaminooxytocinoyl	2.45	0.28				Takashima and du Vigneaud (1970)
N-Glycyl			2.35			Barth *et al.* (1981)
N-(2-Nitro-5-azido)benzoylglycyl						Stadel *et al.* (1978)
N-Acetyl 2-Tyr(Ac)	30:1[a]					Smyth (1967)
N-Acetyl 2-Tyr(Me)	10:1[a]					Jost and Sorm (1971b)
N-Bromoacetyl 2-Tyr(Me)	pA_2 7.33		inactive		inhibitor	Krojidlo *et al.* (1975)
N-Carbamoyl 2-Phe	40:1[a]					Smyth (1970)
N-Carbamoyl 2-Phe(4-Et)	15:1[a]					Smyth (1970)
N-Carbamoyl 2-Phe(4-Me)	9:1[a]					Smyth (1970)
N-Carbamoyl 2-Tyr(Ac)	30:1[a]					Smyth (1970)

Analogs extended at the NH-Terminus							
		Rat Uterus	Avian Vasodepressor	Milk Ejecting	Antidiuretic	Pressor	Ref.
N-Carbamoyl	2-Tyr(Carbamoyl)	25:1[a]					Smyth (1970)
N-Carbamoyl	2-Tyr(Et)	126:1[a]					Smyth (1970)
N-Carbamoyl	2-Tyr(Me)	pA_2 6.4-7.1	inhibitor	inhibitor		inhibitor	Bisset *et al.* (1970)
N-Glycyl	2-Tyr(Me)	pA_2 6.85		0.64		inhibitor	Krojidlo *et al.* (1975)
N-Maleoylglycyl	2-Tyr(Me)	pA_2 7.27		inactive	0.001		Krojidlo *et al.* (1976)
N-Methanesulfonyl	2-Tyr(Me)	pA_2 7.82		inactive		inhibitor	Krojidlo *et al.* (1975)
N-Pivaloyl	2-Tyr(Me)	pA_2 7.26		1.13		inhibitor	Korjidlo *et al.* (1975)
N-Sarcosine	2-Tyr(Me)	pA_2 6.52		inactive		inhibitor	Korjidlo *et al.* (1975)
N-Acetyl	8-Lys						Schillinger *et al.* (1972)
N-Glycyl	8-Lys						Schillinger *et al.* (1972)

Analogs extended at the COOH-Terminus						
	Rat Uterus	Avian Vasodepressor	Milk Ejecting	Antidiuretic	Pressor	Ref.
Oxytocinoyl-glycinamide	~2					Aoyagi *et al.* (1968)

[a]Ratio of analog to oxytocin required to achieve 50% inhibition of agonist response.

extension. The preparation of the oxytocin and deamino-oxytocin dimers with oxytocin constitute an example of both a C-terminal extension of oxytocin via the Gly^9 carboxylate group and the N-terminal via the Cys^1 amino group. As shown in Table III, both of these compounds have very low agonist activity. A very interesting recent case in the N-terminal extended analog of lysine vasopressin (LVP), Gly-Gly-Gly-LVP (Table XIII), which was found to be about 12 times more potent than LVP in the rat uterus assay. These and earlier results suggest that N-terminal extension of neurohypophyseal hormone analogs can result in interesting and unique properties.

3. Monosubstituted Analogs of the Neurohypophyseal Hormones

Early in structure–activity studies of the neurohypophyseal hormones, it was recognized that essentially all of the amino acid residues were involved in determining potency at the uterine receptor. Thus, cleavage of the 20-membered ring at the Cys^1-Try^2 bond by tyrosinase greatly reduced or eliminated the biological activity of the hormone. The importance of having an intact 20-membered ring on the biological activity of the hormone was investigated by the preparation of the reduced sulfhydryl form of oxytocin, H-Cys(SH)-Tyr-Ile-Gin-Asn-Cys(SH)-Pro-Leu-Gly-NH_2 (Yamashiro *et al.*, 1966). This compound was found to possess greatly reduced potency and, perhaps more interesting, the onset of activity was slow, suggesting that perhaps part or all of it was due to oxidation back to the cyclic disulfide. In this regard the pseudoisosteric linear analog of oxytocin, [Ala^1, Ala^6]-oxytocin (Jost et al., 1964) was found to be essentially devoid of biological activity. Perhaps a more realist pseudoisosteric analog is [Ser^1, Ser^6]OT (see Table V). This analog has mixed agonist–antagonist activities. These results and those previously discussed suggest that all the transduction message for oxytocin (at least at the uterine receptor system) is found in the intact 20-membered ring of the hormone.

a. Position 1. As pointed out above, substitution of the N^α-amino group with many substituents leads to weak antagonist analogs (Table III). However, if amino acids are added, a hormonogen with weak agonist potency generally is obtained. A small group such as a carbamoyl group can be added (Table III), but a drop in potency of over 1000-fold results. The uterine receptor clearly prefers a very small group at the α-amino position. Indeed, replacement of the α-amino group with a hydrogen to give [1-β-mercaptopropionic acid]OT ([Mpr^1]OT deamino-oxytocin) (Table IV) gives an oxytocin analog with increased potency at the uterine receptor. Even more interesting, when the α-amino group is replaced with a hydroxyl group ([LHmp^1]OT), an analog with more than double the potency of OT in the oxytocic assay system is obtained. Thus it appears that the

Table IV. Analogs of Oxytocin with a Substitution in a Single Position

Position One Substitutions						
	Rat Uterus	Avian Vasodepressor	Milk Ejecting	Antidiuretic	Pressor	Ref.
L-Hmp	1275			16.6	14.7	Manning *et al.* (1976) Walti and Hope (1974)
D-Hmp	27.3					Walti and Hope (1974)
Maa	25	4		0.1		Jarvis and du Vigneaud (1967)
$(CH_3)_2$Maa	pA_2 6.16	pA_2 7.21				Vavrek *et al.* (1972)
Mpr (deaminooxytocin)	803	975	541	19	1.44	Ferrier *et al.* (1965)
2,2-Dimethyl-3-mercaptopropanoic acid	pA_2 6.60	pA_2 7.41				Vavrek *et al.* (1972)
$(CH_3)_2$Mpr	pA_2 6.94	pA_2 7.88			pA_2 6.27	Nestor *et al.* (1975b)
$(C_2H_5)_2$Mpr	pA_2 7.24	pA_2 8.11			pA_2 6.24	Nestor *et al.* (1975b)
$(CH_2)_5$Mpr	pA_2 7.43	pA_2 8.30			weak	Nestor *et al.* (1975b)
11-mercaptoundecanoic acid	0	0				Fraefel and du Vigneaud (1970b)
5-Mercaptovaleric acid	0	0				Fraefel and du Vigneaud (1970a)
Pen	pA_2 6.86	pA_2 7.50				Vavrek *et al.* (1972)
D-Pen	pA_2 6.32	pA_2 6.78				Vavrek *et al.* (1972)
3-Selenopropanoic acid	560	613	248	24.0	1.7	Walter and Chan (1967)

(continued)

Table IV. *(continued)*

Position Two Substitutions						
	Rat Uterus	Avian Vasodepressor	Milk Ejecting	Antidiuretic	Pressor	Ref.
Ala (deamino)	<0.01	0.02		<0.001	<0.01	Walter and Schwartz (1968)
4-(Bromoacetylamino)phenylalanine (deamino)	pA_2 6.73				inhibitor	Pliska and Marbach (1978)
Cycloleucine	4.9	-	-	-	inactive	Hruby *et al.* (1983)
3,5-Dibromotyrosine (deamino)	pA_2 7.05	pA_2 7.44			0	Lundell and Ferger (1975b)
3,4-Dihydroxyphenylalanine	26	54				Ferrier and Branda (1967a)
2-hydroxyphenylalanine						Midura-Nowaczek and Drabarek (1978)
4-Fluorophenylalanine	97					Marbach and Rudinger (1974b)
4-Fluorophenylalanine (deamino)	57					Marbach and Rudinger (1974b)
Ile	20	48.3	50		0.1	Branda *et al.* (1967)
Ile (deamino)	22	81.2	80		0.1	Branda *et al.* (1967)
3-Iodotyrosine	pA_2 7.2			slight	inhibitor	Marbach and Rudinger (1974a)
Leu	<0.5	3	7	slight	0.05	Hruby and du Vigneaud (1969a)
Pentafluorophenylalanine	low	-	-	-	-	Kaurov *et al.* (1973b)
D-Pentafluorophenylalanine	inhibitor	-	-	-	-	Kaurov *et al.* (1973b)

Position Two Substitutions						
	Rat Uterus	Avian Vasodepressor	Milk Ejecting	Antidiuretic	Pressor	Ref.
D-Phe	inhibitor	-	-	-	-	Kaurov *et al.* (1973b)
D-Phe (4-Et)	pA_2 8.15		1.22	1.5	-	Lebl *et al.* (1985a)
D-Phe (4-Et) (deamino)	pA_2 8.06		0.3	-	pA_2 6.82	Lebl *et al.* (1985a)
4-(Propionylamino)phenylalanine	pA_2 6.65	-	-	-	inhibitor	Pliska and Marbach (1978)
D-Tyr	6.6	34	34	0.02	0.02	Drabarek and du Vigneaud (1695)
D-Tyr	partial agonist	-	-	inactive	inactive	Hruby *et al.* (1979)
D-Tyr (deamino)	<1	30	16	0.07	0.04	Drabarek and du Vigneaud (1965)
D-Tyr(Et)	3.3	-	-	-	-	Kaurov *et al.* (1972)
Tyr(3-Me)	pA_2 6.79	-	-	0.01	inactive	Pliska *et al.* (1977)
Tyr(3-nitro)	1.1	-	-	-	-	Hope and Walti (1973)
Trp	81.2	-	77.5	-	-	Ferrier and Branda (1976b)
D-Trp	inhibitor	-	-	-	0	Kaurov *et al.* (1972)
Val	2.1	6.8	15.6	<0.01	0.05	Hruby and du Vigneaud (1969a)

(continued)

Table IV. *(continued)*

Position Three Substitutions						
	Rat Uterus	Avian Vasodepressor	Milk Ejecting	Antidiuretic	Pressor	Ref.
D-Ala	inhibitor	inhibitor	-	<0.0005	inhibitor	Roy *et al.* (1975)
3-Cyclohexylalanine	1.6	2.8	-	0.3	0.1	Walter *et al.* (1976a)
2-Cyclohexylglycine	8.2	8.5	40	0.25	0.7	Eisler *et al.* (1966)
D-2-Cyclohexylglycine	0.03	0.025	0.1	0.03	-	Eisler *et al.* (1966)
2-Cyclopentylglycine	252	226	380	29.3	7.4	Eisler *et al.* (1966)
3,3-Diethylalanine	305	313	400	1.4	1.2	Eisler *et al.* (1966)
Leu	5	7.1	35	0.03	1.0	Eisler *et al.* (1966)
Phe	~20	~30	~65	~30	~3	Berde *et al.* (1957)
Phe	39	73	-	34	5.1	Walter *et al.* (1976a)
Pro	<0.02	0	-	0.003	0.005	Roy *et al.* (1975)
Val	51	39	67	0.6	0.04	Eisler *et al.* (1966)
Val	59	58	206	~0.1	~0.8	Berde *et al.* (1961)

Position Four Substitutions						
	Rat Uterus	Avian Vasodepressor	Milk Ejecting	Antidiuretic	Pressor	Ref.
Ala	36	65	240	<0.01	<0.01	Guttmann and Boissonnas (1963)
D-Ala	0.05	0.07	-	<0.0001	0	Roy *et al.* (1975)
1-Aminocyclopentanecarboxylic acid	3	4	-	-	-	Ferger (1974)
Asp	-	-	-	inhibitor	-	Datta and Chaudhury (1968) Datta and Chaudhury (1970)
Asp(NH-CH_3)	41	119	-	0.045	<0.5	Havran *et al.* (1969a)
Asp(NH-CH_3) (deamino)	59	186	-	0.012	<0.4	Havran *et al.* (1969a)
Asn	107	201	298	0.044	0.13	Berde and Boissonnas (1968)
Asn	158	-	382	0.17	0.12	Turan *et al.* (1977)
Glu	1.5	0.5	11	<0.01	<0.01	Photaki and du Vigneaud (1965)
Glu (deamino)	13.3	-	34.2	-	-	Ferrier and Branda (1975)
Glu (OMe)	10.5	42	101	-	inactive	Barth *et al.* (1981) Photaki *et al.* (1979)
Glu (OMe) (deamino)	21.4	82	271	-	inactive	Photaki *et al.* (1979) Barth *et al.* (1981)

(continued)

Table IV. *(continued)*

Position Four Substitutions							
		Rat Uterus	Avian Vasodepressor	Milk Ejecting	Antidiuretic	Pressor	Ref.
D-Gln		3.3	<0.05	0.09	0.05	<0.1	Hruby *et al.* (1972)
Des-Gln		3	-	-	-	-	Sawyer and Manning (1973)
Des-Gln	(deamino)	0.5	-	-	-	-	Sawyer and Manning (1973)
Hse		125	-	357	0.24	0.021	Turan *et al.* (1977)
Ile		37	81	184	0.02	<0.05	Hruby *et al.* (1969)
Ile	(deamino)	67	307	156	<0.05	0.06	Takashima *et al.* (1970)
Leu		13	44	66	inhibitor	depressor	Hruby *et al.* (1969) Chan and duVigneaud (1970)
Leu	(deamino)	37	141	143	0.015	<0.05	Takashima *et al.* (1970)
D-Leu		inhibitor	inhibitor	-	0	0.004	Roy *et al.* (1975)
Lys		6	-	-	-	-	Kaurov *et al.* (1971)
Met		15	-	-	-	-	Lozhkina *et al.* (1973)
Nle		20.5	51	87	<0.04	0.015	Flouret and du Vigneaud (1969)
Nle	(deamino)	12.8	53	50	<0.005	0.025	Flouret and du Vigneaud (1969)
Nva		61	99	182	0.04	<0.001	Flouret and du Vigneaud (1969)

Position Four Substitutions							
		Rat Uterus	Avian Vasodepressor	Milk Ejecting	Antidiuretic	Pressor	Ref.
Nva	(deamino)	56	128	134	0.03	<0.001	Flouret and du Vigneaud (1969)
Orn		58	163	127	0.029	<0.1	Havran *et al.* (1969b)
Phe		0.75	pA_2 6.93	-	0	inhibitor	Nestor *et al.* (1975a)
Pro		0.007	-	0.04	low	-	Sawyer *et al.* (1969)
Ser		197	-	423	2.4	0.028	Turan *et al.* (1977)
Ser	(deamino)	37	-	-	-	-	Sawyer and Manning (1973)
2-Thienylalanine		0.51	pA_2 7.44	-	<0.05	<0.05	Smith *et al.* (1978a)
Thr		900	1480	540	1.8	0.43	Manning *et al.* (1970)
Thr	(deamino)	149	781	385	0.9	<0.01	Manning *et al.* (1971b)
aThr		170	-	430	-	<0.2	Bankowski *et al.* (1983)
Val		139	230	-	-	<0.005	du Vigneaud *et al.* (1966)
Val		-	-	419	0.05	-	Hruby *et al.* (1969)

(*continued*)

Table IV. *(continued)*

Position Five Substitutions						
	Rat Uterus	Avian Vasodepressor	Milk Ejecting	Antidiuretic	Pressor	Ref.
Asp	20.3	41	-	0.14	-	Walter *et al.* (1978)
2-Aza-asparagine	-	-	-	-	-	Niedrich (1972)
3-Cyanoalanine	1.33	-	9.8	0.02	inactive	Roy *et al.* (1983a)
Dbu	0.03	0.03	-	<0.001	<0.01	Hase *et al.* (1972c)
3-Malamidic acid	45.1	-	203	0.32	0.37	Roy *et al.* (1982)
Orn	0.29	0.07	0.19	0.002	0.04	Havran *et al.* (1969b)
Val	0.3	0.1	-	<0.001	0.01	Walter and Schwartz (1966)
Val (deamino)	0.7	0.3	-	<0.001	0.01	Walter and Schwartz (1966)
Position Six Substitutions						
	Rat Uterus	Avian Vasodepressor	Milk Ejecting	Antidiuretic	Pressor	Ref.
Hcy	12	0.3	-	-	0	Smith and Ferger (1976)
Pen	2.7	inhibitor	-	-	-	Schulz and du Vigneaud (1967)
Sec	405	385	398	3.4	3.8	Walter and Chan (1967)
Sec (deamino)	492	622	397	15.5	1.1	Walter and Chan (1967)

Position Seven Substitutions						
	Rat Uterus	Avian Vasodepressor	Milk Ejecting	Antidiuretic	Pressor	Ref.
Ala	22	4.8	-	0.08	0.4	Walter *et al.* (1976b)
Ala (deamino)	123	9.8	-	0.17	0.5	Walter *et al.* (1976b)
1-Aminocyclopropane-1-carboxylic acid	9.8	-	62	-	0.2	Prochazka *et al.* (1984)
Azetidine-2-carboxylic acid	43	-	66	0.1	0.25	Barber *et al.* (1979)
3,4-Dehydroproline	1071	548	-	5.9	-	Moore *et al.* (1977)
3,4-Dehydroproline (deamino)	1065	1008	-	23.3	-	Moore *et al.* (1977)
Gly	68	1.5	235	-	-	Bespalova *et al.* (1968)
Gly	65	5.3	-	<0.01	0.3	Walter *et al.* (1976b)
Gly	93	-	392	0.0056	<.01	Lowbridge *et al.* (1977a)
Gly (deamino)	355	17	-	0.062	0.5	Walter *et al.* (1976b)
3-Hyp	~33	-	-	-	-	Bespalova *et al.* (1967)
D-Leu	0.61	0	0.51	-	0	Bespalova *et al.* (1968)
MeAla	62	-	402	0.12	inactive	Grzonka *et al.* (1983b)
MeAla (deamino)	98	-	157	-	inactive	Grzonka *et al.* (1983b)
Nva	-	-	-	-	-	Kaurov *et al.* (1973a)
Pip	~220	-	-	-	-	Martynov *et al.* (1967)
D-Pro	13	<0.005	3.9	1	<0.005	Ferraro and du Vigneaud (1966)
D-Pro (deamino)	45	0	8	-	0	Ferraro and du Vigneaud (1966)
Sar	459	-	564	0.18	inactive	Grzonka *et al.* (1983b)
Sar (deamino)	745	-	498	-	inactive	Grzonka *et al.* (1983b)
Thiazolidine-4-carboxylic acid	1180	720	-	-	3.46	Rosamond and Ferger (1976)
Thiazolidine-4-carboxylic acid (deamino)	1538	2039	-	-	1.69	Rosamond and Ferger (1976)

(*continued*)

Table IV. *(continued)*

Position Eight Substitutions						
	Rat Uterus	Avian Vasodepressor	Milk Ejecting	Antidiuretic	Pressor	Ref.
Ala	141	135	208	0.08	11.6	Jaquenoud (1961)
Ala	166	240	-	2.5	13	Walter and du Vigneaud (1966)
Ala (deamino)	314	453	-	2	8	Walter and du Vigneaud (1966)
2-Hydroxyisocaproic acid	158	-	63	0.06	0.36	Roy *et al.* (1980a)
Gln	58	-	256	5.9	34	Baxter *et al.* (1969b)
Gln (deamino)	-	-	-	-	-	Schillinger *et al.* (1972)
Ile	289	498	328	1.1	6.3	Jaguenoud and Boissonnas (1961)
Laminine	-	-	-	-	-	Kaurov *et al.* (1973c)
MeLeu	45	-	46	-	0.27	Hlavacek *et al.* (1977)
MeLeu (deamino)	101	-	45.2	-	3.15	Roy *et al.* (1983b)
Lys	100	-	-	-	-	Snell and Smyth (1977)
Lys (carbamoyl)	450	-	-	-	-	Snell and Smyth (1977)
Lys(3-carboxypropionyl)	55	-	-	-	-	Snell and Smyth (1977)
Lys(3-carboxybutyryl)	50	-	-	-	-	Snell and Smyth (1977)
Met	200	193	-	0.4	27	Bodanszky *et al.* (1979)
Ser	145	-	-	-	-	Sawyer *et al.* (1970)
Phe	108	191	341	1.16	1	Baxter *et al.* (1969a)
Thr	-	-	-	-	-	Sawyer and Manning (1973)
Tert-leucine	45	-	508	0.7	2.7	Lebl *et al.* (1982b)
Tert-leucine (deamino)	184	-	120	-	1.4	Lebl *et al.* (1982b)
Trp	55	-	-	-	<0.25	Bodanszky *et al.* (1980)
Val	199	278	308	0.8	9	Jaquenoud and Boissonnas (1961)

Position Nine Substitutions						
	Rat Uterus	Avian Vasodepressor	Milk Ejecting	Antidiuretic	Pressor	Ref.
Ala	0.25	-	-	-	-	Dutta *et al.* (1966b)
β-Ala	0.80	0.18	-	-	-	Dutta *et al.* (1966b)
Aminoacetone	2.8	-	2.1	0.2	-	Hlavacek *et al.* (1982)
Methylamine (decarboxamido)	7.2	<0.02	3.8	0.2	0	Branda and du Vigneaud (1966)
Gln	0.05	<0.01	-	0.01	0	Grzonka *et al.* (1974)
Gly-OH	1.3	0.2	12	0.02	0	Ferrier and du Vigneaud (1966)
Gly-OH (deamino)	<0.2	8.6	-	-	-	Takashima and du Vigneaud (1970)
Gly-$NHCH_3$	-	-	-	-	-	Beyerman *et al.* (1968)
Gly-$NHCH_3$ (deamino)	29	0	-	-	-	Takashima *et al.* (1969)
Gly-$N(CH_3)_2$	~15	<0.07	-	<0.01	0.096	Ting *et al.* (1980)
Gly-$N(CH_3)_2$ (deamino)	16	0	-	-	-	Takashima *et al.* (1969)
Glycolic amide (deamino)	134	-	108	2.5	0.35	Roy *et al.* (1980c)
Hydrazino acetic acid (azaglycine)	-	-	-	-	-	Niedrich (1972)
Leu	-	inhibitor	-	-	-	Grzonka *et al.* (1974)
Leu	~144	-	-	-	-	Aoyagi *et al.* (1968)
Met	-	inhibitor	-	-	-	Grzonka *et al.* (1974)
Pro	0.11	inhibitor	-	0.03	0	Grzonka *et al.* (1974)
Thioglycine (deamino)	46	14.9	-	-	-	Jones *et al.* (1973)

the α-amino group actually may have a slightly repulsive effect at the uterine receptor. In addition, reaction of the amino group in the 1 position with acetone [to give the oxazolidenone (Hruby *et al.*, 1969)] leads to a large decrease in potency, as does the N-isopropyl substitution (Hruby and du Vigneaud, 1969b). Clearly there are very stringent steric requirements for the α-amino group in the 1 position.

The stereochemical requirements at the 1 position also are very severe, since [DCys[1]]OT has less than one-thousandth the potency of OT in the oxytocic assay (Jost *et al.*, 1963; Hope *et al.*, 1963).

There also is a stringent ring-size requirement for neurohypophyseal hormone interaction with the uterine receptor. Increasing the ring size one methylene unit ([Hcy[1]]OT) (Jarvis *et al.*, 1961) leads to nearly a 1000-fold decrease in potency in the uterine assay. In like manner, decreasing the ring size by one methylene group ([Maa[1]]OT) (Table IV) leads to a 20-fold decrease in potency in the oxytocic assay. Effort to increase the ring size further ([5-mercaptovaleric acid[1]]OT and [11-mercaptoundecanoic acid[1]]OT) leads to inactive analogs (Table IV).

Substitution of the β-methylene of Cys[1] with one methyl group to give [β-mercaptobutyric acid[1]]OT leads to an analog with weak uterotonic activity (Schulz and duVigneaud, 1966b), but substitution of two methyl groups to give [1-penicillamine]OT leads to an antagonist (Table IV) analog (Schulz and du Vigneaud, 1966a). We will return to this later.

Efforts to modify the disulfide bridge of OT and the other neurohypophyseal hormones have led to mixed results. On the one hand, replacement of the disulfide group with an amide group, cyclo[Asp[1], Dpr[6]]OT, leads to a 30-fold decrease in potency. However, both the carba-1 analogs (the sulfur of the Cys[1] residue is replaced by a CH_2 group) and the carba-6 analogs (Table V) are more potent than the native hormone at the uterine receptor. It would appear that the appropriate van der Walls interaction between the cystine residue of OT and the uterine receptor is important for binding. Interestingly, the bis-carba analog (Table V) has *considerably* less potency than the monocarba analog or even than OT. These results suggest very stringent steric requirements for hormone–uterine receptor interaction in the vicinity of the cystine bridge of the hormone. This is further amplified by a number of other observations. For example, the specific stereoelectronic requirements are illustrated by the surprisingly low potency of [β-mercaptoacetic acid[1],homocysteine[6]]OT, which has the same ring size as OT but has the disulfide bridge shifted over one methylene group toward the amino terminal (Table V). It may be that this structural change causes a large change in the conformation relative to OT, but a more likely cause is the rather stringent stereoelectronic requirements for the correct position of the disulfide bridge in the oxytocin–uterine receptor interaction. In view of the high potency of the carba-1 and carba-6 analogs, it is somewhat surprising that the homolanthionine analog has only one-tenth the potency of OT; on the other hand, in view of the low potency of all analogs with modified ring sizes, the low potency of the deamino-[Lan^6]OT analog (Table V) is not surprising, since a CH_2 group is formally removed from either the 1 or the 6 position. The very stringent stereoelectronic requirements in the vicinity of the disulfide bridge are again illustrated by the low potency of the deamino-1-carba[$S^+(CH_3)I^-$] analog and the dramatically reduced potency of [deamino-1-carba sulfoxide]OT (Table V). It is hard to explain why the 1-carba sulfoxide analog (Table V) should have such low potency, though it is interesting that the deamino-6-carba sulfone analog (Table V) has high potency (about one-half that of the deamino-6-carba analog in the rat uterus assay). One possible explanation is that the sulfide moiety is interacting with the side chain group of Tyr^2 (or some other residue) in an asymmetric manner such that substitution at one of the sulfur atoms affects potency much more than at the other sulfur atom.

In summary, these results suggest that all aspects of the structure of the Cys^1 residue including the stereochemical arrangement of the α-carbon atom are critically involved in interacting with the uterine receptor, and the inhibitory activity of β,β-disubstituted analogs such as [Pen^1]OT suggests that the 1 position may also be involved in transduction, either directly, or indirectly by its transannular conformational effects.

b. Position 2. The Try^2 residue appears to be a critical residue for oxytocin binding and transduction. Indeed, there is much evidence to suggest that its correct stereochemistry, conformational properties, and dynamic behavior are critical for both agonist (Walter, 1977) and antagonist (Meraldi *et al.*, 1977; Hruby, 1981a,b) activities. The partial agonist activity of [DTyr^2]OT (Table IV) (Hruby et al., 1979) has been argued to be consistent with the "cooperative model" (Walter, 1977) of oxytocin activity at the uterine receptor. On the other hand, it was recently shown that [*c*Leu^2]OT (Table IV) is a weak full agonist with the ability to fully stimulate the uterine receptor in a dose–response assay. Since this residue has no aromatic ring and cannot possess either the conformational or stereoelectronic properties proposed for the cooperative model, it is suggested that certain aspects of this model may need revision. Interestingly, replacement of the *p*-hydroxy group of Tyr^2 with a hydrogen leads to an analog, [Phe^2]OT (Bodanszky and du Vigneaud, 1959; Jaquenoud and Boissonnas, 1959), which is about one-fifteenth as potent as the native hormone, and the hydroxyl group can be replaced with a fluorine, [Phe(*p*-F)^2]OT, or an addition hydroxyl can be group added, [Phe(3,4-OH_2)^2]OT (Table IV), with maintenance of fairly high potency. In addition. [Trp^2]OT (Table IV) has good potency in the uterine assay. However, substition of bulky halogens at the 3′ and 3′,5′ positions of Tyr^2 leads to antagonists (Table IV). O-Alkylation of the Tyr^2 hydroxyl group (Rudinger and Krejci, 1968) (Table IV) generally leads to antagonist or mixed

Table V. Disubstituted Analogs of Oxytocin with Modifications at Position One

Substitutions at Positions One and Two							
1-Position	2-Position	Rat Uterus	Avian Vasodepressor	Milk Ejecting	Antidiuretic	Pressor	Ref.
Ace-	tone		0.5				Yamashiro *et al.* (1965)
Pen	Leu	pA_2 7.14	-	inhibitor	-	-	Hruby *et al.* (1979b)
Pen	Phe(4-Et)	pA_2 7.78		0.45	0.025	pA_2 7.8	Lebl *et al.* (1985a)
Pen	D-Phe(4-Et)	pA_2 8.09		pA_2 7.7	0.0007	pA_2 7.4	Lebl *et al.* (1985a)
$(CH_3)_2$Mpr	Phe	pA_2 7.78	-	agonist	0.09	pA_2 6.57	Sawyer *et al.* (1980)
$(CH_3)_2$Mpr	Tyr(Me)	pA_2 7.76	-	pA_2 6.94	0.02	pA_2 7.59	Sawyer *et al.* (1980)
$(C_2H_5)_2$Mpr	3,5-dibromotyrosine	pA_2 7.08	pA_2 7.38	-	-	inactive	Lundell *et al.* (1975)
Substitutions at Positions One and Four							
1-Position	4-Position	Rat Uterus	Avian Vasodepressor	Milk Ejecting	Antidiuretic	Pressor	Ref.
Hmp	Thr	4179	-	808	5.3	4.92	Manning *et al.* (1976)
Pen	Thr	pA_2 7.55	-	inhibitor	-	-	Hruby *et al.* (1980)
$(C_2H_5)_2$Mpr	Leu	pA_2 7.02	pA_2 7.34	-	-	pA_2 6.08	Dyckes *et al.* (1974b)
$(CH_3)_2$Mpr	Thr	pA_2 7.52	-	4	0.02	pA_2 6.67	Manning *et al.* (1978)
$(C_2H_5)_2$Mpr	Thr	pA_2 7.72	-	-	-	pA_2 6.30	Lowbridge *et al.* (1979)
$(CH_2)_5$Mpr	Thr	pA_2 7.91	-	-	-	pA_2 5.86	Lowbridge *et al.* (1979)
Substitutions at Positions One and Five							
1-Position	5-Position	Rat Uterus	Avian Vasodepressor	Milk Ejecting	Antidiuretic	Pressor	Ref.
Maa	isoasparagine	0.098	-	inactive	inactive	inactive	Roy *et al.* (1980b)

Substitutions at Positions One and Six							
1- or 1,6-Position	6-Position	Rat Uterus	Avian Vasodepressor	Milk Ejecting	Antidiuretic	Pressor	Ref.
β-Ala	Asp (cyclo)	1.13	inhibitor	2.93	0.01	inactive	Hase *et al.* (1972b)
Aminoazelaic acid		-	-	-	-	-	Morikawa and Sakakibara (1970)
α-Aminopimelic acid		2.6	inactive	-	0.05	-	Jost and Sorm (1971c)
α-Aminopimelic acid		2.2	0.1	7.7	-	0.01	Barth and Kluh (1974)
Asp	Dpr (cyclo)	16	6.6	-	5.6	-	Smith *et al.* (1978c)
1-Carba		734	175	177	9.3	2.95	Jost *et al.* (1973)
1-Carba (N-glycyl)		-	-	2.35	-	-	Barth *et al.* (1981)
1-Carba sulfoxide		0.26	-	0.02	-	-	Lebl *et al.* (1978b)
Deamino-1-carba		1898	1127	562	24	17.5	Barth *et al.* (1973)
Deamino-6-carba sulfoxide		455.5	-	216.2	-	-	Lebl *et al.* (1978b)
Deamino-1-carba($S^+(CH_3)I^-$)		20.0	-	47.0	-	-	Lebl *et al.* (1979a)
$(CH_3)_2$Mpr	6-carba	16.2	-	6.9	4.81	pA_2 6.82	Lebl and Barth (1983)

(continued)

Table V. *(continued)*

Substitutions at Positions One and Six							
1- or 1,6-Position	6-Position	Rat Uterus	Avian Vasodepressor	Milk Ejecting	Antidiuretic	Pressor	Ref.
Deamino-6-carba		929	571	456	115	1.5	Jost and Sorm (1971a)
Deamino-1-carba sulfone		0.05	-	0.03	-	-	Lebl *et al.* (1978b)
Deamino-1-carba sulfoxide		13.1	-	91.0	-	-	Lebl *et al.* (1978b)
N^{ε}-carbonyllysine		3.9	pA_2 7.27	-	-	<0.05	Bodanszky *et al.* (1978)
1,6-Dicarba		5	-	-	-	<0.1	Keller and Rudinger (1974)
Deamino-1,6-dicarba		100	44	140	4.7	0.1	Yamanaka *et al.* (1970)
Hcy	Hcy	pA_2 6.02	pA_2 6.48	-	-	inactive	Smith and Ferger (1976)
Homolanthionine	(deamino)	60	28	943	0.34	0.02	Barth *et al.* (1973)
Lanthionine	(deamino)	9.2	6.8	42	0.17	inactive	Jost and Sorm (1971)
Maa	Hcy	37	0.3	-	-	inactive	Smith and Ferger (1976)
4-Mercaptobutyric acid	Hcy	pA_2 6.74	pA_2 7.34	-	-	pA_2 5.17	Smith and Ferger (1976)
3-Mercaptopropionic acid	Hcy	22	pA_2 7.57	-	-	inactive	Smith and Ferger (1976)
Ser	Ser	inhibitor + agonist	-	agonist	-	-	Polacek *et al.* (1970)

Substitutions at Positions One and Seven							
1-Position	7-Position	Rat Uterus	Avian Vasodepressor	Milk Ejecting	Antidiuretic	Pressor	Ref.
Hmp	3,4-dehydroproline	880	1295	-	76.7	-	Moore *et al.* (1977)
Substitutions at Positions One and Eight							
1-Position	8-Position	Rat Uterus	Avian Vasodepressor	Milk Ejecting	Antidiuretic	Pressor	Ref.
$(CH_3)_2$Mpr	2-hydroxyisocaproic acid	pA_2 7.37	-	0.025	0.04	inactive	Gazis *et al.* (1983)
$(CH_3)_2$Mpr	Orn	pA_2 7.89	-	agonist	0.87	pA_2 6.95	Sawyer *et al.* (1980)
$(CH_2)_5$Mpr	Orn	pA_2 7.69	-	pA_2 6.85	0.006	pA_2 7.26	Bankowski *et al.* (1980)
Substitutions at Positions One and Nine							
1-Position	9-Position	Rat Uterus	Avian Vasodepressor	Milk Ejecting	Antidiuretic	Pressor	Ref.
1-9 cyclooxytocin		inactive	-	inactive	-	inactive	Zaoral and Krchniac (1977)

agonist–antagonist analogs. Indeed, this was one of the earliest observations in the development of neurohypophyseal hormone antagonist analogs made in the laboratories of du Vigneaud and co-workers and Rudinger and co-workers (reviewed by Rudinger and Krejci, 1968). As may have been expected, similar results were obtained when *p*-alkyl-substituted Phe^2 analogs of the neurohypophyseal hormones were prepared. We will come back to this in our discussion of antagonists. All analogs with an aliphatic amino acid in position 2 have greatly reduced potency, with [Ala^2]OT being essentially inactive. Even [Leu^2]OT has less than one-thousandth the potency of oxytocin, although [Ile^2]OT has about one-twenty-fifth the potency of the native hormone (Table IV) (Berde and Boissonnas, 1968), with the [Val^2]OT analog intermediate in potency (Table IV). Although AVP and LVP analogs are much less potent in the oxytocic assay, examination of similarly substituted Tyr^2 analogs (Tables XVII and XXII) gave results consistent with the above discussion.

These data clearly show that the Tyr^2 residue of the neurohypophyseal hormones is critical for both binding and transduction at the uterine receptor. Continued exploration of the structural (and conformational) features that are controlling both binding and transduction at this residue will be fruitful in developing analogs with specific biological properties.

c. Position 3. A large number of early studies were made in which the Ile^3 residue of OT was substituted with a variety of different aliphatic amino acid residues. These studies have been reviewed (Rudinger *et al.*, 1972; Berde and Boissonnas, 1968). It was found that full activity was maintained, but that potency was reduced in the uterine assay. It appears that there is a specific steric requirement for the 3 position, presumably some sort of lipophilic pocket, for interaction of OT with its receptor on the uterus. More recent studies (Table IV) support this conclusion. One unresolved question relates to the stereospecific requirements of the 3-position substituent. [DIle^3]OT apparently has not been made or tested. There has been a report that [DAla^3]OT has inhibitor activity in the oxytocic assay system (Table IV), but the D-cyclohexylglycine^3 structure, which in some ways is pseudoisosteric to Ile^3, is a weak agonist (Table IV) with no partial antagonist activities. The 3 position does not appear to be directly involved in receptor transduction at the uterine receptor, but it is close to the 2 position and presumably could indirectly affect transduction. An examination of [DIle^3]OT biological activity might be quite interesting.

d. Position 4. The Gln^4 residue in OT has been replaced by a wide series of lipophilic and hydroxyl-containing side-chain residues. With the exception of the single report that [DLeu^4]OT is an antagonist (Table IV), all of these analogs (Table IV) (Berde and Boissonnas, 1968) have been reported to be full agonists in the uterine assay system. [DGln^4]OT was found to be a weak but full agonist at the uterine receptor, so the report that the [DLeu^4]OT is an antagonist is somewhat surprising and worthy of further investigation. A most interesting series of analogs is made up of the [Ser^4]-, [Hse^4]-, and [Thr^4]oxytocin analogs (Table IV). The Thr^4 analog is about two times more potent than OT in the rat uterus assay. Apparently the hydrogen bond donating and/or accepting properties of the Thr^4 hydroxyl group are important for the interaction of the analog with the uterine receptor, because when the pseudoisosteric but lipophilic amino acid Val (which has no hydrogen bond donating or accepting properties) replaced Thr in the 4 position, a sevenfold decrease in potency occurred. Replacement of the slightly polar but neutral glutamine side-chain group with the acid side-chain group of glutamic acid (an essentially isosteric replacement) gives [Glu^4]OT, which exhibits a large, 400-fold, reduction in biological potency (Table IV). This is somewhat reversed by methyl esterification to give [Glu(OMe)^4]OT (Table IV), but in this case there is a significantly greater steric requirement at the uterine receptor than with glutamine. Although the uterine receptor does not accommodate an acid amino acid residue well, it does accommodate a basic amino acid somewhat better. [Orn^4]OT (Table IV) is still a fairly potent analog in the rat uterine assay; the Lys^4 analog is considerably less potent. While the 4-phenylglycine analog has not been examined, substitution of an aromatic amino acid into the 4 position of oxytocin causes a large decrease in potency. For example, [Phe^4]OT (Table IV) is about 700-fold less potent than OT, and [2-thienylalanine^4]OT about 1000-fold less potent than OT in the uterine assay. It has been reported (Table XXII) that the Leu^4 analog of LVP, [Leu^4,Lys^8]VP (VP, Vasopressin), possesses antagonist activities at the uterine receptor. This has not been further examined. In conclusion, it would appear that the 4 position in the neurohypophyseal hormones is important for binding to the uterine receptor, but generally is not critical for transduction.

e. Position 5. The structure–function relationships of this position have not been widely examined. Presumably this is because early studies indicated that changes in structure at this position led to a drastic reduction in biological potency and efficacy (Berde and Boissonnas, 1968). For example, it was discovered early that 5-decarboxamide oxytocin ([5-Ala]OT) (du Vigneaud *et al.*, 1963) was a very weak partial agonist. These and related results led to the assumption that the Asn^5 residue was critical for transduction (Walter, 1977; Hruby, 1981a) at the uterine receptor. This may indeed be the case, but the more recent findings that [Asp^5]OT and the 3-malamidic acid analog [Mal^5]OT both have considerable potency suggest that a more extensive exploration of structure–function relationships at this residue would likely turn up new insights.

f. Position 6. In view of the disulfide nature of the neurohypophyseal hormones and the importance of an intact disulfide bridge, it is perhaps to be expected that the Cys residue in position 6 appears to have many of the same structural, stereochemical, and topological requirements as the Cys[1] residue. On the other hand, the Cys[6] residue is not N-terminal as is the Cys[1] residue, so some differences are likely to be important, and we will emphasize these in our discussion here. First, however, it should be noted that, as with Cys[1], there is a severe stereochemical requirement for Cys[6], since [DCys[6]]OT (Manning and du Vigneaud, 1965) has only about one nine-hundredth the potency of OT in the uterine-receptor assay. In addition, the deamino-6-carba analog (the sulfur of the half-Cys[6] residue is replaced by a CH_2) is only about twofold less potent than the corresponding deamino-1-carba analog (Table V), and both selenocystine (Sec) analogs with Sec in position 1 or position 6 are essentially equipotent (Table IV). Interestingly, the 1,6-dicarba analog and even the deamino-1,6-dicarba analog have greatly reduced potency in the rat uterus assay system relative to the monocarba analogs. Perhaps most interestingly, when the homocysteine is in the 1 position (Jarvis *et al.*, 1961) a drastic reduction in potency (about one eight-hundredth) is observed, while when it is in the 6 position (Table IV) the reduction in potency is only about one forty-fifth. However, when the homocysteine residue is placed in both the 1 and 6 positions, a weak antagonist is obtained.

g. Position 7. As discussed previously, the Pro[7] residue is not required for full biological activity of neurophypophyseal hormones in the rat uterine assay. This residue can be replaced by a variety of residues and still maintain high potency. In several cases analogs have been obtained that are significantly more potent then oxytocin. For example, both [$\Delta^{3,4}$Pro[7]]OT and the thiazolidine-4-carboxylic acid analog, [Thi[4]]OT (Table IV), are about twice as potent as oxytocin in the uterine assay, while the sarcosine analog, [Sar[7]]OT, is essentially equipotent to the native hormone. Indeed, even the *N*-methylalanine analog, [N-MeAla[7]]OT, and the glycine analog, [Gly[7]]OT, retain relatively high potency (Table IV). In addition, though the stereochemical requirements of the 7 position are still severe and [DPro[7]]OT (Table IV) is less than one-fortieth as potent as OT, this stereochemical requirement is not as severe as most of the residues in the 20-membered ring of OT. Interestingly, when the highly conformationally constrained amino acid 1-aminocyclopropane-1-carboxylic acid is substituted into position 7 of OT (Table IV), an analog with greatly reduced potency is obtained.

h. Position 8. As with position 7, position 8 of OT does not appear to be critical to the transduction message of the hormone at the uterine receptor. Thus it is not surprising that the 8 position of OT can accommodate a large number of aliphatic amino acids without a large drop in potency (Table IV). More bulky aliphatic amino acids appear to be less acceptable than less bulky substitutents, but even aromatic amino acids such as Phe and Trp can be substituted in the 8 position with retention of reasonably high potencies in the rat uterus assay. As previously noted, arginine vasotocin, [Arg^8]OT, the naturally occurring neurohypophyseal hormone, has reasonably high potency at the oxytocin receptor system. In addition, both [Lys^8]OT (Table IV) and [Orn^8]OT (Huguenin and Boissonnas, 1963) maintain fairly high potency in the oxytocin assay. Particularly interesting is the N^{ϵ}-carbamoyl analog [Lys(N$^{\epsilon}$-Car)^8]OT, which is essentially equipotent to OT (Table IV). It therefore is not surprising that several analogs of AVP and LVP (see Tables XV–XXI) have quite potent activities in the uterine assay. These results suggest that the uterine receptor can accept a positive charge in the 8 position and still be capable of binding to the uterine oxytocin receptor with high affinity, although it might be presumed that a hydrophobic pocket on the receptor is involved in the hormone–receptor interaction with the 8 position of OT (a leucine). We will see later how this has important consequences for the development of oxytocin antagonist analogs. Finally, the stereochemical requirement for the 8 position would appear to be the least stringent of any of the chiral amino acid residues in OT, since [DLeu^8]OT (Schneider and du Vigneaud, 1962) has about 20 U/mg of oxytocin activity.

i. Position 9. Although, as discussed previously, the glycinamide residue *per se* is not critical for biological activity of neurohypophyseal hormones in the uterine assay system, and the 20-membered ring of oxytocin itself (especially tocinamide; Table II) has considerable biological activity, virtually all efforts to modify the glycinamide-9 residue of OT have lead to a drastic loss in potency to levels much lower than those obtained for tocinamide or even tocinoic acid (Table IV). The glycolic amide and thioglycine analogs can hardly be considered exceptions to the rule, and we have already noted the very low activity of even the deamido analog, oxytocinoic acid (Tables II and IV). The exceptions then are the decarboxamide analog ([methylamine^9]OT) and [Gly-N(CH$_3$)^9]OT and [Gly-N(CH$_3$)$_2$^{9}]OT (Table IV), but even these analogs are 20–80 times less potent than the corresponding oxytocin analog. A reasonable explanation for this has not been suggested, but it would appear that there are severe steric constraints that must be accommodated for correct oxytocin–receptor interaction involving the 9 position of the hormone.

4. Disubstituted Analogs of the Neurohypophyseal Hormones

A fairly large number of agonist analogs of OT with substitutions at two positions in the hormone have been made. Nature has done this (Table I) in aspartocin, glumitocin, and isotocin. The activities of these compounds in the rat uterus assay suggest that the substitution of Gln^4 for an Asn^4 and Ser^4 residue is compatible with high potency, but the substitution of Leu^8 with a Gln^8 residue is not. The disubstituted analogs that have been prepared by synthesis are summarized in Tables V, VI, VII, and VIII. We have already commented extensively

Table VI. Disubstituted Analogs of Oxytocin with Modifications at Position Two

Substitutions at Positions Two and Four							
2-Position	4-Position	Rat Uterus	Avian Vasodepressor	Milk Ejecting	Antidiuretic	Pressor	Ref.
Ile	Ile	0.5	3	-	diuretic	0.05	Hruby and du Vigneaud (1970)
Ile	Leu	inhibitor	-	-	diuretic	0.5	Hruby *et al.* (1973)
Leu	Leu	250:1[a]	-	-	diuretic	<0.2	Hruby and Chan (1971)
Phe	Leu	inhibitor	-	-	inactive	0.5	Hruby *et al.* (1973)
Substitutions at Positions Two and Eight							
2-Position	8-Position	Rat Uterus	Avian Vasodepressor	Milk Ejecting	Antidiuretic	Pressor	Ref.
Phe	Orn	1	3	7	0.55	120	Huguenin (1964)
Phe	Orn (deamino)	2	18	10	4	100	Huguenin (1966)
Phe	*tert*-Leu (deamino)	2.9	-	17.1	6	3.2	Lebl *et al.* (1982b)

[a] Ratio of analog to oxytocin required for 50% inhibition.

Table VII. Disubstituted Analogs of Oxytocin with Modifications at Position Three

Substitutions at Positions Three and Four							
3-Position	4-Position	Rat Uterus	Avian Vasodepressor	Milk Ejecting	Antidiuretic	Pressor	Ref.
Leu	Leu	1	2	-	diuretic	<0.005	Wille *et al.* (1972)
Leu	Thr	26	54	-	-	<0.1	Sawyer and Manning (1971)
Phe	Leu	1.1	2	-	0.1	-	Smith *et al.* (1978b)
Pro	Gly	-	-	-	-	-	Wyssbrod *et al.* (1975)

Substitutions at Positions Three and Eight							
3-Position	8-Position	Rat Uterus	Avian Vasodepressor	Milk Ejecting	Antidiuretic	Pressor	Ref.
Phe	Ala	15	38	-	-	21	Walter and du Vigneaud (1966)
Phe	Ala (deamino)	25	47	-	-	4.6	Walter and du Vigneaud (1966)

Substitutions at Positions Three and Nine							
3-Position	9-Position	Rat Uterus	Avian Vasodepressor	Milk Ejecting	Antidiuretic	Pressor	Ref.
Phe	β-Ala	0.19	-	-	inactive	0.2	Anagnostaras *et al.* (1981)

Table VIII. Disubstituted Analogs of Oxytocin: Positions Four through Nine

Substitutions at Positions Four and Seven							
4-Position	7-Position	Rat Uterus	Avian Vasodepressor	Milk Ejecting	Antidiuretic	Pressor	Ref.
Thr	Gly	166	-	802	0.002	<0.01	Lowbridge *et al.* (1977a)
Thr	MeAla	115	-	243	0.047	<0.6	Grzonka *et al.* (1983a)
Thr	Sar	1173	-	731	0.037	<0.5	Grzonka *et al.* (1983a)
Substitutions at Positions Four and Eight							
4-Position	8-Position	Rat Uterus	Avian Vasodepressor	Milk Ejecting	Antidiuretic	Pressor	Ref.
Leu	Ile	4.4	32	4.9	inactive	-	Rudinger *et al.* (1969)
Lys	Ile	1.2	11	-	0.004	<0.1	Sawyer and Manning (1971)
Pro	Gln	<0.006	-	0.02	inactive	0.001	Sawyer and Manning (1971)
Pro	Ile	<0.006	-	0.03	0.002	0.0016	Rudinger *et al.* (1969)
Ser	Gln	10	-	53	0.41	0.35	Sawyer *et al.* (1969)
Ser	Ile	120	382	530	0.29	-	Rudinger *et al.* (1969)
Ser	Ile	145	-	375	0.75	0.05	Sawyer *et al.* (1969)
Thr	Arg	127	493	-	231	160	Manning *et al.* (1973c)
Thr	Ile	520	1545	-	2.6	1.08	Sawyer and Manning (1971)
Thr	Ile (deamino)	128	1113	251	2.1	<0.5	Manning *et al.* (1971b)

Substitutions at Positions Five and Six							
5-Position	6-Position	Rat Uterus	Avian Vasodepressor	Milk Ejecting	Antidiuretic	Pressor	Ref.
Cys	Asn	<0.002	inactive	-	<0.001	inactive	Theodoropoulos *et al.* (1976)
Substitutions at Positions Seven and Eight							
7-Position	8-Position	Rat Uterus	Avian Vasodepressor	Milk Ejecting	Antidiuretic	Pressor	Ref.
Leu	Pro	27	-	45	-	-	Blaha *et al.* (1981)
2-(1-(2-Oxo-3-aminopyrrolidinyl))-4-methylpentanoic acid	Same as 7 position	1.6	-	8	-	0.2	Prochazka *et al.* (1984)
Substitutions at Positions Eight and Nine							
5-Position	6-Position	Rat Uterus	Avian Vasodepressor	Milk Ejecting	Antidiuretic	Pressor	Ref.
4-(amino)-benzamide	des-Gly	4.2	-	10.2	0.2	0.3	Hlavacek *et al.* (1982)
2-Hydroxyiso-caproic acid	glycolic amide	28	-	19.3	0.153	inactive	Roy *et al.* (1983b)
Lys	glycinemethyl-amide	11.5	inhibitor	-	3.44	0.05	Grzonka *et al.* (1974)

on substitution made in the Cys[1] and Cys[6] position, since these two residues are inextricably "tied together" via the common disulfide bond they share. Thus we will not further comment on these changes again. Many of the other analogs do not provide new insights into structure–function relationships for the uterus system. However, there are a few additional insights that have been obtained in these studies and deserve mention.

First, it should be noted that cyclization of OT by the $N\alpha$-amino group of Cys[1] and the C-terminal of Gly[9] to give a bicyclic compound ($S \rightarrow S$ bond $1 \rightarrow 6$, carboxamide bond $1 \rightarrow 9$) leads to an inactive compound that apparently does not interact with the uterine receptor (Table VI). Nearly as drastic is the effect of replacing β-Mpr in position 1 of deamino oxytocin with α-mercaptoacetic acid and the Asn[5] with isoasparagine (Ias). An analog with about one ten-thousandth the potency of deamino-oxytocin in the uterus assy is obtained (Table V). These results demonstrate the very stringent steric requirements at the N- and C-terminal positions of OT and the importance of the precise placement of the Asn carboxamide group with respect to the 20-membered disulfide ring for interaction of the hormone with the rat uterine receptor. The former is further demonstrated by the very low potency of [Phe[4],β-Ala[9]]OT (Table VII), and the latter by the nil activity when the Asn[5] and Cys[6] residues are interchanged to give [Cys[5],Asn[6]]OT, which has a 17-membered disulfide-containing ring (Table VIII).

As for higher or lower potency than expected based on "additivity rules," a problem with such "additivity rules" as an approach to structure–function analysis is that they generally fail to consider conformational and topological effects. Since oxytocin is conformationally a highly flexible molecule in solution (Brewster and Hruby, 1973), it is not at all surprising that such "additivity rules" will often break down in structure–function studies. With these caveats in mind, it is nonetheless interesting that when α-hydroxy-β-mercaptopropionic acid was substituted in position 1, with the Thr[4] substitution, a highly potent analog was obtained (Table V), much more potent than would be predicted from the potency of [Thr[4]]OT and [Hmp[1]]OT. Similarly, the potency of the [Thr[4],Sar[7]]OT in the uterine assay (Table VIII) is very impressive. On the other hand, use of the Hmp[1] substitution in conjunction with $\Delta^{3,4}$-Pro[7] substitution provides an analog [Hmp[1],$\Delta^{3,4}$-Pro[7]]OT (Table V) that is less potent than either of the monosubstituted analogs. Careful examination of such differences when they occur can provide insight into the relationships of side-chain moieties to one another (topology) that might be important to biological activity.

5. Multisubstituted Analogs of Oxytocin

A number of multisubstituted analogs of OT have been synthesized, and the results of biological studies of these analogs are summarized in Tables IX, X, and XI. Many of these analogs were prepared in conjunction with studies on OT antagonist analogs and will be discussed later. One agonist that deserves comment is [Hmp^1,Thr^4,Gly^7]OT (Table X), which has a disappointingly low potency in the oxytocic assay. On the other hand, the very high potency of deamino-1-carba-[Phe(4-F)^2]OT and the very low potency of deamino-6-carba-[Phe(4-Dma)^2]-OT and of deamino-6-carba-[Phe(4-NO_2)]OT provide interesting insights into the stereoelectronic requirements of the uterine receptor for interacting with the side-chain group in the 2 position. It is also interesting that several of the deamino-6-carba analogs with O-alkyl groups on the phenol hydroxy group of Tyr^2 are actually weak agonists, although based on previous studies one might expect them to be antagonists. The generally low activities of a very extensive series of deamino-1-carba analogs with a Glu(COR)^4 substitutent as well (Table IX) is somewhat disappointing.

6. Summary Comments and Conclusions

The studies to date on structure–function relationships in the uterine assay system suggest the following.

1. The 20-membered disulfide-containing ring system is sufficient for full transduction at the uterine receptor.

2. In the intact hormone there is a cooperativity that occurs between the 20-membered disulfide-containing ring and the tripeptide side-chain that is critical for maximal binding interaction with the uterine receptor. In certain cases it appears that disruption of this cooperativity by "improper" binding can prevent the transduction process.

3. It appears that the acyclic tripeptide moiety in OT is important for binding but not for transduction in the oxytocin–uterine receptor interaction.

4. The $Cys^{1(6)}$, Tyr^2, and Asn^5 residues in the ring appear to be particularly critical for receptor recognition (binding) and especially for transduction. Moreover, these three residues appear to be involved in a cooperative manner in transduction.

5. The uterine receptor for the neurohypophyseal hormone appears to have at least three separate topological sites for oxytocin–receptor interaction (binding). A lipophilic site is associated with residues 1–3 in oxytocin that can accommodate an —H or an —OH more readily than the α-amino group at the α-carbon of position 1, and where a Tyr^2 hydroxyl group is also critical. A more hydrophilic surface with hydrogen bond accepting and donating properties encompasses primarily residues 4 and 5. A more lipophilic site encompasses positions 7–9 of the hormone, but with a possibility for a positive charge at 8 position.

6. While residues 1(6), 2, and 5 appear to be most critical for transduction, conformational restriction is necessary to investigate more precisely the topological relationships that are critical for transduction.

Table IX. Trisubstituted Analogs of Oxytocin with Modifications at Position One and Six

Substitutions at Positions One, Two, and Six							
1,6-Position	2-Position	Rat Uterus	Avian Vasodepressor	Milk Ejecting	Antidiuretic	Pressor	Ref.
Deamino-1-carba	Ile	347	171	478	2	-	Fric *et al.* (1974)
Deamino-1-carba	Phe	233	71	515	0.6	-	Fric *et al.* (1974)
Deamino-1-carba	Phe (4-F)	400	-	319	-	-	Prochazka *et al.* (1981)
Deamino-1-carba (sulfoxide 1)	Phe (4-F)	0.11	-	0.15	-	-	Prochazka *et al.* (1981)
Deamino-1-carba (sulfoxide 2)	Phe (4-F)	6.9	-	34.7	-	-	Prochazka *et al.* (1981)
Deamino-1-carba	tert-Leu	3.3	-	73.6	-	<0.2	Lebl *et al.* (1982b)
Deamino-1-carba	Tyr (Me)	10	-	67	0.9	-	Fric *et al.* (1974)
Deamino-6-carba	Ile	3.1	-	90.0	-	-	Lebl *et al.* (1980b)
Deamino-6-carba (sulfoxide)	Ile	8.44	-	88.1	-	-	Lebl *et al.* (1980b)
Deamino-6-carba	Met	4.74	-	7.3	-	-	Lebl *et al.* (1980b)
$(CH_3)_2Mpr^1$-6-carba	Tyr(Me)	pA_2 8.42	-	0	0.52	pA_2 7.43	Lebl and Barth (1983)

Substitutions at Positions One, Two, and Six							
1,6-Position	2-Position	Rat Uterus	Avian Vasodepressor	Milk Ejecting	Antidiuretic	Pressor	Ref.
Deamino-6-carba (sulfoxide)	Met (O)	0.13	-	<0.05	-	-	Lebl *et al.* (1980b)
Deamino-6-carba	Phe	70	-	170	1.5	0.9	Lebl *et al.* (1982a)
Deamino-6-carba	D-Phe	pA_2 8.06		1.25	0.04	-	Lebl *et al.* (1985a)
Deamino-6-carba	Phe (4-amino)	13	-	6.9	2	inactive	Lebl *et al.* (1982a)
Deamino-6-carba	Phe (4-aminobenzyloxy-carbonyl)	0.07	-	-	0.002	inactive	Lebl *et al.* (1982a)
Deamino-6-carba	Phe (4-Cl)	35	-	140	33	inactive	Lebl *et al.* (1982a)
Deamino-6-carba	Phe (4-dimethylamino)	0.05	-	4.5	4	inactive	Lebl *et al.* (1982a)
Deamino-6-carba	Phe (4-Et)	27	-	6.3	14.3	inhibitor	Lebl *et al.* (1982a)
Deamino-6-carba (sulfoxide)	Phe (4-Et)	17	-	-	0.9	inactive	Lebl *et al.* (1982a)
Deamino-6-carba	D-Phe(4-Et)	pA_2 8.73		pA_2 5.82	0.08	pA_2 6.32	Lebl *et al.* (1985a)
Deamino-6-carba (sulfoxide)	D-Phe(4-Et)	pA_2 7.91		pA_2 6.8	0.002	pA_2 5.75	Lebl *et al.* (1985a)
Deamino-6-carba	Phe (4-Me)	127	-	37.4	31	1.0	Lebl *et al.* (1982a)
Deamino-6-carba	D-Phe(4-Me)	pA_2 8.73		10.3	0.17	-	Lebl *et al.* (1985)
Deamino-6-carba	Phe (4-nitro)	0.001	-	10.4	0.2	inactive	Lebl *et al.* (1982a)

(*continued*)

Table IX. *(continued)*

1,6-Position	2-Position	Rat Uterus	Avian Vasodepressor	Milk Ejecting	Antidiuretic	Pressor	Ref.
Substitutions at Positions One, Two, and Six							
Deamino-6-carba	tert-Leu	7.3	-	44.6	-	-	Lebl *et al.* (1982b)
Deamino-6-carba	D-Tyr	28		40.3	0.12	4	Lebl *et al.* (1985)
Deamino-6-carba	Tyr (Et)	0.001	-	5.6	0.005	inactive	Lebl *et al.* (1982a)
Deamino-6-carba	D-Tyr(Et)	pA_2 7.45		0	-	0	Lebl *et al.* (1985)
Deamino-6-carba	Tyr (Me)	3.1	-	18.0	-	inhibitor	Lebl *et al.* (1982a)
Deamino-6-carba (sulfoxide)	Tyr (Me)	5.25	-	10.1	-	-	Lebl *et al.* (1980b)
Homolanthionine (deamino)	Ile	0.13	-	<0.05	-	-	Lebl *et al.* (1980b)
Homolanthionine (deamino)	Tyr (Me)	0.24	-	0.74	-	-	Lebl *et al.* (1980b)
Substitutions at Positions One, Three, and Six							
1,6-Position	3-Position	Rat Uterus	Avian Vasodepressor	Milk Ejecting	Antidiuretic	Pressor	Ref.
Deamino-1,6-dicarba	Phe	21	-	-	-	<0.1	Yamanaka & Sakakibara (1974)
Deamino-1,6-dicarba	Val	15	-	-	-	<0.25	Yamanaka & Sakakibara (1974)

Substitutions at Positions One, Four, and Six							
1,6-Position	4-Position	Rat Uterus	Avian Vasodepressor	Milk Ejecting	Antidiuretic	Pressor	Ref.
deamino-1-carba	Ile	58.7	-	295.5	3.0	2.8	Lebl *et al.* (1980c)
deamino-1-carba	Glu	1	-	-	-	-	Lebl and Jost (1978)
deamino-1-carba	Glu	1.0	-	39.6	0.04	<0.2	Lebl *et al.* (1980c)
deamino-1-carba	Glu-R						
deamino-1-carba	R = OMe	26.0	-	1307	-	-	Lebl *et al.* (1979a)
deamino-1-carba	R = OEt	5.7	-	132.7	-	-	Lebl *et al.* (1979a)
deamino-1-carba	R = $NHCH_3$	40.7	-	12.2	-	-	Lebl *et al.* (1979a)
deamino-1-carba	R = $N(CH_3)_2$	4.26	-	34.5	-	-	Lebl *et al.* (1979a)
deamino-1-carba	R = NH-1-adamantyl	9.2	-	8.2	-	-	Lebl *et al.* (1979a)
deamino-1-carba	R = 4-methylanalide	21.2	-	100	-	-	Lebl *et al.* (1979a)
deamino-1-carba	R = NHNHBoc	3.24	-	21.6	-	-	Lebl *et al.* (1979a)
deamino-1-carba	R = $NHNH_2$	19.4	-	13.8	-	-	Lebl *et al.* (1979a)

(continued)

Table IX. *(continued)*

Substitutions at Positions One, Four, and Six							
1,6-Position	4-Position	Rat Uterus	Avian Vasodepressor	Milk Ejecting	Antidiuretic	Pressor	Ref.
deamino-1-carba	R = Gly-NH_2	5.3	-	36.5	-	-	Lebl *et al.* (1979a)
deamino-1-carba	R = Gly-OMe	10.2	-	98.0	-	-	Lebl *et al.* (1979a)
deamino-1-carba	R = Pro-Leu-Gly-NH_2	0.15	-	0.29	-	-	Lebl *et al.* (1979a)
deamino-1-carba	R = $NH(CH_2)_5COOBu^t$	2.36	-	23.3	-	-	Lebl *et al.* (1979a)
deamino-1-carba	R = $NH(CH_2)_5COOH$	4.78	-	42.7	-	-	Lebl *et al.* (1979a)
deamino-1-carba	R^a = Cys-OMe	1.8	-	-	-	-	Lebl *et al.* (1978c)
deamino-1-carba	R^a = OCOO-iBu^t	16	-	-	-	-	Lebl *et al.* (1978c)
deamino-1-carba	R^a = 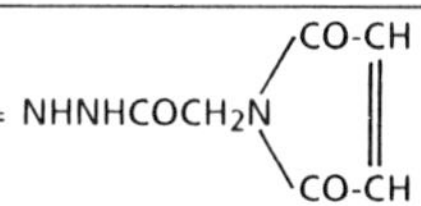	5	-	-	-	-	Lebl *et al.* (1978c)
deamino-1-carba	R^a = $NHNHCOCH_2Br$	2	-	-	-	-	Lebl *et al.* (1978c)
deamino-1-carba	R^a = azide	0.6	-	-	-	-	Lebl *et al.* (1978c)
deamino-1-carba	R^a = 1-hydroxybenzotria-zole ester	160	-	-	-	-	Lebl *et al.* (1978c)
deamino-1-carba	R^a = p-nitrophenyl ester	5	-	-	-	-	Lebl *et al.* (1979b)
deamino-1-carba	R^a = $NH(CH_2)_5CO$-p-nitrophenyl ester	-	-	-	-	-	Lebl *et al.* (1979b)

Substitutions at Positions One, Four, and Six							
1,6-Position	4-Position	Rat Uterus	Avian Vasodepressor	Milk Ejecting	Antidiuretic	Pressor	Ref.
deamino-1-carba (sulfoxide)	Glu (OMe)	0.2	-	11.3	-	-	Lebl *et al.* (1979c)
deamino-1-carba	Leu	0.4	-	0.5	<0.15	<0.2	Lebl *et al.* (1980c)
deamino-6-carba	Thr	695		692		1.1	Lebl *et al.* (1985)
deamino-1-carba	Val	2.8	-	10.9	0.003	<0.2	Lebl *et al.* (1980c)
deamino-1-carba	Thr	272		409		1.3	Lebl *et al.* (1985)
Substitutions at Positions One, Six, and Seven							
1,6-Position	7-Position	Rat Uterus	Avian Vasodepressor	Milk Ejecting	Antidiuretic	Pressor	Ref.
Deamino-1,6-dicarba	Gly	245	4.6	-	0.02	weak	Yamanaka and Sakakibara (1974)
Deaminodicarba	Gly	206	2.9	235	0.01	inactive	Walter *et al.* (1974)

[a]R is reactive group and compounds are noncompetative uterotonic inhibitors *in vitro* (Lebl *et al.* 1978c).

Table X. Other Trisubstituted Analogs of Oxytocin

Substitutions at Positions One, Two, and Four								
1-Position	2-Position	4-Position	Rat Uterus	Avian Vasodepressor	Milk Ejecting	Antidiuretic	Pressor	Ref.
$(CH_3)_2$Mpr	Phe	Thr	pA_2 7.49	-	pA_2 6.82	0.01	pA_2 6.63	Sawyer *et al.* (1980)
$(CH_3)_2$Mpr	Tyr (Me)	Thr	pA_2 7.64	-	pA_2 ~7	0.04	pA_2 7.32	Sawyer *et al.* (1980)
Pen	Phe	Thr	pA_2 7.67	-	inhibitor	-	inhibitor	Hruby *et al.* (1980)
Substitutions at Positions One, Two, and Eight								
1-Position	2-Position	8-Position	Rat Uterus	Avian Vasodepressor	Milk Ejecting	Antidiuretic	Pressor	Ref.
$(CH_3)_2$Mpr	Phe	Orn	pA_2 7.20	-	pA_2 6.65	0.04	pA_2 6.57	Sawyer *et al.* (1980)
$(CH_3)_2$Mpr	Tyr (Me)	Orn	pA_2 7.70	-	pA_2 7.17	0.08	pA_2 7.62	Sawyer *et al.* (1980)
$(C_2H_5)_2$Mpr	Tyr (Me)	Orn	pA_2 8.91	-	pA_2 7.11	0.04	pA_2 8.02	Bankowski *et al.* (1980)
$(C_2H_5)_2$Mpr	Tyr(Et)	Orn	pA_2 8.50	-	pA_2 7.10	0.18	pA_2 8.11	Bankowski *et al.* (1980)
$(CH_2)_5$Mpr	Tyr(Me)	Orn	pA_2 8.52	-	pA_2 7.16	0.009	pA_2 7.96	Bankowski *et al.* (1980)
$(CH_2)_5$Mpr	Tyr(Et)	Orn	pA_2 8.25	-	pA_2 7.30	0.005	pA_2 7.84	Bankowski *et al.* (1980)

Substitutions at Positions One, Four, and Seven								
1-Position	2-Position	7-Position	Rat Uterus	Avian Vasodepressor	Milk Ejecting	Antidiuretic	Pressor	Ref.
Hmp	Thr	Gly	218	-	959	0.004	<0.01	Lowbridge *et al.* (1979a)
Substitutions at Positions Two, Four, and Eight								
2-Position	4-Position	8-Position	Rat Uterus	Avian Vasodepressor	Milk Ejecting	Antidiuretic	Pressor	Ref.
Tyr(Et)	Val	Orn(deamino)	pA_2 7.6	-	-	-	-	Melin *et al.* (1983)
Tyr(Et)	Thr	Orn(deamino)	pA_2 8.9	-	-	-	-	Melin *et al.* (1983)
D-Tyr(Et)	Val	Orn(deamino)	pA_2 8.2	-	-	-	-	Melin *et al.* (1983)
D-Tyr(Et)	Thr	Orn(deamino)	pA_2 8.3	-	-	-	-	Melin *et al.* (1983)
D-Trp	Val	Arg(deamino)	pA_2 8.4	-	-	-	-	Melin *et al.* (1983)
Substitutions at Positions Three, Four, and Eight								
3-Position	4-Position	8-Position	Rat Uterus	Avian Vasodepressor	Milk Ejecting	Antidiuretic	Pressor	Ref.
Phe	Leu	Met	0.62	1.05	-	0.33	0.02	Smith *et al.* (1977)
Phe	Phe	Met	agonist	pA_2 6.29	-	0.43		Smith *et al.* (1977)

Table XI. Analogs of Oxytocin with Modification in Four Positions

Substitutions at Positions One, Two, Four, and Six								
1,6-Position	2-Position	4-Position	Rat Uterus	Avian Vasodepressor	Milk Ejecting	Antidiuretic	Pressor	Ref.
Deamino-1-carba	Tyr (Me)	Glu	0.1	-	0.5	-	-	Lebl *et al.* (1979a)
Deamino-1-carba (sulfoxide)	Tyr (Me)	Glu	-	-	-	-	-	Lebl *et al.* (1979a)
Deamino-1-carba	Tyr (Me)	Glu (OMe)	0.2	-	2.0	-	-	Lebl *et al.* (1979a)
Deamino-1-carba	Tyr (Me)	Glu (HBT ester)	0[a]	-	-	-	-	Lebl *et al.* (1978c)
Deamino-1-carba	Tyr (Me)	Glu (ONp)	[a]	-	-	-	-	Lebl *et al.* (1979b)

Substitutions at Positions One, Two, Four, and Eight									
1-Position	2-Position	4-Position	8-Position	Rat Uterus	Avian Vasodepressor	Milk Ejecting	Antidiuretic	Pressor	Ref.
$(CH_3)_2$Mpr	Phe	Asn	Orn	pA_2 7.98	-	pA_2 6.63	0.014	pA_2 7.39	Sawyer *et al.* (1980)

[a]Noncompetitive inhibitors.

B. At the Avian Vasodepressor Receptor

The avian vasodepressor (AVD) assay is the standard assay for oxytocin in the "U.S. Pharmacopeia," and although it is not as widely used today, many structure–function studies of oxytocin analogs have utilized this assay. Because it is no longer as widely used, a discussion of structure–function relationships as comprehensive as that presented for the uterine assay system is not possible.

A cursory inspection of the structure–activity relationships of the neurohypophyseal hormones in the avian vasodepressor assay (Table I) reveals that the AVD receptor system has quite different structural requirements for binding and transduction than does the uterine receptor. Of particular note is the observation that the AVD system readily accepts a basic residue in position 8. Hence arginine vasotocin is essentially equipotent to OT in the AVD assay system. In addition, AVP and LVP are both about 10-fold more potent in the AVD assay system than in the uterine assay. Thus it will not be surprising to find LVP and AVP analogs with potent activities in the AVD assay.

1. Fragment Analogs of the Neurohypophyseal Hormones

The data of Table II suggest that in most ways the avian vasodepressor receptor is much more sensitive than the rat uterine receptor to the removal of one or more of the amino acids from the C-terminal tripeptide sequence of oxytocin. Although the data is not as extensive (Table XII), this also appears to be the case for vasopressin analogs. Simple modification of the carboxamide to a carboxylate in OT leads to a 2500-fold decrease in AVD potency. Removal of the C-terminal tripeptide in either oxytocin or vasopressin appears to essentially eliminate agonist activity, although pressinamide appears to have about one two-thousandth the potency of oxytocin. Even [des-Leu^8,Gly-NH_2^{9}]OT has minimal or no AVD activity (Table II).

2. Structurally Extended Analogs of the Neurohypophyseal Hormones

Only a few neurohypophyseal hormone analogs have been prepared in which the structures have been extended at C-terminal, and even fewer have been examined in the AVD assay system. In a sense, oxytocinoyl-oxytocin (Table II) can be thought of as primarily a C-terminal extended oxytocin, although of course, it is extended at the N-terminal from the standpoint of the "second" oxytocin structure. In any case, this analog has only about 1/1200 times the potency of OT in the AVD assay. Earlier work (Rudinger and Krejci, 1968) with

Table XII. Fragments of Vasopressin

Compound	Rat Uterus	Avian Vasodepressor	Milk Ejecting	Antidiuretic	Pressor	Ref.
[9-des-Gly]arginine vasopressin	inactive	-	-	-	depressor	Popsuevich and Cipens (1970)
[9-des-Gly]arginine vasopressin	-	-	-	164	pA_2 6.05	Manning *et al.* (1984)
[9-des-Gly-NH_2]arginine vasopressin	-	-	-	5.6	<0.02	Manning *et al.* (1984)
[8-D-Arg, 9-des-Gly-NH_2]vasopressin	0.07	-	0.12	4.5	<0.2	Brtnik *et al.* (1983)
[8-D-Arg, 9-des-Gly-NH_2]vasopressin (deamino)	0.08	-	0.01	14.5	<0.2	Brtnik *et al.* (1983)
[8-Orn, 9-des-Gly-NH_2] vasopressin	0.09	-	0.02	0.3	<0.2	Brtnik *et al.* (1983)
[Deamino-1-carba, 8-D-Arg, 9-des-Gly-NH_2] vasopressin	0.09	-	-	0.8	inhibitor	Brtnik *et al.* (1983)
[des-(8-Arg, 9-Gly)] vasopressin	<0.01	-	-	0.29	<0.1	Barth and Kluh (1974)
[1-$(CH_2)_5$Mpr, 2-D-Tyr(Et), 4-Val, 7-Pro-NH$(CH_2)_5NH_2$, des-(8-Arg, 9-Gly-NH_2)]vasopressin	-	-	-	27[a]	-	Huffman *et al.* (1985)
[1-$(CH_2)_5$Mpr, 2-D-Tyr(Et), 4-Val, 7-Lys, des-(8-Arg, 9-Gly)]-vasopressin	-	-	-	59[a]	-	Huffman *et al.* (1985)
Pressinamide	0.24	inactive	-	-	inactive	Ferger *et al.* (1972)
Pressinamide	0.26	0.3	0.7	-	inactive	Zaoral and Flegel (1972a)
Pressinamide (deamino)	0.5	inactive	-	-	inactive	Ferger *et al.* (1972)
Pressinamide (deamino)	1.03	-	1.21	-	-	Zaorel and Flegel (1972b)
Pressinoic acid	inactive	inactive	-	-	inactive	Ferger *et al.* (1972)
Pressinoic acid	-	-	-	0.03	-	Chan *et al.* (1973)
Pressinoic acid (deamino)	0.05	inactive	-	-	inactive	Ferger *et al.* (1972)

Compound	Rat Uterus	Avian Vasodepressor	Milk Ejecting	Antidiuretic	Pressor	Ref.
Pressinoic acid (deamino)	-	-	-	0.6	-	Chan *et al.* (1973)
[1-$(CH_2)_5$Mpr, 2-D-Tyr(Et), 4-Val]pressinoic acid cadoverineamide	-	-	-	94[a]	-	Huffman *et al.* (1985)
Phe-Gln-Asn-Cys(Cys)-Pro-Arg-Gly-NH_2	-	-	-	-	-	van Nispen *et al.* (1984)
<Glu-Asn-Cys(Cys)-Pro-Arg-Gly-NH_2	-	-	-	-	-	van Nispen *et al.* (1984)
<Glu-Asn-Cys(Cys)-Pro-Arg	-	-	-	-	-	van Nispen *et al.* (1984)
Asn-Cys(Cys)-Pro-Arg-Gly-NH_2	-	-	-	-	-	van Nispen *et al.* (1984)
<Glu-Asn-Cys-Pro-Arg-Gly-NH_2 | <Glu-Asn-Cys-Pro-Arg-Gly-NH_2	-	-	-	-	-	van Nispen *et al.* (1984)
<Glu-Asn-Cys-Pro-Arg | <Glu-Asn-Cys-Pro-Arg	-	-	-	-	-	van Nispen *et al.* (1984)

[a] Dose of peptide delivered i.p. (μg/kg) which results in reduction of urine osmolality from hydropenic levels to 300 m-Osmoles/kg H_2O.

N-terminal extended analogs of OT such as N-Leu-OT, N-Pro-OT, N-Gly-Gly-OT, N-Leu-Leu-OT, etc. gave very interesting results. On the one hand, the analogs were almost all weak agonists that showed a slow onset of biological activity but a prolonged weak agonist effect. Interestingly, however, some of these analogs acted as partial antagonists, in that the biological effect of oxytocin given shortly after the onset of the response of the extended analogs was greatly reduced even after the agonist effects of the extended analogs had faded. The precise explanation for this unique type of antagonism is not known, though an explanation has been proposed (Rudinger and Krejci, 1968), but unfortunately these very interesting observations have not been examined further.

3. Monosubstituted Analogs of the Neurohypophyseal Hormones

Although the avian vasodepressor assay has not been utilized as extensively as in the past, considerable work has been done so that it is possible to obtain insights into the structural features of neurohypophyseal hormones important for binding and transduction at this receptor. These studies indicate that there are some important differences but also many similarities between the uterus and avian vasodepressor receptor systems.

a. Position 1 Substitutions. As in the uterine assay, replacement of the α-amino groups of Cys^1 with a proton leads to a more potent analog (Table IV) in the AVD assay, and replacement of the sulfur with a selenium atom (Table IV) also gives a highly potent analog. Unfortunately, the analog in which the α-amino group was replaced by a hydroxyl group, [LHmp^1]OT (Table IV), was not tested for its AVD activity. However many of the carba-1 analogs were examined in the AVD assay (Table V). In all cases, especially the carba-1 analogs, there is a significant decrease in potency relative to the potency in the uterine assay system and often relative to the corresponding sulfur-containing analog. Analogs with reduced ring size, such as the lanthionine analog (Table V) and the mercaptoacetic acid analog [Maa^1]OT (Jarvis *et al.*, 1961, Berde and Boissonnas, 1968), either have little or no activity or else are antagonists (e.g., [Hcy^1,Hcy^6]OT and [β-Mpa^1,Hcy^6]OT). We return to modification at the β-carbon atom that leads to antagonist analogs later in this chapter. Unfortunately, the open-chain pseudoisosteric analog of OT in which the Cys^1 and Cys^6 residues are replaced with Ser (Table V) was not examined in the AVD assay. Interestingly, however, the cyclic analogs in which the Cys^1 is replaced by β-Ala and Cys^6 by Asp and the ring closed via an amide bond, [β-Ala^1,Asp^6]OT (Table V), is an inhibitor in the AVD assay system but a weak agonist in the other oxytocin assay systems being reviewed here. Very interesting is the observation that [Maa^1,Hcy^6]OT, which has a 20-membered disulfide-containing ring but in which the disulfide bond is frame-shifted by one CH_2 group, has only about one two-thousandth the potency of OT in the AVD assay. These studies indicate that the Cys^1 (in conjunction with the Cys^6) position is very important to both binding and transduction at the AVD receptor. This is further born out by the observation that [DCys^1]OT is a very weak agonist with < 1/2500 times the potency of OT (Hope *et al.*, 1963; Berde and Boissonnas, 1968).

b. Position 2 Substitutions. It has been known for some time that replacement of the Tyr residue in position 2 with any amino acid, even Phe, leads to a dramatic decrease in potency in the avian vasodepressor assay (Berde and Boissonnas, 1968). Generally the decrease in potency is not as large as in the oxytocic assay. The more recent studies summarized in Table IV are consistent with these general conclusions. Nonetheless, there clearly is a rather stringent requirement for a bulky aromatic or aliphatic group in position 2 for high potency at the AVD receptor system. It also appears that substitutions that lead to antagonists in the uterine receptor also lead to antagonists at the avian vasodepressor assay, but unfortunately this has not been examined in many cases.

c. Position 3 Substitutions. Replacement of the Ile^3 in OT with other aliphatic amino acids generally leads to a large reduction in biological potency. Potencies at the uterine and avian vasodepressor assay are generally quite similar (Table IV) (Berde and Boissonnas, 1968; Rudinger *et al.*, 1972). The diastereoisomer analog of OT, [DIle^3]OT, apparently has never been prepared. Interestingly, however, [DAla^3]OT (Table IV) is reported to be a weak inhibitor, suggesting that the stereochemistry of the amino acid at position 3 can be important not only for receptor recognition but also for transduction. Interestingly, the most potent 3-substituted analog has the 3-methyl group of Ile^3 replaced by an ethyl group to give the 3,3-diethylalanine analog, while the corresponding pseudoisosteric side-chain-constrained analog [2-cyclopentylglycine^3]oxytocin has slightly reduced potency (Table IV). It is interesting to note that the analog in which the β-ethyl group is replaced by a methyl group, [Val^3]OT, has greatly reduced potency, and further extended analog with the CH_2 group, [3-cyclohexylalanine^3]OT, has an even further reduced potency. In the light of these findings, it is interesting that [Phe^3]OT (oxypressin) has quite high AVD potency (Table IV), while [Try^3]OT has only about 1/15,000 times the activity of OT in this assay (Berde *et al.*, 1961a).

d. Position 4 Substitutions. Replacement of the Gln^4 by a glutamic acid residue (Table IV) lead to a large, 1000-fold, reduction in potency in the AVD assay. Clearly this receptor cannot readily tolerate a negative charge in this position. In like manner, replacement of the LGln^4 with DGln^4 leads to greater than a 10,000-fold reduction in potency. In this regard it is interesting that

[DAla^{4}]OT is a very weak agonist in the AVD assay (about one six-thousandth that of OT; Table IV), while the [DLeu^{4}]OT and [DPhe^{4}]OT analogs are weak inhibitors. On the other hand, the 4 position of OT can be substituted with nearly any aliphatic and hydroxyl-containing amino acid and still retain rather high potency. Particularly noteworthy is the high potency of [Thr^{4}]OT (Table IV), which is almost three times more potent than OT in the AVD assay. Since the AVD assay is an *in vivo* assay, it is not clear whether this increased potency is due to greater efficacy, greater binding interaction, or greater *in vivo* stability than OT, or a combination of all three. In view of the other structure–function results, it is believed to be due to a greater binding interaction with the AVD receptor. It is interesting to note in this regard that the deamino Thr^{4} analog (Table IV) is actually *less* potent than [Thr^{4}]OT in the AVD assay. Since the deamino analogs of OT are generally believed to be more "stable" in the biological milieu, these results would seem to indicate that the increased potency is a receptor-related effect, especially since in general deamino analogs of OT are more potent than amino-containing analogs. Indeed, this is true for most other 4-substituted analogs of OT (Table IV). Finally, it is interesting to note the [1-aminocyclopentane^{4}]OT (Table II) has less than one-hundredth the potency of OT, considerably less than the Leu^{4} and Ile^{4} analogs (Table IV). Since such α-substituted derivatives are expected to occupy a narrow range of conformations in the vicinity of the α helix and the 4_{10} helix, it would appear that an α-helical conformation in the vincinity of the 4 position is not compatible with high potency for OT analogs.

e. Position 5 Substitutions. As discussed previously, there has been little done to examine structure–function relationships regarding the 5 position of OT. Recent studies summarized in Table IV indicate that changes in this position generally are detrimental to potency in the AVD assay, although interestingly none of the analogs prepared to date are antagonists. As with most other amino acids, replacement of Asn with DAsn gave a diastereoisomeric analog of OT with only about 1/1200 times the potency of OT in the AVD assay (Dutta *et al.*, 1966a). The one exception to the conclusions given above is [Asp^{5}]OT (Table IV), which has about one-tenth the potency of OT in the AVD assay. Thus the AVD receptor can accept a negative charge in this site without a drastic loss in potency. Hopefully more work will be done to examine the importance of this residue in the agonist series. Although it has long been believed that the Asn^{5} side-chain is critical of transduction at the AVD receptor, there is still little hard evidence in support of this idea. Clearly, further work is needed.

f. Position 6 Substitutions. Since the 1 and 6 positions are so intimately interrelated due to their covalent attachment via the disulfide bond, we have already examined several aspects of structure–function relationships for this position in our discussion of position 1. Replacement of LCys^6 with DCys^6 led to an analog with essentially no activity in the AVD assay [< 1/50,000 times that of OT (Manning and du Vigneaud, 1965)]. Replacement of the Cys^6 sulfur with the more bulky selenium atom is compatible with high potency (Table IV), but addition of a methylene group to give [Hcy^6]OT led to a drastic loss of potency (about 1600-fold) in the AVD assay. This is a much more dramatic loss than seen at the uterine receptor. This seems to be true for most other 6-substituted analogs of OT in the agonist series (Table V). It would thus appear that the stereo and structural relation of position 6 to the rest of the OT structure is critical to high potency in the AVD assay.

g. Position 7 Substitutions. A number of analogs of OT substituted in the 7 position have been prepared, but only a small portion of these have been examined in the AVD assay (Table IV). Nonetheless, a few comments can be made. There is a strict stereochemical requirement for Pro^7 and for interaction of oxytocin with the AVD receptor. As opposed to the uterine receptor, where considerable activity is retained, essentially all activity is lost for [DPro^7]OT in the AVD assay (Table IV). On the other hand, the 3,4-dehydroproline^7 and thiazolidine-4-carboxylate^7 analogs are more potent than OT in the AVD assay. However, the increased potency is not as great as at the uterine receptor. This is seen more dramatically for other analogs, where 7-substituted OT analogs tend to be at least 10-fold less potent in the AVD assay then in the rat uterus assay (Table IV). It would appear that the AVD receptor is more sensitive to structural change at the 7 position of the hormone than is the uterine receptor.

h. Position 8 Substitutions. As already noted, natural hormones substituted at the 8 position (Table I) generally retain higher potency in the AVD assay relative to the rat uterus assay. This is especially dramatic when basic amino acid residues are found in the position, since these are as much as 10-fold more potent in the AVD assay. As shown in Table IV, similar trends are seen when aliphatic amino acids are substituted in position 8. Generally a wide range of substitutions are compatible with high potency, including aromatic amino acid, e.g., [Phe^8]OT (Table IV), and generally the relative potencies in the AVD assay are greater than in the uterine assay (Table IV) (Berde and Boissonnas, 1968). Even [DLeu^8]OT retains relatively high potency in the AVD assay (Schneider and du Vigneaud, 1962). Clearly, there is considerable flexibility in structural requirements for the 8 position and strong interaction with the AVD receptor.

i. Position 9 Substitutions. As compared to the 8 position, there appear to be surprisingly stringent requirements for the 9 position. Nearly all modifications of the 9 position lead to drastic reductions in potency (Table IV). Just modifying the C-terminal from carboxamide to carboxylate leads to over 2000-fold reduction in potency. Clearly the AVD receptor cannot readily accommodate a negative charge in this region of the receptor, but the *N,N'*-dimethyl carboxamide analog is equally inactive, suggesting that a well-defined steric requirement also may be of considerable importance. Interestingly, [des-Gly^9]OT is a weak inhibitor (Berde and Boissonnas, 1968) in the AVD assay system. In addition, substitution of several aliphatic amino acids in the 9 position led to weak inhibitors (Table IV). These results suggest that the glycinamide residue (perhaps in conjunction with residues in the ring of OT) may be critical for transduction at the AVD receptor.

4. Multiple Substitutions in OT

Examination of Tables V–XI shows that most multiply substituted analogs of oxytocin have not been examined in the avian vasodepressor assay. However, several that have been deserve comment. A long-standing problem, the question of the structure of the product that results from reaction of oxytocin with acetone to give a product with about one-thousandth the potency of oxytocin in the AVD assay, was answered by determining that an imidazolidinone forms by reaction of acetone with the N-terminal amino group and the Tyr^2 amino groups (Hruby *et al.*, 1968; Hruby and du Vigneaud, 1969b). [Hmp,$\Delta^{3,4}Pro^7$]OT (Table V) was found to be a very potent analog; unfortunately, [Hmp^1,Thr^4]OT was not examined in the AVD assay. The high potency of [Thr^4,Arg^8]OT in the AVD assay (Table VIII) again illustrates the ease with which the AVD receptor accepts a basic residue in position 8. More interesting is the very high potency of [Thr^4,Ile^8]OT, which in fact appears to be slightly more potent than [Thr^7]OT (Table IV), although OT and IT (Table I) have essentially identical potencies in the AVD assay. Again, the deamino analog (Table VIII) is slightly less potent than the amino-containing analog.

5. Substitutions in Vasopressin Analogs

Examination of Tables XII–XXIII shows that although AVP and LVP analogs generally have not been examined in the AVD assay, it is not surprising that in several cases that have been examined they have been found to have high potency. Some notable examples include (a) [Mpa^1,Ile^3,Orn^8]OT (Table XV), which has about 400 U/mg of AVD activity—essentially equivalent to OT; (b) [Thr^4]AVP (Table XVII), which has about 70% the potency of OT; (c) [Ile^3]LVP (Table XXII), which has about 50% the potency of OT; and (d) [Ile^3,Thr^4]AVP, which was found to be about 60% more potent than OT in the AVD assay—interestingly, the deamino analog (Table XIX) was essentially equipotent to OT.

Table XIII. Extended Analogs of Vasopressin

NH_2 Terminal Extensions							
Arginine Vasopressin Analogs							
NH_2- Extension	Substitutions at other Positions	Rat Uterus	Avian Vasodepressor	Milk Ejecting	Antidiuretic	Pressor	Ref.
Ala-Gly-		-	-	-	-	-	Chan and Smith (1980)
Val-Asp-		-	-	-	-	-	Chan *et al.* (1981)
Pro-Arg-Val-		-	-	-	-	0.3	Moore (1978)
Dansyl	2-(4-Azido)phenylalanine	-	-	-	active	-	Fahrenholz *et al.* (1980)
Ac	2-Tyr (me)	pA_2 7.29	-	-	0.026	pA_2 7.18	Jones and Sawyer (1980)

(continued)

Table XIII. *(continued)*

NH_2 Terminal Extensions							
Lysine Vasopressin Analogs							
NH_2- Extension	Substitutions at other Positions	Rat Uterus	Avian Vasodepressor	Milk Ejecting	Antidiuretic	Pressor	Ref.
acetone						7	Yamashiro *et al.* (1967)
4-Aminobenzoyl-		-	-	-	-	-	Geiser *et al.* (1968)
Leu-		-	-	-	-	-	Kasafirek *et al.* (1966)
Phe-		-	-	-	-	-	Kasafirek *et al.* (1966)
Tyr-		-	-	-	-	-	Kasafirek *et al.* (1966)
Trp-		-	-	-	-	-	Kasafirek *et al.* (1966)
Pro-		-	-	-	-	-	Kasafirek *et al.* (1966)
Gly-Gly-		-	-	-	-	-	Kasafirek *et al.* (1966)
Leu-Leu-		-	-	-	-	-	Kasafirek *et al.* (1966)
Gly-Pro-		-	-	-	-	-	Kasafirek *et al.* (1966)
Sar-Gly-		-	-	-	-	-	Kasafirek *et al.* (1966)
Gly-Gly-Gly-		60	-	45	~2	~2.5	Prochazka *et al.* (1978)
Acetyl-	3-Ile	-	-	-	-	-	Schillinger *et al.* (1972)
Gly-	2-Tyr(Me)	-	-	-	~2	1.5-3	Zaoral and Sorm (1965)
Gly-	3-Ile	-	-	-	-	-	Schillinger *et al.* (1972)

NH_2 Terminal Extensions							
Lysine Vasopressin Analogs							
NH_2- Extension	Substitutions at other Positions	Rat Uterus	Avian Vasodepressor	Milk Ejecting	Antidiuretic	Pressor	Ref.
Gly-Gly-Gly-	8-Orn	-	-	-	~0.6	~6	Lebl *et al.* (1978)
Gly-Gly-Gly-	9-des-Gly-NH_2	0.1	-	inactive	10	220	Prochazka *et al.* (1978)
COOH- Terminal Extensions							
Arginine Vasopressin Analogs							
NH_2- Extension	Substitutions at other Positions	Rat Uterus	Avian Vasodepressor	Milk Ejecting	Antidiuretic	Pressor	Ref.
-Ala-Met-Ala-NH_2		-	-	-	-	0.2	Moore (1978)

Table XIV. Vasopressin Analogs with Substitutions in Position Eight

8-Position		Rat Uterus	Avian Vasodepressor	Milk Ejecting	Antidiuretic	Pressor	Ref.
L-2-Aminoheptanoic acid					~8	~19	Fink and Bodanszky (1973)
L-2-Amino-3-glycylaminopropionic acid	(deamino)	0.2-0.3	-	-	435	5.75	Zaoral *et al.* (1979c)
L-2-Amino-3-guanidinoproprionic acid	(deamino)	1.5	-	-	95[a]	49.5	Zaoral *et al.* (1979b)
D-2-Amino-3-guanidinopropionic acid	(deamino)	0.4	-	-	830[a]	2.7	Zaoral *et al.* (1979b)
D-Arg		1.07	-	-	257	1.08	Lowbridge *et al.* (1977)
D-Arg	(deamino)	14	-	-	955	0.5	Manning *et al.* (1973a)
Dbu		3.5	-	-	120	149	Zaoral and Sorm (1966a)
Dbu	(deamino)	24.7	-	134	1373	450	Zaoral *et al.* (1976b)
D-Dbu		0.1	-	-	90-150	3.6	Zaoral and Sorm (1966b)
D-Dbu	(deamino)	0.39	-	-	360	2.05	Zaoral *et al.* (1967b)
Dpr		0.45	-	-	40-60	109	Zaoral *et al.* (1970)
Dpr	(deamino)	0.9	-	-	1079	14	Krchnak *et al.* (1979)
D-Dpr		0.53	-	-	150-170	21	Zaoral *et al.* (1970)
D-Dpr	(deamino)	1.72	-	-	1000	22.8	Krchnak *et al.* (1979)
Gln		0.4	-	-	-	2	Papsuevich *et al.* (1972)
Gly		inactive	-	-	-	0.01	Papsuevich *et al.* (1972)
Har		13	-	-	427	250	Lindeberg *et al.* (1974)
Har	(deamino)	20	-	-	460	96	Lindeberg *et al.* (1974)
D-Har		-	-	-	84[b]	0.83	Lindeberg *et al.* (1976)
D-Har	(deamino)	1.6	-	-	17	0.02	Zaoral and Brtnik (1975)
D-Har	(deamino)	-	-	-	196[b]	0.06	Lindeberg *et al.* (1976)
Homolysine		-	-	-	159	267	Lindeberg *et al.* (1974)
Homolysine	(deamino)	-	-	-	634	153	Lindeberg *et al.* (1974)

8-Position		Rat Uterus	Avian Vasodepressor	Milk Ejecting	Antidiuretic	Pressor	Ref.
D-Homolysine		-	-	-	207[b]	1.8	Lindeberg (1975)
D-Homolysine	(deamino)	-	-	-	170[b]	<0.5	Lindeberg (1975)
L-Norarginine	(deamino)	7.4	-	-	1910[a]	1500	Zaoral *et al.* (1979a)
D-Norarginine	(deamino)	16.2	-	-	190[a]	16.3	Zaoral *et al.* (1979a)
D-Lys	(deamino)	0.07	-	-	3.8	1.1	Zaoral *et al.* (1967b)
Lys (D-Ala)		-	-	-	-	-	Cort *et al.* (1982)
Lys (Gly)	(deamino)	-	-	-	-	-	Cort *et al.* (1982)
Lys (Gly-Gly)	(deamino)	-	-	-	-	-	Cort *et al.* (1982)
Lys (Gly-Gly-Gly)	(deamino)	-	-	-	-	20	Cort *et al.* (1982)
Lys (Tos)	(deamino)	-	-	-	-	negligible	Havran (1973)
Orn		10	21	50	88	360	Huguenin and Boissonnas (1963)
Orn	(deamino)	15	43	43	202	355	Huguenin (1966)
D-Orn		0.2	-	-	50-70	0.24	Zaoral *et al.* (1970)
D-Orn	(deamino)	1.9	-	3.1	29	4.8	Zaoral *et al.* (1970b)
D-Orn	(deamino)	-	-	-	590[b]	-	Zaoral *et al.* (1970b)
Thl		5	-	-	400	55	Hermann *et al.* (1965)

[a] Results in references were reported as percent of a standard batch of DDAVP. For purposes of comparison DDAVP is taken as possessing 955 units/mg of antidiuretic activity.

[b] Used modified Burn (Burn *et al.*, 1950) assay technique (Cort *et al.*, 1975) that may give high results for analogs with prolonged duration of action.

Table XV. Disubstituted Vasopressin Analogs with Modifications at Position Eight

Substitutions at Positions Two and Eight							
2-Position	8-Position	Rat Uterus	Avian Vasodepressor	Milk Ejecting	Antidiuretic	Pressor	Ref.
Phe	Orn	0.5	<0.01	3	16	153	Huguenin (1964)
Phe	Orn (deamino)	<1	weak pressor	2.6	13	45	Huguenin (1966)
Tyr (Me)	D-Arg	inactive	-	-	10	inactive	Krchnak *et al.* (1979)
Substitutions at Positions Three and Eight							
3-Position	8-Position	Rat Uterus	Avian Vasodepressor	Milk Ejecting	Antidiuretic	Pressor	Ref.
Ile	Cit (deamino)	-	-	-	-	-	Cheesman *et al.* (1983)
Ile	Gln (deamino)	-	-	-	-	-	Cheesman *et al.* (1983)
Ile	Orn (deamino)	80	400	300	70	270	Huguenin (1966)
Ile	Orn (Ac) (deamino)	-	-	-	-	-	Cheesman *et al.* (1983)
Ile	Orn (HCO) (deamino)	-	-	-	-	-	Cheesman *et al.* (1983)
Ile	Lys (4-carboxybutyryl)	50	-	-	-	-	Snell and Smyth (1977)
Ile	Lys (3-carboxypropionyl)	55	-	-	-	-	Snell and Smyth (1977)
Ile	Hyl (deamino)	-	-	-	-	-	Cheesman *et al.* (1983)

Substitutions at Positions Four and Eight							
4-Position	8-Position	Rat Uterus	Avian Vasodepressor	Milk Ejecting	Antidiuretic	Pressor	Ref.
Val	D-Har (deamino)	-	-	-	31[a]	<0.01	Lindeberg *et al.* (1977)
Val	D-Homolysine (deamino)	-	-	-	21[a]	0.01	Lindeberg *et al.* (1977)
Substitutions at Positions Seven and Eight							
7-Position	8-Position	Rat Uterus	Avian Vasodepressor	Milk Ejecting	Antidiuretic	Pressor	Ref.
Arg	Pro	20	-	1.6	<1	3.2	Blaha *et al.* (1981)
Gly	Orn	0.45	-	7	1.2	0.78	Lebl *et al.* (1980b)
Gly	Orn (deamino)	1.6	-	7	0.61	0.63	Lebl *et al.* (1980b)
Gly	Orn (N-Gly-Gly-Gly-)	0.006	-	-	0.032	depressor	Lebl *et al.* (1980b)
Substitutions at Positions Eight and Nine							
8-Position	9-Position	Rat Uterus	Avian Vasodepressor	Milk Ejecting	Antidiuretic	Pressor	Ref.
D-Arg	Ala	-	-	-	70	<0.4	Buku *et al.* (1984)
D-Arg	Ala (deamino)	-	-	-	184	<0.4	Buku *et al.* (1984)

[a] Antidiuretic activity was determined by a modified Burn (Burn *et al.*, 1950) assay (Cort *et al.*, 1975) that can give high values with analogs possessing prolonged action.

Table XVI. Trisubstituted Vasopressin Analogs with Modifications at Position Eight

Substitutions at Positions One, Six, and Eight

1,6-Position	8-Position	Rat Uterus	Avian Vasodepressor	Milk Ejecting	Antidiuretic	Pressor	Ref.
Deamino-6-carba	D-Arg	0.66	-	-	2200[a]	2.9	Jost *et al.* (1974)
Deamino-6-carba	Orn	27	-	-	4618	172	Jost *et al.* (1974)

Substitutions at Positions Two, Three, and Eight

2-Position	3-Position	8-Position	Rat Uterus	Avian Vasodepressor	Milk Ejecting	Antidiuretic	Pressor	Ref.
Phe	Ile	Orn (deamino)	<2	18	10	4	100	Huguenin (1966)

[a] Antidiuretic activity was determined by a modified Burn (Burn *et al.*, 1950) assay (Cont *et al.*, 1955) that can give high values for analogs possessing prolonged duration of action.

Table XVII. Analogs of Arginine Vasopressin with Substitutions in a Single Position

Position One Substitutions						
1-Position	Rat Uterus	Avian Vasodepressor	Milk Ejecting	Antidiuretic	Pressor	Ref.
Hmp	31	96	-	467	549	Lowbridge *et al.* (1977b)
Mpr (deamino arginine vasopressin)	29	160	85	1400	395	Huguenin *et al.* (1965)
$(CH_3)_2$Mpr	pA_2 6.93	-	-	42	pA_2 7.45	Manning *et al.* (1982b)
$(CH_2H_5)_2$Mpr	pA_2 7.30	-	-	0.38	pA_2 8.36	Manning *et al.* (1982b)
$(CH_2)_5$Mpr	pA_2 8.15	-	-	0.033	pA_2 8.35	Kruszynski *et al.* (1980)
Position Two Substitutions						
2-Position	Rat Uterus	Avian Vasodepressor	Milk Ejecting	Antidiuretic	Pressor	Ref.
4-(Azido)phenylalanine	-	-	-	active	-	Fahrenholz *et al.* (1980)
4-(Azido)phenylalanine (N-dansyl)	-	-	-	active	-	Fahrenholz *et al.* (1980)
4-(Bromoacetylamino)phenylalanine	-	-	-	active	-	Fahrenholz *et al.* (1980)
Phe	0.2	<1	3.0	350	122	Huguenin and Boissonnas (1962)
Phe (deamino)	0.3	<1	4	810	29	Huguenin and Boissonnas (1966)
Tyr (Me)	pA_2 7.44	-	-	386	9.7	Bankowski *et al.* (1978)
Tyr (Me) (N-acetyl)	pA_2 7.29	-	-	0.026	pA_2 7.18	Jones and Sawyer (1980)
Tyr (deamino)	pA_2 8.4	-	-	-	-	Melin *et al.* (1983)
D-Tyr	1.5	10	-	207	194	Hruby *et al.* (1979b)
D-Tyr (deamino)	pA_2 8.7	-	-	-	-	Melin *et al.* (1983)

(continued)

Table XVII. *(continued)*

Position Three Substitutions							
3-Position		Rat Uterus	Avian Vasodepressor	Milk Ejecting	Antidiuretic	Pressor	Ref.
Abu	(deamino)	-	-	-	-	-	Cheesman *et al.* (1983)
Ala	(deamino)	-	-	-	-	-	Cheesman *et al.* (1983)
Gly	(deamino)	-	-	-	-	-	Cheesman *et al.* (1983)
Ile	(arginine vasotocin)	75	150	-	-	125	Katsoyannis and du Vigneaud (1958)
Ile	(arginine vasotocin)	115	285	210	250	245	Berde *et al.* (1962)
Ile	(arginine vasotocin)	127	493	-	231	160	Manning *et al.* (1973c)
Ile	(deamino)						Schillinger *et al.* (1972)
alle	(deamino)						Cheesman *et al.* (1983)
Leu	(deamino)						Cheesman *et al.* (1983)
Met	(deamino)						Cheesman *et al.* (1983)
Thr	(deamino)						Cheesman *et al.* (1983)
Trp	(deamino)						Cheesman *et al.* (1983)
Tyr	(deamino)						Cheesman *et al.* (1983)
Tyr (Me)	(deamino)						Cheesman *et al.* (1983)
Val	(deamino)						Cheesman *et al.* (1983)

Position Four Substitutions							
4-Position		Rat Uterus	Avian Vasodepressor	Milk Ejecting	Antidiuretic	Pressor	Ref.
Abu		5	25	-	760	38	Gillessen and du Vigneaud (1970)
Abu	(deamino)	3.3	9.9	-	1020	10.7	Gillessen and du Vigneaud (1970)
Arg		-	-	-	430	168	Rekowski *et al.* (1985)
Arg	(deamino)	-	-	-	214	128	Rekowski *et al.* (1985)
D-Gln		0.11	0.10	-	0.45	0.26	Hruby *et al.* (1982)
Thr		18	321	-	231	104	Manning *et al.* (1973c)
Thr	(deamino)	10.2	195	-	758	30	Manning *et al.* (1973b)
Val		-	-	-	738	32	Sawyer *et al.* (1974)
Val	(deamino)	-	-	-	1150	51	Sawyer *et al.* (1974)
Position Five Substitutions							
5-Position		Rat Uterus	Avian Vasodepressor	Milk Ejecting	Antidiuretic	Pressor	Ref.
Asp		0.38	-	-	86.5	6.93	Smith *et al.* (1980a)

(continued)

Table XVII. *(continued)*

Position Seven Substitutions							
7-Position		Rat Uterus	Avian Vasodepressor	Milk Ejecting	Antidiuretic	Pressor	Ref.
3,4-Dehydroproline		40.6	76.9	-	1260	255	Botos *et al.* (1979)
3,4-Dehydroproline	(deamino)	69	345	-	4134	240	Botos *et al.* (1979)
Gly		0.1	-	-	-	0.01	Papsuevich *et al.* (1972)
MeAla		2.8	-	35	343	10.6	Grzonka *et al.* (1983)
MeAla	(deamino)	1.7	-	35	113	1.3	Grzonka *et al.* (1983)
Sar		9.3	-	110	188	3.6	Grzonka *et al.* (1983)
Sar	(deamino)	38.2	-	193	580	0.84	Grzonka *et al.* (1983)
Position Eight Substitutions							
8-Position		Rat Uterus	Avian Vasodepressor	Milk Ejecting	Antidiuretic	Pressor	Ref.
D-Arg		1.07	-	-	257	1.08	Lowbridge (1977b)
D-Arg		0.4	-	-	114	4	Zaoral *et al.* (1967a)
D-Arg	(deamino)	5.1	-	-	870	11	Zaoral *et al.* (1967a)
D-Arg	(deamino)	14	-	-	955	0.5	Manning *et al.* (1973a)
D-Arg	(deamino, dimer)	0.3	-	-	equals monomer	hypotensive	Zaoral *et al.* (1978)
MeArg		-	-	-	<50	0.5	Hlavacek *et al.* (1977)

Position Nine Substitutions						
9-Position	Rat Uterus	Avian Vasodepressor	Milk Ejecting	Antidiuretic	Pressor	Ref.
deamido (arginine vasopressinoic acid)	low	inactive	-	-	inhibitor	Meienhofer *et al.* (1971)
deamido (arginine vasopressinoic acid)	-	-	-	-	inhibits cyclase	Dousa *et al.* (1970)
deamido (arginine vasopressinoic acid)	-	--	-	4.7	<0.3	Manning *et al.* (1984)
Ala	-	--	-	104	<0.4	Buku *et al.* (1984)
Ala (deamino)	-	-	-	168	<0.4	Buku *et al.* (1984)
D-Ala	-	-	-	inactive	inactive	Cort *et al* (1981)
D-Ala	-	-	-	189	7.6	Buku *et al.* (1984)
D-Ala (deamino)	-	-	-	inactive	inactive	Cort *et al.* (1981)
D-Ala (deamino)	-	-	-	312	0.5	Buku *et al.* (1984)

Table XVIII. Disubstituted Analogs of Arginine Vasopressin with Modifications at Position One

Position One and Two Substitutions							
1-Position	2-Position	Rat Uterus	Avian Vasodepressor	Milk Ejecting	Antidiuretic	Pressor	Ref.
$(CH_3)_2$Mpr	Tyr (Me)	pA_2 7.61	-	-	3.5	pA_2 7.96	Bankowski *et al.* (1978)
$(CH_2)_5$Mpr	D-Ile	pA_2 7.62	-	-	pA_2 6.96	pA_2 7.79	Manning *et al.* (1984a)
$(CH_2)_5$Mpr	D-Phe	pA_2 8.59	-	-	pA_2 7.21	pA_2 8.35	Manning *et al.* (1983)
$(CH_2)_5$Mpr	Tyr (Me)	pA_2 8.13	-	-	0.31	pA_2 8.62	Kruszynski *et al.* (1980)
Position One and Three Substitutions							
1-Position	3-Position	Rat Uterus	Avian Vasodepressor	Milk Ejecting	Antidiuretic	Pressor	Ref.
Hmp	Leu						Cheesman *et al.* (1983)
Position One and Four Substitutions							
1-Position	4-Position	Rat Uterus	Avian Vasodepressor	Milk Ejecting	Antidiuretic	Pressor	Ref.
$(CH_3)_2$Mpr	Val	pA_2 8.09	-	-	312	pA_2 7.97	Manning *et al.* (1982b)
$(C_2H_5)_2$Mpr	Val	pA_2 7.32	-	-	1.50	pA_2 8.29	Manning *et al.* (1982b)
$(CH_2)_5$Mpr	Arg	-	-	-	1.5	28	Rekowski *et al.* (1985)
$(CH_2)_5$Mpr	Val	pA_2 7.34	-	-	0.32	pA_2 7.97	Manning *et al.* (1982b)

Position One and Six Substitutions							
1-Position	6-Position	Rat Uterus	Avian Vasodepressor	Milk Ejecting	Antidiuretic	Pressor	Ref.
Ala	Ala	<0.2	<0.2	-	0.08	0.025	Huguenin and Guttmann (1965)
Ala	Ala (deamino)	<0.2	<0.2	-	0.2	0.041	Huguenin and Guttmann (1965)
β-Ala	Ala	-	-	-	-	-	Jager and Geiger (1973)
Deamino-	1-carba	34	-	65	15,000-23,000[a]	550	Prochazka *et al.* (1978a)
Deamino-	1-carba	15.4	12.8	-	1569	223	Jost *et al.* (1974)
Deamino-	1-carba	15.4	12.8	-	1274	37.7	Hase *et al.* (1974)
Position One and Seven Substitutions							
1-Position	7-Position	Rat Uterus	Avian Vasodepressor	Milk Ejecting	Antidiuretic	Pressor	Ref.
$(CH_3)_2$Mpr	MeAla	pA_2 7.81	-	none	0.007	pA_2 7.01	Gazis *et al.* (1984)
$(CH_3)_2$Mpr	Sar	pA_2 7.08	-	0.015	2.06	pA_2 7.14	Gazis *et al.* (1984)
$(CH_2)_5$Mpr	MeAla	pA_2 7.54	-	pA_2 5.86	0.015	pA_2 8.10	Gazis *et al.* (1984)
$(CH_2)_5$Mpr	Sar	pA_2 6.95	-	none	0.112	pA_2 7.93	Gazis *et al.* (1984)

(continued)

Table XVIII. *(continued)*

Position One and Eight Substitutions							
1-Position	8-Position	Rat Uterus	Avian Vasodepressor	Milk Ejecting	Antidiuretic	Pressor	Ref.
Hmp	D-Arg	4.3	2.6	-	886	0.86	Lowbridge *et al.* (1977b)
$(CH_3)_2$Mpr	D-Arg	pA_2 7.89	-	-	5.8	pA_2 7.45	Manning *et al.* (1982b)
$(C_2H_5)_2$Mpr	D-Arg	pA_2 6.95	-	-	0.067	pA_2 7.96	Manning *et al.* (1982b)
$(CH_2)_5$Mpr	D-Arg	pA_2 6.97	-	-	0.31	pA_2 8.52	Manning *et al.* (1982b)
Position One and Nine Substitutions							
1-Position	9-Position	Rat Uterus	Avian Vasodepressor	Milk Ejecting	Antidiuretic	Pressor	Ref.
$(CH_2)_5$Mpr	des-Gly	-	-	-	0.003	pA_2 8.38	Manning *et al.* (1984b)
$(CH_2)_5$Mpr	des-Gly-NH_2	-	-	-	0.04	pA_2 7.96	Manning *et al.* (1984b)

[a]Used threshold doses.

Table XIX. Other Disubstituted Analogs of Arginine Vasopressin

Position Two and Three Substitutions							
2-Position	3-Position	Rat Uterus	Avian Vasodepressor	Milk Ejecting	Antidiuretic	Pressor	Ref.
Phe	Phe (4-azido)	-	-	-	active	-	Fahrenholz *et al.* (1980)
Phe	Phe (4-bromoacetylamino)	-	-	-	active	-	Fahrenholz *et al.* (1980)
Phe	Ile	2	10	10	109	125	Huguenin & Boissonnas (1962)
Phe	Ile (deamino)	-	-	-	-	-	Cheesman *et al.* (1983)
Tyr (Me)	Ile (deamino)	-	-	-	-	-	Cheesman *et al.* (1983)
Position Two and Four Substitutions							
2-Position	4-Position	Rat Uterus	Avian Vasodepressor	Milk Ejecting	Antidiuretic	Pressor	Ref.
Tyr(3-I)	Val	-	-	-	-	-	Moore *et al.* (1985)
$Tyr(3,5\text{-}I_2)$	Val	-	-	-	-	-	Moore *et al.* (1985)
Position Two and Seven Substitutions							
2-Position	7-Position	Rat Uterus	Avian Vasodepressor	Milk Ejecting	Antidiuretic	Pressor	Ref.
Phe	3,4-dehydroproline	0.80	4.2	-	1541	63.8	Botos *et al.* (1979)
Phe	3,4-dehydroproline (deamino)	6.0	2.9	-	13,000	inactive	Botos *et al.* (1979)

(continued)

Table XIX. *(continued)*

Position Three and Four Substitutions							
3-Position	4-Position	Rat Uterus	Avian Vasodepressor	Milk Ejecting	Antidiuretic	Pressor	Ref.
Ile	Abu (deamino)						Cheesman *et al.* (1983)
Ile	Ala (deamino)						Cheesman *et al.* (1983)
Ile	Asn (deamino)						Cheesman *et al.* (1983)
Ile	Glu (deamino)						Cheesman *et al.* (1983)
Ile	Glu(NH-CH_3) (deamino)						Cheesman *et al.* (1983)
Ile	Gly (deamino)						Cheesman *et al.* (1983)
Ile	His (deamino)						Cheesman *et al.* (1983)
Ile	Ile (deamino)						Cheesman *et al.* (1983)
Ile	Leu	inhibitor	-	-	3.2	6	Pliska *et al.* (1973)
Ile	Leu (deamino)						Straub *et al.* (1973)
Ile	Lys (deamino)						Cheesman *et al.* (1983)
Ile	Orn (deamino)						Cheesman *et al.* (1983)
Ile	Ser	66	311	-	-	20.2	Chauvet *et al.* (1972)
Ile	Thr	201	831	-	279	106	Manning *et al.* (1973c)
Ile	Thr (deamino)	63	786	-	348	33	Manning *et al.* (1973b)

Position Three and Four Substitutions							
3-Position	4-Position	Rat Uterus	Avian Vasodepressor	Milk Ejecting	Antidiuretic	Pressor	Ref.
Ile	Val (deamino)						Cheesman *et al.* (1983)
Leu	Leu						Straub *et al.* (1973)
Leu	Leu (deamino)						Straub *et al.* (1973)

Position Three and Five Substitutions							
3-Position	5-Position	Rat Uterus	Avian Vasodepressor	Milk Ejecting	Antidiuretic	Pressor	Ref.
Ile	Asp (deamino)						Cheesman *et al.* (1983)
Ile	Asp(NH-CH_3) (deamino)						Cheesman *et al.* (1983)
Ile	Gln (deamino)						Cheesman *et al.* (1983)

Position Three and Seven Substitutions							
3-Position	7-Position	Rat Uterus	Avian Vasodepressor	Milk Ejecting	Antidiuretic	Pressor	Ref.
Ile	Ala (deamino)						Cheesman *et al.* (1983)
Ile	Gly (deamino)						Cheesman *et al.* (1983)
Ile	Hyp (deamino)						Cheesman *et al.* (1983)
Ile	Thiazolidine-4-carboxylic acid (deamino)						Cheesman *et al.* (1983)

(continued)

Table XIX. *(continued)*

Position Three and Eight Substitutions							
3-Position	8-Position	Rat Uterus	Avian Vasodepressor	Milk Ejecting	Antidiuretic	Pressor	Ref.
Ile	D-Arg	-	-	-	-	-	Schillinger *et al.* (1972)
Ile	Har	320	316		340[a]	300	Larsson *et al.* (1977)
Ile	Har (deamino)	149	325	-	125[a]	149	Larsson *et al.* (1977)
Position Three and Nine Substitutions							
3-Position	9-Position	Rat Uterus	Avian Vasodepressor	Milk Ejecting	Antidiuretic	Pressor	Ref.
Ile	Ala (deamino)						Cheesman *et al.* (1983)
Ile	des-Gly (deamino)						Cheesman *et al.* (1983)
Ile	Pro (deamino)						Cheesman *et al.* (1983)
Ile	Ser (deamino)						Cheesman *et al.* (1983)

Position Four and Eight Substitutions								
4-Position	8-Position		Rat Uterus	Avian Vasodepressor	Milk Ejecting	Antidiuretic	Pressor	Ref.
Abu	D-Arg	(deamino)	-	-	-	620[a,b]	inhibitor	Cort *et al.* (1976)
Ala	D-Arg	(deamino)	-	-	-	460[a,b]	-	Cort *et al.* (1976)
Asn	D-Arg	(deamino)	0.007	-	inactive	10750	inactive	Zaoral and Blaha (1977)
Gly	D-Arg	(deamino)	-	-	-	10[a,b]	-	Cort *et al.* (1976)
Ile	D-Arg	(deamino)	-	-	-	470[a,b]	inhibitor	Cort *et al.* (1976)
Leu	D-Arg	(deamino)	-	-	-	230[a,b]	inhibitor	Cort *et al.* (1976)
Val	D-Arg		0.6	-	-	653	0.037	Sawyer *et al.* (1974)
Val	D-Arg	(deamino)	8	inactive	-	1230	<0.01	Manning *et al.* (1973a)

a Used modified Burn (Burn *et al.*, 1950) assay technique (Cort *et al.*, 1975) that may give high results for analogs with prolonged duration of action.
b Values reported by Cort *et al.*, (1976) were relative to deamino (8-D-Arg) vasopressin. In this table the antidiuretic potency of deamino (8-D-Arg) vasopressin of 955 reported by Manning *et al.*, (1973a) was used to calculate the potencies shown above.

Table XX. Trisubstituted Analogs of Arginine Vasopressin

Substitutions at Positions One, Two, and Four								
1-Position	2-Position	4-Position	Rat Uterus	Avian Vasodepressor	Milk Ejecting	Antidiuretic	Pressor	Ref.
$(CH_2)_5$Mpr	D-Ala	Val	pA_2 6.16	-	-	agonist	pA_2 5.79	Manning *et al.* (1982a)
$(CH_2)_5$Mpr	D-Arg	Val	pA_2 <5.0	-	-	pA_2 <5.9	pA_2 5.4	Manning *et al.* (1982a)
$(CH_2)_5$Mpr	Gly	Val	pA_2 5.54	-	-	agonist	agonist	Manning *et al.* (1982a)
$(CH_2)_5$Mpr	D-Ile	Abu	pA_2 7.49	-	-	pA_2 8.22	pA_2 6.73	Manning *et al.* (1984a)
$(CH_2)_5$Mpr	D-Ile	Ala	pA_2 7.50	-	-	pA_2 7.76	pA_2 6.03	Manning *et al.* (1984a)
$(CH_2)_5$Mpr	D-Ile	Asn	pA_2 7.93	-	-	pA_2 6.51	pA_2 7.26	Manning *et al.* (1984a)
$(CH_2)_5$Mpr	D-Ile	Cha	pA_2 7.44	-	-	pA_2 6.64	pA_2 6.56	Manning *et al.* (1984a)
$(CH_2)_5$Mpr	D-Ile	Ile	pA_2 7.47	-	-	pA_2 8.04	pA_2 6.42	Manning *et al.* (1984a)
$(CH_2)_5$Mpr	D-Ile	Leu	pA_2 7.84	-	-	pA_2 6.80	pA_2 6.75	Manning *et al.* (1984a)
$(CH_2)_5$Mpr	D-Ile	Lys	pA_2 7.10	-	-	pA_2 6.76	mixed	Manning *et al.* (1984a)
$(CH_2)_5$Mpr	D-Ile	Nva	pA_2 7.83	-	-	pA_2 7.01	pA_2 6.96	Manning *et al.* (1984a)
$(CH_2)_5$Mpr	D-Ile	Orn	pA_2 7.57	-	-	pA_2 6.46	pA_2 6.76	Manning *et al.* (1984a)
$(CH_2)_5$Mpr	D-Ile	Phe	pA_2 7.13	-	-	mixed	mixed	Manning *et al.* (1984a)
$(CH_2)_5$Mpr	D-Ile	Ser	pA_2 7.28	-	-	pA_2 7.26	pA_2 6.21	Manning *et al.* (1984a)
$(CH_2)_5$Mpr	D-Ile	Thr	pA_2 7.91	-	-	pA_2 7.91	pA_2 6.83	Manning *et al.* (1984a)
$(CH_2)_5$Mpr	D-Ile	Val	pA_2 7.98	-	-	pA_2 7.98	pA_2 6.94	Manning *et al.* (1982a)

Substitutions at Positions One, Two, and Four								
1-Position	2-Position	4-Position	Rat Uterus	Avian Vasodepressor	Milk Ejecting	Antidiuretic	Pressor	Ref.
$(CH_2)_5Mpr$	D-Leu	Val	pA_2 7.32	-	-	pA_2 7.79	pA_2 6.45	Manning *et al.* (1982a)
$(CH_2)_5Mpr$	D-Phe	Abu	pA_2 7.73	-	-	pA_2 7.96	pA_2 7.70	Manning (1983)
$(CH_2)_5Mpr$	D-Phe	Ala	pA_2 7.68	-	-	pA_2 7.52	pA_2 7.55	Manning (1983)
$(CH_2)_5Mpr$	D-Phe	Cha	pA_2 7.78	-	-	pA_2 7.19	pA_2 6.70	Manning (1983)
$(CH_2)_5Mpr$	D-Phe	Gly	pA_2 7.76	-	-	pA_2 5.85	pA_2 6.79	Manning (1983)
$(CH_2)_5Mpr$	D-Phe	Ile	pA_2 8.04	-	-	pA_2 8.24	pA_2 7.86	Manning *et al.* (1983)
$(CH_2)_5Mpr$	D-Phe	Leu	pA_2 7.76	-	-	pA_2 6.07	pA_2 7.70	Manning *et al.* (1983)
$(CH_2)_5Mpr$	D-Phe	Lys	pA_2 7.60	-	-	pA_2 7.22	pA_2 6.93	Manning *et al.* (1983)
$(CH_2)_5Mpr$	D-Phe	Nle	pA_2 7.54	-	-	pA_2 7.12	pA_2 7.52	Manning *et al.* (1983)
$(CH_2)_5Mpr$	D-Phe	Nva	pA_2 8.03	-	-	pA_2 6.99	pA_2 7.73	Manning *et al.* (1983)
$(CH_2)_5Mpr$	D-Phe	Orn	pA_2 7.73	-	-	pA_2 6.5	pA_2 7.57	Manning *et al.* (1984a)
$(CH_2)_5Mpr$	D-Phe	Phe	pA_2 7.57	-	-	pA_2 6.07	pA_2 6.45	Manning *et al.* (1983)
$(CH_2)_5Mpr$	D-Phe	Pro	pA_2 5.76	-	-	0.004	pA_2 <0.01	Manning *et al.* (1983)
$(CH_2)_5Mpr$	D-Phe	Ser	pA_2 7.64	-	-	pA_2 7.33	pA_2 7.63	Manning *et al.* (1984a)
$(CH_2)_5Mpr$	D-Phe	Thr	pA_2 8.04	-	-	pA_2 7.62	pA_2 7.38	Manning *et al.* (1983)
$(CH_2)_5Mpr$	D-Phe	Tyr	pA_2 7.68	-	-	pA_2 5.57	pA_2 6.58	Manning *et al.* (1983)

(continued)

Table XX. *(continued)*

Substitutions at Positions One, Two, and Four								
1-Position	2-Position	4-Position	Rat Uterus	Avian Vasodepressor	Milk Ejecting	Antidiuretic	Pressor	Ref.
$(CH_2)_5$Mpr	D-Phe	Val	pA_2 7.44	-	-	pA_2 8.07	pA_2 8.06	Manning *et al.* (1982a)
$(CH_2)_5$Mpr	D-Tyr	Val	pA_2 8.11	-	-	pA_2 7.51 32[a]	pA_2 8.41	Manning *et al.* (1982a) Huffman and Moore (1984)
$(CH_2)_5$Mpr	Tyr(Et)	Val	-	-	-	pA_2 7.57	pA_2 8.16	Sawyer *et al.* (1981)
$(CH_2)_5$Mpr	D-Tyr(Et)	Val	-	-	-	pA_2 7.81 6.5[a]	pA_2 8.22	Manning *et al.* (1982c) Huffman and Moore (1984)
$(CH_2)_5$Mpr	Tyr(3-I)	Val	-	-	-	-	-	Moore *et al.* (1985)
$(CH_2)_5$Mpr	Tyr(3,5-I_2)	Val	-	-	-	-	-	Moore *et al.* (1985)
$(CH_2)_5$Mpr	Tyr(Me)	Val	-	-	-	pA_2 7.35	pA_2 8.32	Sawyer *et al.* (1981)
$(CH_2)_5$Mpr	D-Tyr(Me)	Val	-	-	-	pA_2 7.77	pA_2 8.48	Manning *et al.* 1982c)
$(CH_2)_5$Mpr	Tyr(n-Pr)	Val	-	-	-	pA_2 7.29	pA_2 8.22	Manning *et al.* (1981)
$(CH_2)_5$Mpr	D-Tyr(n-Pr)	Val	-	-	-	pA_2 7.61	pA_2 8.14	Manning *et al.* (1982c)
$(CH_2)_5$Mpr	Tyr(i-Pr)	Val	-	-	-	pA_2 7.32	pA_2 8.36	Manning *et al.* (1981)
$(CH_2)_5$Mpr	D-Tyr(i-Pr)	Val	-	-	-	pA_2 7.66	pA_2 8.40	Manning *et al.* (1982c)
$(CH_2)_5$Mpr	D-Val	Val	pA_2 7.90	-	-	pA_2 7.48	pA_2 6.41	Manning *et al.* (1982a)

Substitutions at Positions One, Three, and Six

1,6-Position	3-Position	Rat Uterus	Avian Vasodepressor	Milk Ejecting	Antidiuretic	Pressor	Ref.
Deaminodicarba	Ile	135	64.2	36.8	313	27.0	Hase *et al.* (1972a)
Deaminodicarba	Phe(4-azido)						Fahrenholz *et al.* (1983)

Substitutions at Positions One, Four, and Eight

1-Position	4-Position	8-Position	Rat Uterus	Avian Vasodepressor	Milk Ejecting	Antidiuretic	Pressor	Ref.
Hmp	Val	D-Arg	1.1	-	-	892	<0.02	Lowbridge *et al.* (1977b)
$(CH_3)_2$Mpr	Val	D-Arg	pA_2 7.23	-	-	123	pA_2 7.82	Manning *et al.* (1977)
$(C_2H_5)_2$Mpr	Val	D-Arg	pA_2 7.29	-	-	0.71	pA_2 8.18	Manning *et al.* (1982b)
$(CH_2)_5$Mpr	Val	D-Arg	pA_2 6.63	-	-	0.10	pA_2 7.68	Manning *et al.* (1982b)

Substitutions at Positions One, Six, and Seven

1,6-Position	7-Position	Rat Uterus	Avian Vasodepressor	Milk Ejecting	Antidiuretic	Pressor	Ref.
deamino-1-carba	Gly	30	-	50	47.5	<1	Brtnik *et al.* (1981)

(continued)

Table XX. *(continued)*

Substitutions at Positions One, Six, and Nine							
1,6-Position	9-Position	Rat Uterus	Avian Vasodepressor	Milk Ejecting	Antidiuretic	Pressor	Ref.
Deamino-6-carba	D-Ala	-	-	-	inactive	inactive	Cort *et al.* (1981)

Substitutions at Positions Three, Four, and Five								
3-Position	4-Position	5-Position	Rat Uterus	Avian Vasodepressor	Milk Ejecting	Antidiuretic	Pressor	Ref.
Ile	Asn	Gln (deamino)	-	-	-	-	-	Cheesman *et al.* (1983)

Substitutions at Positions Three, Four, and Eight								
3-Position	4-Position	8-Position	Rat Uterus	Avian Vasodepressor	Milk Ejecting	Antidiuretic	Pressor	Ref.
Ile	Leu	D-Arg	0.24	-	-	100[a]	0.15	Zaoral *et al.* (1976a)
Ile	Leu	D-Arg (deamino)	0.005	-	-	130[a]	0.17	Zaoral *et al.* (1976a)

[a]Dose of peptide delivered i.p. (μg/kg) which results in reduction of urine osmolality from hydropenic level to 300 m-Osmoles/kg H_2O.

[a]Used modified Burn (Burn *et al.*, 1950) assay technique (Cort *et al.*, 1975) that may give high results for analogs with prolonged duration of aciton.

Table XXI. Tetrasubstituted Analogs of Arginine Vasopressin

Substitutions at Positions One Two, Four, and Eight									
1-Position	2-Position	4-Position	8-Position	Rat Uterus	Avian Vasodepressor	Milk Ejecting	Antidiuretic	Pressor	Ref.
$(CH_2)_5$Mpr	D-Phe	Val	D-Arg	pA_2 7.70			pA_2 7.07	pA_2 7.98	Manning *et al.* (1982a)
$(CH_2)_5$Mpr	D-Tyr	Val	D-Arg	pA_2 7.99			pA_2 7.03	pA_2 8.05	Manning *et al.* (1982a)
$(CH_2)_5$Mpr	Tyr(Et)	Val	D-Arg	-			pA_2 7.10	pA_2 8.31	Sawyer *et al.* (1981)
$(CH_2)_5$Mpr	D-Tyr(Et)	Val	D-Arg	-			pA_2 7.59	pA_2 7.88	Manning *et al.* (1982c)
$(CH_2)_5$Mpr	Tyr(Me)	Val	D-Arg	-			pA_2 6.68	pA_2 8.44	Sawyer *et al.* (1981)
$(CH_2)_5$Mpr	D-Tyr(Me)	Val	D-Arg	-			pA_2 7.19	pA_2 8.10	Manning *et al.* (1982c)
$(CH_2)_5$Mpr	Tyr(n-Pr)	Val	D-Arg	-			pA_2 6.67	pA_2 7.86	Manning *et al.* (1981)
$(CH_2)_5$Mpr	D-Tyr(n-Pr)	Val	D-Arg	-			pA_2 7.60	pA_2 7.92	Manning *et al.* (1982)
$(CH_2)_5$Mpr	Tyr(i-Pr)	Val	D-Arg	-			pA_2 6.88	pA_2 8.41	Manning *et al.* (1981)
$(CH_2)_5$Mpr	D-Tyr(i-Pr)	Val	D-Arg	-			pA_2 7.51	pA_2 8.19	Manning *et al.* (1982c)
Substitutions at Positions One Two, Four, and Nine									
1-Position	2-Position	4-Position	9-Position	Rat Uterus	Avian Vasodepressor	Milk Ejecting	Antidiuretic	Pressor	Ref.
$(CH_2)_5$Mpr	D-Phe	Ile	des-Gly	-			pA_2 8.01	pA_2 7.83	Manning *et al.* (1984b)
$(CH_2)_5$Mpr	D-Phe	Val	des-Gly	-			pA_2 8.06	pA_2 8.15	Manning *et al.* (1984b)
$(CH_2)_5$Mpr	D-Phe	Val	des-Gly-NH_2	-			pA 7.71	pA_2 7.92	Manning *et al.* (1984b)
$(CH_2)_5$Mpr	D-Tyr	Val	des-Gly				63[a]		Huffman and Moore (1984)
$(CH_2)_5$Mpr	D-Tyr	Val	des-Gly-NH_2				156[a]		Huffman and Moore (1984)
$(CH_2)_5$Mpr	Tyr(Et)	Val	des-Gly	-			pA_2 7.83	pA_2 8.17	Manning *et al.* (1984b)
$(CH_2)_5$Mpr	D-Tyr(Et)	Val	des-Gly	-			pA_2 7.57 5.0[a]	pA 8.17	Manning *et al.* (1984b) Huffman and Moore (1984)

[a]Dose of peptide delivered i.p. (μg/kg) which results in reduction of urine osmolality from hydropenic levels to 300 m-Osmoles/kg H_2O.

Table XXII. Analogs of Lysine Vasopressin with Substitutions in a Single Position

Substitutions in Position One						
1-Position	Rat Uterus	Avian Vasodepressor	Milk Ejecting	Antidiuretic	Pressor	Ref.
Maa	3.3	pA_2 8.29	-	-	1.1	Dyckes *et al.* (1974c)
Mba	pA_2 6.74	pA_2 7.89	-	-	1.7	Dyckes *et al.* (1974c)
Mpr (Deamino lysine vasopressin)	12	61	32	301	126	Kimbrough *et al.* (1963)
$(C_2H_5)_2$Mpr	pA_2 6.84	pA_2 7.63	-	-	pA_2 7.15	Dyckes *et al.* (1974a)
Hmp	17.9	-	-	-	431	Walti and Hope (1975)
D-Hmp	0.015	-	-	-	0.91	Walti and Hope (1975)

Substitutions in Position Two						
2-Position	Rat Uterus	Avian Vasodepressor	Milk Ejecting	Antidiuretic	Pressor	Ref.
3,5-Dibromotyrosine (deamino)	pA_2 6.29	pA_2 6.97	-	-	inactive	Lundell and Ferger (1975a)
Ile	<0.005	0.5	-	3.7	1.8	Havran and du Vigneaud (1969)
Ile (deamino)	<0.1	1.0	-	3.0	1.5	Havran and du Vigneaud (1969)
Phe	0.3	0.15	2.5	21	57	Berde *et al.* (1961b)
Tyr (Me)	inhibitor	-	-	1.5-3	79	Zaoral (1965)
Tyr (Et)	inhibitor	-	-	5	inhibitor	Zaoral (1965)
Tyr (Et) (deamino)	-	-	-	0.2-2 (inhibitor)	0.5-3	Larsson *et al.* (1978)
Tyr (n-Pr) (deamino)	-	-	-	0.2-2 (inhibitor)	0.5-3	Larsson *et al.* (1978)
Tyr (n-Bu) (deamino)	-	-	-	0.2-2	0.5-3	Larsson *et al.* (1978)
Tyr (n-Bu) (deamino)	-	-	-	0.2-2 (inhibitor)	0.5-3	Larsson *et al.* (1978)
Tyr (n-hex) (deamino)	-	-	-	0.2-2 (inhibitor)	0.5-3	Larsson *et al.* (1978)

Substitutions in Position Three						
3-Position	Rat Uterus	Avian Vasodepressor	Milk Ejecting	Antidiuretic	Pressor	Ref.
1,4-Cyclohexadienyl-L-alanine	6	-	-	125-130	129-132	Banerjee *et al.* (1979)
Ile	78	210	180	24	130	Huguenin & Boissonas (1962)
Ile (deamino)	-	-	-	-	-	Schillinger *et al.* (1972)
Nle (deamino)	-	-	-	-	-	Schillinger *et al.* (1972)
Pro (deamino)	-	-	-	-	-	Schillinger *et al.* (1972)
β-(2-Thienyl)-L-alanine	19.0	87	-	332	243	Smith *et al.* (1975)
Val (deamino)	-	-	-	-	-	Schillinger *et al.* (1972)

Substitutions in Position Four						
4-Position	Rat Uterus	Avian Vasodepressor	Milk Ejecting	Antidiuretic	Pressor	Ref.
Abu	1.5	13.1	-	707	10.2	Gillessen & du Vigneaud
Abu (deamino)	1.5	12.6	-	729	3.5	Gillessen & du Vigneaud
Asn	3	22	18	24	54	Boissonnas *et al.* (1963)
Asn	4		-	2430	109.5	Pliska & Krejci (1966)
D-Gln	2.5			-	0.38	Duttu *et al.* (1969)
Leu	pA_2 5.99	pA_2 6.31	-	1-2	1.33	Dyckes *et al.* (1973)
Leu (deamino)	pA_2 5.94	4.6	-	5-6	0.55	Dyckes *et al.* (1973)
Thr	13	267	-	155	49	Manning *et al.* (1973c)
Thr (deamino)	2.7	195	-	544	10	Manning *et al.* (1973b)

(continued)

Table XXII. *(continued)*

Substitutions in Position Five						
5-Position	Rat Uterus	Avian Vasodepressor	Milk Ejecting	Antidiuretic	Pressor	Ref.
Ala	<0.001	<0.01	-	0.2	0.15	Gillessen & du Vigneaud (1967)
Ala (deamino)	<0.002	<0.01	-	0.05	0.15	Gillessen & du Vigneaud (1967)
Asp(N(CH_3)$_2$)	<0.05	0.39	-	2.55	5.5	Smith *et al.* (1980b)
Ser	<0.05	<0.1	-	0.04	<0.01	Boissonnas *et al.* (1963)
Substitutions in Position Seven						
7-Position	Rat Uterus	Avian Vasodepressor	Milk Ejecting	Antidiuretic	Pressor	Ref.
Azetidine-2-carboxylic acid	2.0	-	3.5	80	27	Barber *et al.* (1979)
Gly	-	-	-	1.0	0.15	Kolc *et al.* (1967)
Hyp	2.0	-	-	-	0.13	Dutta *et al.* (1969)
D-Pro	0.15	-	-	20	0.05	Kolc *et al.* (1967)
Val	-	-	-	0.7	0.20	Boissonnas *et al.* (1963)

Substitutions in Position Nine						
9-Position	Rat Uterus	Avian Vasodepressor	Milk Ejecting	Antidiuretic	Pressor	Ref.
Ala	0.08	-	-	2	0.05	Zaoral *et al.* (1967c)
D-Ala	0.03	-	-	0.8	0.01	Zaoral *et al.* (1967c)
D-Ala	-	-	-	22	3.7	Buku *et al.* (1984)
D-Ala (deamino)	-	-	-	22.5	0.7	Buku *et al.* (1984)
β-Ala	0.04	-	-		inhibitor	Dutta *et al.* (1969)
des-Gly	<0.01	-	-	3	1.0	Barth and Kluh (1974)
Ethylenediamine	0.02	-	-	2	0.002	Zaoral *et al.* (1967c)
deamido (Lysine vasopressinoic acid)	4.8	-	-	-	1.3	Dutta *et al.* (1969)
Gly-N$(CH_3)_2$	0.4	0.1	-	-	0.06	Glass and du Vigneaud (1972)
Gly-NH-CH_3	0.4	<0.1	-	-	1.5	Glass and du Vigneaud (1972)
Hyp	2.0	-	-	-	0.13	Dutta *et al.* (1969)
Sar	0	<0.01	-	-	0.5	Meienhofer and du Vigneaud (1961)

Table XXIII. Analogs of Lysine Vasopressin with Multiple Substitutions

Substitutions in Position One and Four							
1-Position	4-Position	Rat Uterus	Avian Vasodepressor	Milk Ejecting	Antidiuretic	Pressor	Ref.
$(C_2H_5)_2$Mpr	Leu	pA_2 6.69	pA_2 7.56	-	0.05	pA_2 6.91	Dyckes *et al.* (1974b)
Substitutions in Position One and Six							
1-Position	6-Position	Rat Uterus	Avian Vasodepressor	Milk Ejecting	Antidiuretic	Pressor	Ref.
Ala	Ala	<0.2	<0.1	-	0.05	0.012	Huguenin & Boissannas (1965)
Ala	Ala (deamino)	<0.2	<0.1	-	0.05	0.008	Huguenin & Boissonnas (1965)
Deamino-	dicarba	5.1	4.2	-	7.8	10.4	Sakakibara & Hase (1968)
Deamino-	dicarba	5.3	2.4	3.2	208	6.2	Hase *et al.* (1972a)
Substitutions in Position Two and Three							
2-Position	3-Position	Rat Uterus	Avian Vasodepressor	Milk Ejecting	Antidiuretic	Pressor	Ref.
Tyr (t-Bu)	Ile	-	-	-	-	-	Schillinger *et al.* (1972)
Substitutions in Position Two and Five							
2-Position	5-Position	Rat Uterus	Avian Vasodepressor	Milk Ejecting	Antidiuretic	Pressor	Ref.
Phe	Asp (deamino-5,8-cyclo)	-	-	-	pA_2 6.15	mixed	Skala *et al.* (1984)

Substitutions in Positions Three and Four							
3-Position	4-Position	Rat Uterus	Avian Vasodepressor	Milk Ejecting	Antidiuretic	Pressor	Ref.
Ile	Glu	-	-	-	-	-	Schillinger *et al.* (1972)
Ile	Leu	-	-	-	-	-	Strub *et al.* (1973)
Ile	Leu (deamino)	-	-	-	-	-	Strub *et al.* (1973)
Ile	Ser (deamino)	-	-	-	-	-	Schillinger *et al.* (1972)

Substitutions in Positions One, Three, and Six							
1,6-Position	3-Position	Rat Uterus	Avian Vasodepressor	Milk Ejecting	Antidiuretic	Pressor	Ref.
Deaminodicarba	Ile	63.6	20.5	36.3	34.8	9.6	Hase *et al.* (1972a)

Substitutions in Positions One, Two, Four, and Nine									
1-Position	2-Position	4-Position	9-Position	Rat Uterus	Avian Vasodepressor	Milk Ejecting	Antidiuretic	Pressor	Ref.
$(CH_2)_5$Mpr	D-Tyr(Et)	Val	des-Gly	-	-	-	22[a]	-	Huffman *et al.* (1985)

[a] Dose of peptide delivered i.p. (µg/kg) which results in reduction of urine osmolality from hydropenic levels to 300 m-Osmol/kg H_2O.

6. Summary Comments and Conclusions

Although the avian vasodepressor assay is one of the classical assays for the neurohypophyseal hormones, it has fallen out of favor. This perhaps is understandable, since the assay does not appear to be particularly relevant to the receptors in mammalian systems that are of primary interest to most investigators. Structure–function analysis clearly shows that the receptors for the AVD activity do not correspond to those for the uterus or for other mammalian receptor systems. In a sense, the AVD receptor appears to be a sort of hybrid between oxytocin-like receptors and vasopressin-like receptors, in that it readily accommodates a basic residue in the 8 position for agonist analogs. It also accommodates a Phe^3 substitution reasonably well, although only slightly better than the uterine receptor. On the other hand, it appears that the $Gly\text{-}NH_2{}^9$ residue is much more critical for interaction with the AVD receptor than the uterine receptor, and it may be involved in transduction. It is clear that should further insights into the structure–function relationships be desired, it would be highly desirable to examine more carefully the requirements of the 5 and 9 positions for transduction.

C. Agonist Analogs of the Neurohypophyseal Hormones in the Milk-Ejecting Assay

Although the milk-ejecting activity of oxytocin is the most widely accepted physiological activity of OT, the milk-ejecting assay is not utilized universally for examining structure–function relationships of the neurohypophyseal hormones. This is unfortunate, since there are significant differences in the structural requirements at the milk-ejecting receptor compared to the uterine receptor or other neurohypophyseal hormone receptors. This is immediately obvious when examining the activities of the naturally occurring neurohypophyseal hormones (Table I). First, the range of the milk-ejecting potencies for natural hormones is the narrowest range of all the five assay systems, being only over about 140-fold. It is particularly interesting that aspartocin is almost equipotent to OT and more potent than valitocin (Table I). Also very interesting is the potency of isotocin, which again is nearly identical to OT in the milk ejecting assay and more potent than mesotocin (Table I). Clearly the milk-ejecting receptor is quite accommodating of structural changes in the 4 and 8 position of the neurohypophyseal hormones. On the other hand, the milk-ejecting activities of naturally occurring neurohypophyseal hormones with a basic residue are reduced relative to the avian vasodepressor activities, but nonetheless are more active than observed in the rat uterus assay. The low potency of phenylpressin suggests that the Tyr^2 residue is important for high potency in the milk-ejecting assay.

1. Fragment Analogs of Neurohypophyseal Hormones

Conversion of the carboxamide-terminal group to a carboxylate, which has such a deleterious effect on the activity of OT in the uterine and AVD assays, results in only a 34-fold decrease in potency in the milk-ejecting assay (Table II). While this is still substantial, it suggests that the mammary-gland receptor can more readily accommodate a negative charge at the C-terminal than any of the OT receptor systems. Similarly, the removal of the C-terminal tripeptide still left an analog, tocinamide (Table II), with about one-seventieth the potency of OT in the milk ejecting assay. Pressinamide, the ring of vasopressin (Table XII), also is fully active, although about eightfold less potent than tocinamide in the milk-ejecting assay. It appears that the C-terminal tripeptide, although important for potency in the milk-ejecting assay, is not critical for transduction (see also Berde and Boissonnas, 1968).

2. Structurally Extended Analogs of Neurohypophyseal Hormones

Although [N^{α}-glycyl]OT (Table III) has been reported to be active in the milk-ejecting assay under some conditions, earlier studies [reviewed in Berde and Boissonnas (1968) and Rudinger and Krejci (1968)] showed that it inhibited the milk ejection action of oxytocin in the guinea pig assay. This also was found to be the case for [N-leucyl]oxytocin and for N-Leu-Gly-Gly-OT (Beránková-Ksandrová *et al.*, 1966). Other N-terminal extended analogs of OT and vasopressin examined were found to be agonists. The inhibitory activity reported for the *N*-carbamoyl Tyr(Me)^2 analog (Table III) is interesting. However, none of this inhibitory activity has been examined in more detail in the milk-ejecting assay, nor does it appear that anyone has vigorously pursued this lead to design inhibitor analogs specifically for the mammary gland receptor. Hence it is difficult to interpret these results, although tempting to speculate that the inhibitory activity may be the result of a disruption of the spatial relationship(s) between the N-terminal and the Tyr^2 residues. The results of Table II, however, clearly show that relationships of the N-terminal to the 2 position that lead to antagonist analogs for the rat uterus do not necessarily lead to antagonists for the mammary gland. Indeed, in some cases reasonably potent antagonists of oxytocin in the rat uterus assay are agonists in the milk-ejecting assay.

3. Monosubstituted Analogs of the Neurohypophyseal Hormones

As is the case for activity in the rat uterus and AVD assay systems, deamino-oxytocin ([Mpr^1]OT; Table IV) is more potent than OT in the milk-ejecting assay. In addition, the seleno analog (Table IV) also is a potent analog. The 1-carba analog (Table V) is considerably less potent than OT, and the deamino-1-carba analog is essentially equipotent to deamino oxytocin. Interestingly, the deamino-1,6-dicarba analog has about one-third the potency of OT and one-fourth the potency of deamino-oxytocin, making it relatively more potent in the milk-ejecting assay than in the AVD or uterine assays. Of particular note is the exceptionally high potency of the deamino homolanthionine analog of OT (Table V); even the lanthionine derivative (Table V) retains about one-tenth the potency of OT, which again makes it relatively much more potent in this assay than in the AVD or uterine assays. Apparently the mammary-gland receptor can accept a reduction in ring size much more readily than either the AVD or uterine systems. Unfortunately, the milk-ejecting activity of [Maa^1]OT (Table IV) apparently has not been examined, although it would be most interesting to do so. Another interesting comparison is the 1-carba sulfoxide, which is a very weak agonist (about 1/20,000 times that of OT), and deamino-1-carbasulfoxide, which is a potent agonist (about one-half of OT) in the milk-ejecting assay. It appears that the mammary gland receptor can accommodate one highly polar or formally charged group in the 1 position, but not two; it is interesting that these charges can apparantly be either at the α-carbon or the γ-sulfur. Interestingly, the pseudoisosteric ring-opened analog with Ser residues at positions 1 and 6 (Table V) has weak agonist activity. This suggests that the 20-membered ring apparently is not essential for biological activity in the milk-ejecting assay.

b. Position 2 Substitution. The stereochemical requirements for agonist activity at the 2 position of the neurohypophyseal hormone in the mammary gland do not appear to be as stringent as for the rat uterus, since [DTyr^2]OT has about one-twelfth the potency of OT (Table IV) and appears to be a full agonist. In addition, the substitution of Tyr^2 by Phe^2 provides an analog with about one-third the potency of OT (Berde and Boissonnas, 1968), and the loss in potency appears to be primarily due to replacement of the hydroxyl group with a hydrogen. This appears to be due primarily to a steric factor, since [Phe(p-Me)^4]OT has about 60% the potency of OT (Zhuze *et al.*, 1964), and there apparently is no partial antagonist activity, although this may have been difficult to detect in the *in vivo* milk-ejecting assay. Interestingly, some of these analogs cause tachyphylaxis in the isolated mammary-gland strip (reviewed by Rudinger and Krejci, 1968), and thus it may be that although they have agonist activities they also are partial antagonists at the mammary gland. It would appear that more extensive studies with the *in vitro* mammary gland could provide further insights into the importance of the 2-position to transduction at the mammary gland receptor. Of particular interest would be the [3′,4′-Br_2Tyr^2]OT, [3′-I-Tyr^2]OT, and [D Trp^2]OT analogs. As with the other oxytocin receptor systems, replacement of the aromatic residue in position 2 with aliphatic residues leads to a significant loss in potency, but the loss in potency is generally less severe than in the rat uterus or AVD assays (Table IV) (Berde and Boissonnas, 1968).

c. Position 3 Substitution. The Ile^3 residue can be replaced by a variety of aliphatic amino acids and still retain high potency. The stereochemistry at the α-carbon atom does not appear to be important, since [$aIle^3$]OT retains about one-third the potency of OT in the milk-ejecting assay (Nesvadba *et al.*, 1963). This is also seen from the very high potencies of [Val^3]OT and especially [3,3-Et_2Ala^3]OT and [2-cyclopentylglycine^3]OT (Table IV). Even the Phe^3 substitution is fairly well accommodated (Table IV), but analogs with more bulky aromatic amino acids—e.g., [Tyr^2]OT and [Trp^2]OT—show severe loss in potency in the milk-ejecting assay (Berde and Boissonnas, 1968).

d. Position 4 Substitutions. As with the 3 position of neurohypophyseal hormones, the 4 position can be substituted with a wide variety of amino acids and still retain high potency. [Thr^4]OT (Table IV) is about 30% more potent than OT in the milk-ejecting assay, but this increased potency is not as great as that seen for this analog in the uterine and AVD assays. On the other hand, the very high potencies of [Val^4]OT, [Ser^4]OT, [Hse^4]OT, and [Asu^4]OT, all of which are essentially equipotent to OT, suggest that there is considerable structural variation possible for the side-chain group in position 4 of OT compatible with a strong interaction with mammary-gland receptor. Particularly impressive also are the high potencies of [Ala^4]OT, [Glu(OMe)^4]OT, [Mpr^1,Glu(OMe)^4]OT, [Ile^4]OT, and [Leu^4]OT, all of which are 5–10 times more potent in the milk-ejecting assay than in the uterus assay. On the other hand, the stereochemical requirements are much more stringent for the mammary-gland receptor than for the uterine receptor, since [DGln^4]OT (Table IV) has only about 1/4500 times the potency of OT in the milk-ejecting assay, which is less than one-thirtieth the relative potency of [DGln^4]OT in the rat uterus assay. The fairly severe stereochemical requirements are also evident in the low potency of [Pro^4]OT in the milk-ejecting assay.

e. Position 5 Substitutions. Although the milk-ejecting activities of analogs of the neurohypophyseal hormones substituted in position 5 have not been thoroughly investigated (Table IV) (Berde and Boissonnas, 1968), the high potency of [3-malamidic acid^5]OT, of [3-CN-Ala^5]OT (Table IV), and of [Gln^5]OT (Berde and Boissonnas, 1968) suggests that there is more flexibility in structure possible for the side-chain group in position 5 for interaction with the mammary-gland receptor than for any other receptor for the neurohypophyseal hormones. Further studies with both more bulky and otherwise modified Asn derivatives would be most instructive.

f. Position 6 Substitutions. The replacement of Cys^6 with $\text{D}Cys^6$ leads to an analog with virtually no activity in the milk-ejecting assay (Manning and du Vigneaud, 1965). Thus there appears to be a substantial stereochemical requirement for interaction of OT with the mammary gland receptor. The Cys^6 sulfur can be replaced by selenium with essentially no change in potency (Table IV). Most of the other modifications that affect activity in the milk-ejecting assay related to cystine (Cys^1 and Cys^6) have been discussed already in Section III,C,3,a. One rather interesting observation that deserves special mention is the very large difference in the activities of deamino-6-carba sulfoxide (about one-fourth of OT) and deamino-6-carba sulfone (about 1/14,000 times OT) (Table V).

g. Position 7 Substitutions. The Pro^7 residue of OT can be substituted with a wide variety of amino acids and still retain high potency in the milk-ejecting assay. For example, [$MeAla^7$]OT, [Sar^7]OT, and even [Gly^7]OT are essentially equipotent to OT in the milk-ejecting assay (Table IV). Even the 7-substituted OT analogs containing 1-aminocyclopropane-1-carboxylic acid and azetidine-2-carboxylic acid have greater than one-seventh the potency of the native hormone. The diastereoisomer analog [$\text{D}Pro^7$]OT has only about one-hundredth the potency of the native hormone. This is one of the few positions in the hormone where the milk-ejecting receptor appears to have more stringent stereostructural requirements than is the case for the rat uterus receptor.

h. Position 8 Substitutions. As already mentioned, the milk-ejecting receptor can accommodate many changes in the 8 position of OT without a significant loss in activity (Table IV). In addition to the Ile and Val substitutions already mentioned, the Gln^8 and Ala^8 analogs are within a factor of two of the potency of OT. The [Phe^8]OT analog is nearly as active as OT in the milk-ejecting assay, and the bulky amino acid *tert*-leucine^8 analog of OT is actually about 15% more potent than OT. Clearly, the site of interaction of the side-chain of position 8 of OT with the mammary-gland receptor can accommodate a large lipophilic group. It is thus interesting that the potencies of AVP and LVP in the milk-ejecting assay are quite substantial, and that [Lys^8]OT and [Orn^8]OT have about one-half and one-fourth the potency of OT, respectively, in the milk-ejecting assay (Berde and Boissonnas, 1968). Thus the binding site on the receptor of the mammary gland can apparently accommodate a positive charge, though apparently it prefers not to since [N^ϵ-For-Lys^8]OT is about 30% more potent than [Lys^8]OT (Berde and Boissonnas, 1968).

i. Position 9 Substitutions. Most OT analogs that have been modified in the 9 position have not been examined in the milk-ejecting assay (Table IV). We have noted that the carboxylate terminal derivative of OT, [Gly[OH^9]OT, where the only change is from a C-terminal carboxamide peptide (the native hormone) to a carboxylate C-terminal compound, has only about one thirty-fifth the potency of OT, but actually is (relatively speaking) much more active in the milk-ejecting assay than in the AVD or uterine assays. It would appear that further studies of analogs substituted in the 9 position of OT are needed before much can be said regarding the importance of this residue to binding and transduction in the mammary-gland receptor.

4. Multiple Substitution in OT Analogs

A variety of multiply-substituted analogs of OT have been examined for their activities in the milk-ejecting assay (Tables V–XI) and provide a few additional insights into structure–function relationships for this receptor system. As expected (Table V), [Hmp^1,Thr^4]OT is much more potent (about twofold) than OT in the milk-ejecting assay system, and interestingly the deaminopenicillamine analog [$(CH_3)_2$Mpr^1,Thr^4] is a weak agonist (about one-hundredth the potency of OT) rather than an antagonist. Even [$(CH_3)_2$Mpr^1,Orn^8]OT is an agonist, but [$(CH_2)_5$Mpr^1,Orn^8]OT is a weak antagonist in the milk-ejecting assay (Table V). Many other Thr^4 and Ser^4 analogs of OT are quite potent in the assay—e.g., [Thr^4,Gly^7]OT and [Thr^4,Ser^7]OT (Table VIII)—although it is interesting to note that the former compound is relatively speaking about five times more potent in the milk-ejecting compared to the oxytocic assay, while the latter compound is about 30% less active in the milk-ejecting than in the oxytocic assay. The most potent of the Thr^4-substituted analogs appears to be [Hmp^1,-Thr^4,Gly^4]OT (Table X), with 959 U/mg of milk-ejecting activity. However, the most potent milk-ejecting analog reported to date appears to be the deamino-1-carba analog deamino-1-carba[Glu(OMe)^4]OT, which has 1307 U/mg of milk-ejecting activity (Table IX). These results illustrate that it is possible to substitute alkyl groups at the carboxyl or carboxamide moiety when Glu^4 or Gln^4 is used. On the other hand, when the Tyr^2 group was substituted by Tyr(Me)^2, the potency dropped substantially (Table XI). Nonetheless, it should be possible to utilize these results to design analogs of OT that utilize substitutents in the side-chain moiety of position 4 for side-chain to side-chain covalent attachment (Hruby, 1982) to some other part of the OT molecule, and hence obtain conformationally constrained analogs of OT with unique biological activities. Finally, it should be noted that when a bicyclic oxytocin analog was made by covalently bonding the N-terminal amino group of Cys^1 to the carbonyl group of Gly^9 in OT, the analog obtained, 1-9-*cyclo*-oxytocin (Table V), was inactive.

5. Substitutions in Vasopressin Analogs

In view of the accommodation of the mammary-gland receptor to a basic residue in the 8 position of the neurohypophyseal hormones, it is not surprising that most vasopressin agonist analogs are found to be weak agonists in the milk-ejecting assay. In general, the analogs that have been tested are relatively more potent in the milk-ejecting assay than in the uterus assay (Tables XII–XXIII), although there are some exceptions. Some of the analogs actually have quite high potency, including [Mpr^1,Dbu^8]VP (Table XIV), [Mpr^1,Ile^3,Orn^8]VP (Table XV), [Sar^7]AVP (Table XVII), and [Ile^3]LVP (Table XXII).

6. Summary Comments and Conclusions

Although the milk-ejecting activity of the neurohypophyseal hormones is a well-accepted physiological role for OT in mammals, we are still a long way from having a good understanding of the structure–function relationships of OT at this receptor system. In general there appears to be considerable flexibility for structural modifications in positions 4, 7, and 8 and to a lesser extent in position 3, and in very special ways, in position 1(6). A few highly potent analogs have been prepared, but a truly superpotent analog apparently has still not been obtained. It appears that bicyclic or other conformationally constrained analogs with side-chain to side-chain cyclization should be possible using appropriately designed substituents on side-chain groups in positions 4, 7, and 8 and perhaps others. Such analogs might prove exceptionally useful for understanding conformation–biological activity relationships at the mammary-gland receptor for milk-ejecting activity.

IV. STRUCTURE–ACTIVITY RELATIONSHIPS OF ANTAGONIST ANALOGS OF THE NEUROHYPOPHYSEAL HORMONES THAT COMPETITIVELY BLOCK THE ACTIONS OF THE NATIVE HORMONE AT RECEPTORS FOR OXYTOCIN

The primary focus of this section of the discussion is to examine the structure–function relationships of analogs of the neurohypophyseal hormones that antagonize (inhibit) the agonist activities of OT at oxytocin receptors in the rat uterus, avian vasodepressor, and milk-ejecting assays. Inhibitors of the biological effects of a peptide hormone at its receptors are important for developing a full understanding of the hormone–receptor interaction in relationship to biological information transfer, since competitive inhibitors bind to the receptor but do not

not transduce the biological message. Hence they lack some structural, conformational, and/or dynamic property that is essential for transduction. In essence, antagonists require a separation of those chemical–physical features of the hormone that are necessary for binding from those necessary for transduction. Thus antagonists can provide direct information regarding stereostructural and conformational properties related to the hormone-binding message, and indirect evidence (by comparison with structurally similar agonist analogs) of the conformational and structural features that are critical for transduction. Such considerations have led us to suggest that in general the binding and the transduction processes will require different stereostructural and conformational properties of the hormone (Meraldi *et al.*, 1977; Hruby, 1981a,b, 1984).

Structure–function studies of oxytocin analogs and their activities at several receptors were among the first studies to lead to peptide hormone antagonist analogs in the laboratories of du Vigneaud and of Rudinger (for a review see Rudinger and Krejci, 1968). Since the early work through about 1967 has been thoroughly reviewed, we concentrate our discussion in the following sections primarily on more recent studies, refering back to older studies only when they are critical to the development of ideas regarding structure–function relationships. We draw from other recent reviews on the structure–function and conformation–activity relationships for antagonists of oxytocin in the rat uterus and avian vasodepressor systems (Hruby and Mosberg, 1982; Hruby, 1981b, 1985).

A. Antagonist Analogs of the Neurohypophyseal Hormones in the Uterus Assay

Quite early in the studies of oxytocin structure–function relationships it was found that substitution of the Tyr^2 residue by O-alkyl-substituted tyrosine derivatives and/or O-alkyl-substituted tyrosine derivatives and N^α-substituted or extended analogs of the neurohypophyseal hormones led to analogs with weak antagonist, partial antagonist, or agonist/antagonist activities, depending on assay conditions. In addition, it was found that certain analogs of OT with aliphatic or aromatic amino acids led to analogs with weak partial antagonist or weak inhibitor activities in the uterine assay system. These studies have been thoroughly reviewed (Rudinger and Krejci, 1968; Hruby and Mosberg, 1982). However, the major impetus for the development of potent oxytocin inhibitor analogs came from the discovery of Schulz and du Vigneaud (1966a) that [penicillamine^1]oxytocin ([Pen^1]OT) is a fairly potent antagonist of OT action in the isolated rat uterus assay. This was followed by the development of a series of β,β-dialkylcysteine-1-substituted analogs with increasingly potent antagonist activities. This antagonist activity can be accounted for by conformational restrictions that accompany these structural modifications (Meraldi *et al.*, 1977; Hruby, 1981a,b, 1985; Hruby and Mosberg, 1982), but in this review we analyze antagonist activity on a residue-by-residue basis, with special emphasis on those aspects of structure that appear to produce antagonist analogs.

1. Fragment Analogs of the Neurohypophyseal Hormones

Examination of Tables II and XII shows that few fragment analogs have been prepared that have antagonist activities in the rat uterus system. Although many of these fragment analogs have little or no activity at the rat uterus, only those with β,β-diethyl-β-mercaptopropionic acid—[(C_2H_5)Mpr] or β-cyclopentamethylene-β-mercaptopropionic acid [$(CH_2)_5$Mpr]—form analogs that are antagonists.

2. Extended Analogs of the Neurohypophyseal Analogs

As pointed out above, substitution of an N^{α}-amino group with a variety of amino acid residues or of a variety of acylation agents lead to antagonist analogs in the uterus assay system, but only when used in conjunction with either O-alkylated Tyr^2 residues or *p*-alkyl-substituted Phe^2 residues. Especially potent antagonists are [N^{α}-Ac-Cys^1,Tyr$(Me)^2$]OT, [N^{α}-BrAc-Cys^1,Tyr$(Me)^2$]OT, [N^{α}-Car-Cys^1, Phe$(Et)^2$]OT, [N^{α}-Mes-Cys^1,Tyr$(Me)^2$]OT (Table III), and [N^{α}-Ac-cys^1, Tyr$(Me)^2$]AVP (Table XIII). This inhibitory activity appears to be due to a steric effect, in which the steric relationship of the Cys^1 and Tyr^2 residue necessary for transduction is distorted by the substitutions on the N^{α}-amino group and the *p*-position of the aromatic amino acid in position 2. Interestingly, N^{α}-acylation of the amino group in Pen^1 antagonist analogs of OT has not been examined, and it might be interesting to do so in view of the results seen in these studies.

3. Monosubstituted Antagonist Analogs of the Neurohypophyseal Hormones

a. Position 1 Substitutions. Substitution of Cys^1 with β,β-dialkyl-substituted residues in position 1 leads to antagonist analogs in the rat uterus assay system, and generally the larger the substituents, the more potent the antagonism. Thus [Pen^1]OT ≈ [$(CH_3)_2Mpr^1$]OT < [$(C_2H_5)_2Mpr^1$]OT < [$(CH_2)_5Mpr^1$]OT (Table IV) in antagonist potency as measured by p*A*2* values.

*When possible, inhibitory potencies are expressed as pA_2 values as defined by Schild (1949).

In addition, the 19-membered ring analog $[(CH_3)_2Maa^1]OT$ and the diastereoisomer analog [DPen[1]]OT also are weak antagonist analogs. In this regard, it is most interesting that the homocystine 1(6) analog of OT, $[Hcy^1,Hcy^6]OT$, is a weak antagonist (Table V), and the related deamino analog $[4\text{-}Mba^1,Hcy^6]OT$ is a more potent antagonist (Table V). It also is noteworthy that the open-chain pseudoisosteric Ser^1 analog of OT, $[Ser^1,Ser^6]OT$, is a mixed antagonist/agonist (Table V). It is clear from even these studies that the conformation constraints imposed by the gemminal dialkyl groups in the 1 position, or other extended alkyl groups in the same position, lead to hormone–receptor interactions that prevent transduction. The nature of the conformational constraints that have occurred as a result of these substitutions has been reviewed (Hruby, 1981b, 1985; Hruby and Mosberg, 1982).

b. Position 2 Substitutions. Substitution of the tyrosine in position 2 of OT with more bulky *p*-substituted aromatic amino acid derivatives or 3′- and/or 5′-substituted analogs can lead to antagonist analogs. The partial agonist activity of $[DTyr^2]OT$ (Table IV) and the reasonably potent antagonist activities of $[Mpr^1,3',5'\text{-}Br_2\text{-}Tyr^2]OT$, of $[3'\text{-}I\text{-}Tyr^2]OT$, of $[(\beta\text{-}Me)Tyr^2]OT$, and of $[DTrp^2]OT$ (Table IV) suggest that an improper steric fit of the aromatic side-chain of a neurohypophyseal hormone when it interacts with the uterine receptor leads to a hormone–receptor complex that cannot transduce the message but that can effectively compete for the OT binding site on the receptor.

c. Positions 2, 3, 4, 5, 7, 8, or 9 Substitutions. Single substitution *per se* generally does not lead to antagonist activities of neurohypophyseal hormones in the uterine assay system, although it has been reported (Table IV) that $[DLeu^4]OT$ and $[DAla^3]OT$ are weak antagonists in this assay system. Also interesting is the weak antagonist activity of the 23-membered ring analog of oxytocin $[Tyr^{2-3}]OT$ (Berde and Boissonnas, 1968).

4. Multiple Substitution in Antagonist Analogs of OT

Substitution of β,β-dialkyl derivatives of Cys^1, such as Pen, $\beta,\beta\text{-}Et_2Mpr$, etc., in conjunction with other modifications in the oxytocin structure can lead to potent antagonist (Tables V–XI). For example, $[(CH_3)_2Mpr^1]OT$, $[(CH_2)_5Mpr^1,Thr^4]OT$, and $[(CH_3)_2Mpr^1,Orn^8]OT$ (Table V), $[Pen^1,Phe^2,Thr^4]OT$, $[(C_2H_5)_2Mpr^1,Tyr(Me)^2,Orn^8]OT$, $[(CH_2)_5Mpr^1,Tyr(Me)^2,Orn^8]OT$ (Table X), and $[(CH_3)_2Mpr^1,Phe^2,Asn^4,Orn^8]OT$ (Table XI) are all potent antagonists in the uterus assay system. An interesting aspect of these studies is the observation that substitution of a basic amino acid in position 8, such as Arg^8 or Orn^8, produced antagonist analogs that are more potent antagonists than the corresponding Leu^8.

This is particularly interesting since, as we have already discussed for oxytocin agonists, replacement of the Leu^8 residue with a basic residue such as Arg or Orn leads to agonist analogs that are *less* potent. This suggests that oxytocin antagonists have a different binding mode than that of agonists when they interact with the uterine receptor.

It is interesting to note that [Leu^2,Leu^4]OT, [Ile^2,Leu^4]OT, and [Phe^2,Leu^4]-OT are all weak inhibitors in the rat uterus assay. This weak antagonist activity could be a result of a slight steric perturbation in the vicinity of the 2 position on interaction of the analog with the receptor, which in turn leads to a slightly different binding mode such that transduction does not occur.

5. Substitutions in Vasopressin Analogs

In view of the large increases in potency often seen for oxytocin antagonist analogs that have a basic residue in position 8, it is perhaps not surprising that many vasopressin analogs might be expected to quite potent antagonists, especially if substituted in position 1 with β,β-dialkyl-substituted cysteine derivatives. Indeed, as seen in Tables XVII–XXII, quite a few analogs prepared primarily as antagonists for the pressor and antidiuretic receptors also have strong antagonist activities in the rat uterus system. For example, the simple substitution of $(CH_2)_5$Mpr into AVP to give [$(CH_2)_5Mpa^1$]AVP produced a good inhibitor at the rat uterus. In addition, [$(CH_2)_5Mpa^1$,DPhe^2]OT is a very potent antagonist analog (Table XVIII), showing that while a D amino acid in position 2 leads to a large decrease in potency in agonist analogs, such amino acids are very compatible with high potency in antagonist analogs. Again, this suggests that agonist and antagonist analogs bind differently to the uterine receptor. Other highly potent disubstituted vasopressin antagonist analogs include [$(CH_3)_2Mpr^1$,Val^4]AVP, [$(CH_3)_2Mpr^1$,$MeAla^7$]AVP, and [$(CH_3)_2Mpr^1$,DArg^8]AVP (Table XVIII). It is particularly interesting that the β,β-dimethyl-substituted analogs generally are more potent antagonists than the β,β-diethyl or β,β-cyclopentamethylene analogs; this is the opposite of what is seen in the monosubstituted series. Nonetheless, highly potent antagonist analogs can be obtained for $(CH_2)_5$Mpr analogs when the proper substituents are utilized at the 2, 4, and 8 positions. Particularly potent analogs include [$(CH_2)_5Mpr^1$,DIle^2,Asn^4]AVP, [$(CH_2)_5Mpr^1$,DIle^2,Thr^4] AVP, [$CH_2)_5Mpr^1$,DIle^2,Val^4]AVP, [$(CH_2)_5Mpr^1$,DPhe^2,Ile^4]AVP, [$(CH_2)_5$ Mpr^1,DPhe^2,Nva^4]AVP, [$(CH_2)_5Mpr^1$,DPhe^2,Ile^4]AVP, [$(CH_2)_5Mpr^1$,DTyr^2, Val^4]AVP, [$(CH_2)_5Mpr^1$,DVal^2,Val^4]AVP (Table XX), and [$(CH_2)_5Mpr^1$,D Tyr^2,Val^4]AVP, all of which have pA_2 values of about 8. Again, the potencies of these analogs do not follow those expected from structure–function studies of agonist analogs. Finally, it should be noted that though they are fairly weak antagonist analogs, [Mba^1]LVP (Table XXII), [Tyr(Me)^2]LVP (Table XXII), and [Leu^4]LVP (Table XXII) all have antagonist activities in the rat uterus assay.

Interestingly, [$(C_2H_5)_2$Mpr[1],Leu[4]]LVP (Table XXIII) is only seven or eight times more potent than [Leu[4]]LVP (Table XXII) as an oxytocin antagonist.

6. Summary Comments and Conclusions

Highly potent antagonist analogs of oxytocin in the rat uterine receptor system have been prepared. Especially potent are those analogs with a combination of β,β-dialkyl substituents at the Cys[1] position (derivatives of penicillamine), lipophilic D amino acids in position 2 that are either aliphatic or aromatic, lipophilic amino acids at position 4, and aliphatic or basic amino acids at position 8. The structure–function results strongly support the conclusion that agonist and antagonist analogs, although competitive inhibitors, utilize different stereostructural and conformational properties for interaction with the uterine receptor (Hruby 1981a,b, 1984; Hruby and Mosberg, 1982; Meraldi *et al.*, 1977). It has also been shown that an additional property of some of these antagonists is their ability to inhibit oxytocin-mediated prostaglandin release at the myometrium (Chan *et al.*, 1981). In preliminary studies, Chan and Hruby have shown (1983) that antagonist analogs with these properties can be quite effective in the prevention of labor and preterm labor in animals. In preliminary studies it has also been reported (Hruby, 1984) that some of these analogs can have prolonged *in vivo* activities. The development of all of these properties for potent, long-acting tocolytic agents seem imminent.

B. Antagonist Analogs of the Neurohypophyseal Hormones in the Avian Vasodepressor Assay

The early structure–function studies that led to the discovery of antagonist analogs of the neurohypophyseal hormones at the uterine receptor also provided antagonist analogs for the avian vasodepressor (AVD) system. Although the structure–function analysis of these antagonist analogs has not been as extensive, potent antagonists of the AVD system have been developed. From these studies it has been possible to arrive at some tentative conclusions regarding inhibition of the transduction message at the AVD receptor. Since the AVD assay is an *in vivo* assay, caution in interpretation of absolute inhibitor potency numbers is needed. Nonetheless, the results clearly show that the structure–function relationships that provide inhibitory activity at the oxytocic and avian vasodepressor system are different.

1. Fragment Analogs of the Neurohypophyseal Hormones

As with the effects on the uterine receptor, substitution of the β-hydrogens of Cys[1] with alkyl groups in the tocin ring leads to antagonists in the AVD assay system as well. Both [$(C_2H_5)_2$Mpr[1]]tocinamide and [$(CH_2)_5$Mpr[1]]tocinoic acid are inhibitor analogs, and the former compound is quite potent with a pA_2 = 7.42 (Table II). In view of this rather high antagonist potency, it might be possible to develop considerably more potent antagonists in the 2 and 4 positions of the 20-membered ring and possibly in other positions as well. No analogs of the pressin ring have been reported that have antagonist activities in the AVD assay system. However, the truncated analogs of OT—[des-Gly-NH_2^9]OT, [des-Leu^8]OT, and [des-Pro^7]OT—all have been reported to be weak antagonists in the AVD assay (reviewed by Hruby and Mosberg, 1982).

2. Extended Analogs of the Neurohypophyseal Hormones

Early studies with extended analogs of oxytocin in which amino acids and small peptides were added to the N^α-amino groups showed that several of these compounds had inhibitory properties toward oxytocin in the AVD assay [for reviews see Rudinger and Krejci (1968) and Berde and Boissonnas (1968)]. Even N^α-acetyloxytocin was a weak inhibitor in the AVD assay system (Berde and Boissonnas, 1968), and similarly [N^α-acetyllysine] vasopressin also had inhibitory effects in this assay (Cash and Smith, 1963). Furthermore, as shown in Table III, [N^α-Car-Cys^1,$Tyr(Me)^2$]OT is an inhibitor in the AVD assay system. Unfortunately, very little subsequent work has been done to follow up on these findings.

3. Monosubstituted-Antagonist Analogs of the Neurohypophyseal Hormones

a. Position 1 Substitutions. The substitution of Cys^1 for Pen^1 and other β,β-disubstituted cysteine derivatives led to potent antagonists in the AVD assay. In this assay system the order of potency increased with increasing bulk of the β,β-substituents, with [$(CH_3)_2$Mpr[1]]OT < [$C_2H_5)_2$Mpr[1]]OT < [$(CH_2)_5$Mpr[1]]OT in antagonist potency. In addition, these analogs are almost an order of magnitude more potent as antagonists than in the uterine assay system, with [$(CH_2)_5$Mpr[1]]OT having a p*A*2 of 8.30 (Table IV), which is one of the highest antagonist potencies reported for a neurohypophyseal hormone analog in this assay system. Again, the [DPen[1]] analog (Table IV) is also a reasonably potent antagonist analog, as is [$(CH_3)_2$Maa[1]]OT, the 19-membered ring analog.

b. Position 2 Substitutions. Substitution of the Tyr^2 residue for more bulky and lipophilic amino acid residues can lead to antagonist analogs in the AVD assay system much as they do for the uterine system. Although many analogs have not been tested (Table IV), [$Tyr(Et)^2$]OT, [Mpr^1,$Tyr(Me)^2$]OT (Hruby and Mosberg, 1982), and [3′,5′-Br_2Tyr^2]OT (Table IV) are inhibitors in the AVD assay.

c. Positions 3, 4, and 5 Substitutions. As with the uterine system, [DAla^3]OT and [DLeu^4]OT appear to be weak antagonists in the avian vasodepressor assay system. Even more interesting, however, are the strong antagonist activities reported for [4(β-(2-thienyl)alanine)]oxytocin (Table IV). No monosubstituted analogs in which the Asn^5 residue has been modified have been reported with antagonist activities in the AVD assay.

d. Position 6 Substitutions. Modification of position 6 in OT can readily lead to antagonist analogs. Very interesting is the observation that whereas [Pen^6]OT (Table IV) is an agonist in the uterine assay, it is an antagonist in the AVD assay. In addition, a number of other substitutions in this position can provide antagonist analogs. For example, the 22-membered ring system [Hcy^1,Hcy^6]OT is an AVD antagonist analog, as are its deamino analog [Mba^1,Hcy^6]OT and the related 21-membered system [Mpr^1,Hcy^6]OT. Interestingly, the 20-membered amide bridged cyclic analog *c*-[β-Ala^1,Asp^6]OT is an antagonist in the AVD system but an agonist in the uterine and milk-ejecting assays (Table V). Another interesting antagonist analog is the cyclic N^ϵ-carbonyllysine residue (Table V). From these studies it is clear that a disruption of the interaction of the Cys^6 residue with the AVD receptor can lead to antagonist activities.

e. Position 7, 8, and 9 Analogs. The weak antagonist activities of the des-Gly-NH_2^{2}, des-Leu^8, and des-Pro^7 analogs of OT in the AVD assay have already been noted (*vide supra*). Other than that, no monosubstituted analogs in which the 7, 8, and 9 residues have been modified have been reported to be antagonists in the AVD assay.

4. *Multiply-Substituted Analogs Including Analogs of the Vasopressins*

Very few studies have been reported that examine the potential antagonist activities of multiply-substituted neurohypophyseal hormone analogs. In cases where such studies have been made, antagonists are found for analogs that are antagonists in the uterine assay system. Generally the analogs are more potent in the AVD assay system. For example, [$(C_2H_5)_2$Mpr^1,Leu^4]OT (Table V) is about three times more potent an inhibitor in the AVD than the uterus assay. Where differences are seen, it is for some compounds that are antagonists in the AVD assay but agonists in the oxytocic assay. For example, [Lys^8,Gly-NHMe^9]OT, an agonist in the uterine assay, is an antagonist in the AVD assay (Table VIII). In like manner, [Phe^3,Phe^4,Met^8]OT is an agonist in the oxytocic assay and an antagonist in the AVD assay. Exceptions to the above comments are some studies done by Dyckes *et al.*, on a series of LVP analogues (Tables XXII and XXIII). In these studies they found a series of rather simple analogs that were quite potent antagonists. Examples include [Maa^1]LVP, a 19-membered ring analog; [Mba^1]LVP, a 21-membered ring analog; and [$(C_2H_5)_2$Mpr^1]LVP, [Leu^4]LVP (Table XXII), and [$(C_2H_5)_2$Mpr^1,Leu^4]LVP (Table XXIII). A follow-up on these studies could undoubtedly lead to more potent antagonist analogs.

5. Summary Comments and Conclusions

The results show that the AVD receptor is much more sensitive to modifications in structure of the neurohypophyseal hormones that lead to antagonist activity. Appropriate modifications of positions 1, 2, 3, 4, and 6 can lead to antagonist analogs, and the different potency characteristics of agonist and antagonist analogs substituted with a basic residue in position 8 suggest that the side-chain in this residue interacts differently with the AVD receptor in agonist and antagonist states. These and other results show that agonist and antagonist analogs use different aspects of primary structure and of conformational properties for interaction with their receptors. Further work with the AVD assay system would certainly provide more evidence for this particularly where parallel agonist and antagonist series were investigated.

C. Antagonist Analogs of the Neurohypophyseal Hormones in the Milk-Ejecting Assay

Thus far, very few studies have been done to examine or develop inhibitory analogs of neurohypophyseal hormone in the mammary-gland milk-ejecting system. This is rather strange considering that the milk-ejecting activity is the classical physiological system for OT. In any case, some early studies did show that the N^α-amino group extended analogs of OT—in particular, [N^α-Gly]OT, [N^α-Leu]OT, and [N^α-Leu-Gly-Gly]OT—were weak antagonists in the milk-ejecting assay (Berde and Boissonnas, 1968; Rudinger and Krejci, 1968). A cursory examination of Tables I–XXIII shows that only a few studies have been made since that time for possible inhibitors in the milk-ejecting assay. Some notable exceptions can be found in Tables III, V, X, XI, and XVIII. Some of the more potent antagonist analogs include [$(CH_3)_2$Mpr^1,Tyr(Me)^2]OT (Table V), [$(CH_3)_2$Mpr^1,Tyr(Me)^2,Thr^4]OT (Table X), [$(CH_3)_2$Mpr^1,Phe^2,Asn^4,Orn^8]OT (Table XI), and E(CH_2)5Mpr^1,Tyr(E+)^2,Orn^8]OT (Table X), of which the latter appears to be the most potent antagonist analog at the mammary-gland receptor reported thus far. However, it still is an order of magnitude less potent as an antagonist in the milk-ejecting assay ($pA_2 = 7.36$) (Table X) than in the rat uterus assay (Table III) ($pA_2 = 8.25$). A good deal more work is needed to obtain insight into the structural features that lead to potent antagonist activities for neurohypophyseal hormone analogs at the mammary-gland receptor.

V. STRUCTURE–ACTIVITY RELATIONSHIPS OF VASOPRESSIN AGONISTS

A. At the Pressor Receptor

The pressor activity of neurohypophyseal peptides has been a well-established biological activity of these compounds from the earliest times, although there is still discussion regarding the circumstances in which this activity is of physiological significance. Nonetheless, the rat pressor assay has been a standard assay especially for the "vasopressin-like" activities of the neurohypophyseal hormones. Examination of the pressor potencies of the naturally occurring neurohypophyseal hormones (Table I) suggests that the pressor receptor is the most sensitive to the structural changes that occur in the native peptides. There is essentially a 10,000-fold difference in the potencies of arginine vasopressin and isotocin. A basic residue in position 8 is critical for high pressor potency, since all neurohypophyseal hormones without a basic residue in position 8 have drastically reduced potencies in the pressor assay. However, equally important for potency is position 4 (especially $Gln^4 \rightarrow Ser^4$), since isotocin is about 100-fold less potent than mesotocin and the only difference in structure between the two compounds is that the former has Ser^4 and the latter Gln^4. This is quite different from what is seen in other neurohypophyseal hormone assay system, where the Ser^4 analog is more active or nearly equipotent to the Gln^4 analog (compare MT and IT in Table I).

1. Fragment Analogs of the Neurohypophyseal Hormones

The 20-membered ring analogs of the neurohypophyseal hormones—e.g., tocinamide, pressinoic acid, etc.—all appear to be inactive (Tables II and XII). Interestingly, [des-Gly^9]AVP and the corresponding $\text{D}Arg^8$ carba analog (Table XII) are both analogs with antagonist activities. It is not clear why the pressin ring is not an antagonist. As expected, [$(C_2H_5)_2Mpr^1$]tocinamide and [$(CH_2)_5Mpr^1$]tocinoic acid are both weak inhibitors in the pressor assay system.

2. Structurally Extended Analogs of the Neurohypophyseal Hormones

Although early work with N-terminal extended analogs of the neurohypophyseal hormones generally showed agonist activities in the pressor assay system (Berde and Boissonnas, 1968; Rudinger and Krejci, 1968), there were a few analogs that were shown to be weak antagonists in this assay. However, subsequent studies (Table III and XIII) have shown that certain N-terminal extended analogs, especially those with a $Tyr(Me)^2$ substitution and N^{α}-Ac, N^{α}-

BrAc, N^{α}-Gly, and other simple acylated analogs, could be weak inhibitors. Clearly, work is needed to fully explore the use of N^{α}-extended analogs for agonist activities at pressor receptor systems, since many of these compounds are strong agonists.

3. Monosubstituted Analogs of the Neurohypophyseal Hormones

It is commonly believed that the 8 position of the neurohypophyseal hormones is the most important in determining pressor activity, and, indeed, an extraordinary number of vasopressin analogs with various substitutions in the 8 position of the hormones has been made (Table XIV). However, many analogs of the neurohypophyseal hormones have been made with substitutions at other positions, and we will make a more systematic examination of structure–function relationships.

a. Position 1. As we have seen earlier, replacement of the α-amino group of Cys^1 with a hydrogen to give the Mpr^1 analogs generally leads to analogs with greater potency. This is not the case for pressor activities. As can be seen in Tables XVII and XXII, both deamino-LVP and deamino-AVP are less potent than the parent compound in the pressor assay system, and the same is found for [Mpr^1]OT (Table IV). On the other hand, replacement of the N^{α}-amino group with an N^{α}-hydroxyl group does lead to more potent analogs of AVP (Table XVII), LVP (Table XXII), and oxytocin (Table IV). In fact, in the OT analog, the Hmp derivative is almost five times more potent than OT. In the 1-carba (and 6-carba) series of analogs (Tables XVII, XVIII, and XXIII), the very substantial increases seen in potency at all other neurohypophyseal hormone assay systems is seldom seen in the pressor assay system. Apparently no pressor agonist analog of vasopressin has been prepared and tested that is substantially (more than five times) more potent than the native hormones. On the other hand, very potent antagonist analogs for the pressor–receptor system have been prepared. It is interesting that whereas the acyclic analogs of oxytocin are generally inactive or weak antagonists at the oxytocic receptor system, similar vasopressin analogs such as [Ala^1,Ala^6]AVP (Table XVIII) and [Ala^1,Ala^6]LVP (Table XXIII) have some weak (about 1/20,000 times the potency of the native hormones) activity in the pressor assay system.

b. Position 2. There do not appear to be severe stereochemical requirements for the 2 position of the neurohypophyseal hormones, since [DTyr^2]AVP has about one-half the potency of AVP in the pressor assay (Table XVII). In addition, the phenol hydroxyl group does not appear to be too critical either, since both [Phe^2]AVP (Table XVII) and [Phe^2]LVP (Table XXII) are quite potent.

Although [Tyr(Me)^2]LVP (Table XXII) appears to be quite potent, [Tyr(Me)^2]AVP (Table XVII) and especially O-alkylated analogs with two or more carbon atoms (Tables XVII and XXII) have low potency, and some are antagonists.

c. Position 3. Replacement of the Phe^3 of AVP (or LVP) with an Ile^3 residue to give AVT is consistent with high biological potency in the pressor assay system, and the 2-thienyl-L-alanine analog of LVP (Table XXII) is also highly potent. However, little else has been done to explore structure–activity relationships of this position in the pressor assay.

d. Position 4. Substitution of Gln^4 by D-Gln leads to a drastic loss of potency (about 100 times) for both AVP (Table XVII) and LVP (Table XXII). Substitution of the Gln^4 for Thr^4, which was effective in increasing potency at the "oxytocin-like" assay systems, leads to a modest loss (about one-fourth to one-fifth) in potency in the pressor assay. Also, the α-aminobutyric^4 (Abu) and Val^4 substitutions (which are very effective for increasing potency in the antidiuretic assay) do not effect increases in potency in the pressor assay, but instead lead to large decreases in potency (Tables XVII and XXII).

e. Position 5. As for all other neurohypophyseal hormone receptors, the Asn^5 appears to be a critical residue for both binding and presumably transduction in the rat pressor assay system. It appears that Asp^5 and Asp[$N(CH_3)_2$] can be substituted for Asn^5 with maintenance of activity (Tables XVII and XXII), although with a substantial decrease in potency (45-fold to 70-fold). Other structural modifications appear to give analogs with less than one-thousandth the potency of the native hormone at the receptor, but studies thus far are limited.

f. Position 6. As already mentioned, the ring-opened analogs [Ala^1,Ala^6]LVP and -AVP are weak agonists. The deamino-[6-carba]AVP analog is quite potent, but the low potencies of the deamino-dicarba analogs (Tables XVIII and XXIII) of AVP and LVP suggest that there are quite stringent steric and perhaps stereoelectronic requirements for vasopressin binding at the pressor receptor vis-à-vis the cystine bridge.

g. Position 7. The stereoelectronic properties at the 7 position seem to be quite stringent, since [DPro^7]LVP has only about 1/5000 times the potency of LVP in the pressor assay. In addition, there is a severe reduction in potency when Pro^7 is substituted by a Gly^7, Sar^7, or Hyp^7 (Tables XVII and XXII). On the basis of these limited studies, it would appear that the conformational structure properties of proline are quite important for the strong hormone–receptor interaction at the pressor receptor. In this regard, it is interesting that the $\Delta^{3,4}$Pro^7

analog of AVP does not lead to the large potency increase, as it does for some other receptors of the neurohypophyseal hormones; nonetheless, it is still quite potent at the pressor receptor.

h. Position 8. As might be expected, an extraordinarily large number of analogs of vasopressin have been prepared with various substitutions in position 8 (Tables XIV and XVII). The pressor assay system is sensitive to structural change. The stereochemical requirement for the 8 position of the neurohypophyseal hormones at the pressor receptor are severe. [DArg^8]VP, [DLys^8]VP, and [D Dbu^7]VP, for example (Table XIV), are all considerably less potent than the all L analogs. Basic residues that will be accepted at the 8 position and still retain a high potency in the pressor assay include the α,ω-diaminobutyric acid (Dbu^8), homolysine^8, Orn^8, and Har^8 (Table XIV). All have pressor activities within a factor of two to four times that of AVP or LVP. Most interesting is the high potency of [Mpr^1,NArg^8]VP (Table XIV) (1500 U/mg), which appears to be the most potent analog reported in the pressor assay. Since the side-chain in position 8 is important to potency in the pressor assay, it is not surprising that [Gly^8]VP (Table XIV) is inactive.

i. Position 9. Quite a few derivatives of the vasopressins substituted in the 9 position have been prepared, and most of them have very low potency in the pressor assay (Tables XVII and XXII). It is clear from these studies that the Gly-$NH_2{}^9$ residue is a critical residue for both binding and transduction. Whether substituted with Sar, DAla, LAla, Gly($NHCH_3$), or by elimination, analogs with low or no potency are obtained. Most interesting are the deamido (Table XVII) and the β-Ala analogs (Table XXII), both of which appear to be weak inhibitors in the pressor assay. It would appear that the Gly-$NH_2{}^9$ residue either directly or indirectly is critical for transduction at the pressor receptor. Further studies could provide some new insights into binding and transduction at the pressor receptor.

4. Multiply Substituted Analogs of the Neurohypophyseal Hormones

The most extensive work on multiply substituted analogs of the vasopressins generally has been made with antagonist analogs. However, there are a few agonist analogs that are multiply substituted aside from the 1- and 6-substituted analogs that have been discussed. For example, the high potency of [Phe^2,Orn^8]VP (Table XV) and [Ile^3,Har^8]VP (Table XIX) are noteworthy, and it is especially interesting that in both cases the deamino analogs are considerably less potent. On the other hand, it is not clear why the analog [Phe^2,$\Delta^{3,4}$Pro^7]AVP has lost considerable potency and the corresponding deamino analog appears to be inactive (Table XIX). Of considerable interest are the mixed agonist–antagonist activities of the bicylic LVP analog Mpr-Phe-Phe-Gln-Asp-Cys-Pro-Lys-Gly-NH_2 (Table XXIII). Apparently, modification of the conformational relationships of the Asn^5 and the Lys^8 side-chain groups has led to an analog that has difficulty transducing at the pressor receptor. Further examination of these conformational relationships should provide insight into the conformational requirements for transduction at this receptor.

Most oxytocin analogs that have been examined for their pressor activity are either inactive, weakly active, or weak antagonists (Tables IV–XII). A few notable exceptions are $[Phe^2,Orn^8]$OT (Table VI), $[Thr^4,Arg^8]$OT (Table VII), and perhaps $[Phe^3,Ala^8]$OT (Table VII). However, these analogs have either a basic residue at the 8 position or a Phe residue at position 3. Hence the low potency or even inactive properties of oxytocin analogs are expected based on the activities of the native neurohypophyseal hormones. We will discuss the properties of the inhibitors later.

5. Summary Comments and Conclusions

The pressor assay is the most sensitive assay to structural modification for any of the classical neurohypophyseal-hormone assays. It appears that the topological relationships of the side-chain moieties of the 5 and 8 positions, the stereostructural integrity of the 7 position, and the structural properties of the $Gly\text{-}NH_2{}^9$ and of the 20-membered disulfide ring are all critical for obtaining a potent agonist analog in the pressor assay system.

B. At the Kidney Antidiuretic Receptor System

The antidiuretic activity of the vasopressins is universally accepted as a critical physiological role for these hormones (vasopressin is often referred to as antidiuretic hormone, ADH). In view of this well-accepted physiological role of vasopressin, a good deal of studies have been made on the properties of the vasopressin receptor, including membrane preparations [see for example, C. Roy *et al.* (1975) and Hechter *et al.* (1978)] and the antidiuretic assay that will be discussed here. Based on these and on biophysical studies, Walter and coworkers (1977) made a number of suggestions regarding vasopressin structure–function relationships at the kidney antidiuretic receptor. They suggested that the key "binding elements" for agonist activity at the vasopressin receptor in the kidney were the side-chain groups of the Phe^3, Gln^4, Pro^7, and Arg^8 (or Lys^8) residues, while the key "active elements" for transduction were the Asn^5 carboxamide group and the basic moiety on the side-chain of Arg^8 (or Lys^8). The picture for antagonists was complicated by the fact that full *in vivo* antidiuretic responses are achieved at very low receptor occupancies, thus masking partial agonists in this assay. However, receptor antagonists for the kidney antidiuretic receptor have been obtained that must be interpreted in terms of this model, or we must change some aspects of it.

Examination of the antidiuretic activities of the naturally occurring neurohypophyseal hormones (Table I) shows that for high potency, a basic residue is needed in position 8. Furthermore, even though oxytocin exhibits full *in vivo* agonist activity with low potency, it is not capable of fully stimulating vasopressin-sensitive renal adenylate cyclase to the levels observed for vasopressin (C. Roy *et al.*, 1975), thus indicating a possible role for the basic moiety on the side-chain of position 8 in transduction.

1. Fragment Analogs of the Neurohypophyseal Hormones

As can be seen in Tables II and XII, most of the fragment analogs of the neurohypophyseal hormones, especially the 20-membered ring fragment analogs, have not been examined for their antidiuretic activity. Where they have been, they have very low potency. Interestingly, the des-Gly^9 analog of AVP (Table XII) is quite potent in the antidiuretic assay, suggesting that as compared with all other neurohypophyseal hormone assay systems examined in this chapter, this residue is not important for binding or transduction. However, when both the Gly^9 and Arg^8 residues are removed from the C-terminal, a very weak agonist is obtained (Table XII).

2. Structurally Extended Analogs of the Neurohypophyseal Hormones

The earliest studies with N^{α}-amino extended analogs of the neurohypophyseal hormones (Berde and Boissonnas, 1968; Walter *et al.*, 1967) indicated that these kinds of analogs, when tested in the antidiuretic assay, were all weak agonists in this assay, even though in some of the neurohypophyseal hormone–receptor systems these extended analogs were antagonists or partial antagonists (*vide supra*). Unfortunately, many of the more recently prepared extended analogs of the neurohypophyseal hormones (Tables III and XIII) have not been tested for their antidiuretic biological activities.

3. Monosubstituted Analogs of the Neurohypophyseal Hormones

a. Position 1. As for oxytocin analogs, replacement of the α-amino group of Cys^1 with a hydrogen gave [Mpr^1]AVP with very potent antidiuretic activity. It is possible that some of this increased potency is a result of the greater stability of the Mpr analogs against enzymatic breakdown *in vivo,* since the antidiuretic assay is an *in vivo* assay. On the other hand, binding experiments with kidney membrane (C. Roy *et al.*, 1975; Hechter *et al.*, 1978) support the idea that the increased potency of these compounds is due to increased binding activity at the kidney receptors. The α-hydroxyl-β-mercaptopropronic acid (Hmp) analogs of AVP (Table XVII) and LVP (Table XXII) have high potency in the antidiuretic assay. Removal of the disulfide bridge of AVP (Table XVIII) and LVP (Table XXIII) leads to analogs with highly reduced potency (about 1/5000 times that of the native hormone). The replacement of one of the sulfur atoms of the disulfide bridge with a methylene group in deamino-1-carba-AVP (Table XVIII) gives an analog about 30 times more potent than AVP in the antidiuretic assay, making this analog the most potent ADH compound known. Interestingly, the deamino-dicarba analogs of AVP (Table XVIII) and LVP (Table XXIII) are considerably less potent and, in the case of LVP, less potent than the native hormone. Finally, it should be noted that whereas substitution of Cys^1 by $(CH_3)_2$Mpr and especially $(C_2H_5)_2$Mpr and $(CH_2)_5$Mpr generally leads to antagonist analogs for the other neurohypophyseal hormone–receptor systems, all of these compounds are agonists in the AVP series (Table XVII), although [$(CH_2)_5Mpr^1$]AVP has only about 1/15,000 times the potency of AVP. These results clearly show the importance of position 1 of vasopressin for binding to the antidiuretic hormone receptor in the kidney.

b. Position 2. While position 2 is critical for transduction in the oxytocin receptor system (*vide supra*), there is much more latitude for substitution in this position, particularly in singly substituted analogs, in the antiduretic assay. For example, the phenol hydroxyl group of Tyr^2 can be replaced by a hydrogen and the analog still retains high potency (Tables XVII and XXII), especially in the arginine vasopressin series. In addition, the phenol hydroxyl group can be alkylated with a methyl group and the vasopressin analog can be quite potent, especially in the AVP series. [$Tyr(Me)^2$]AVP has been reported to be almost as potent as AVP (Table XVII). In this regard, however, it is interesting to note that N^α-acetylation of this analog to give [N^α-Ac-Cys^1,$Tyr(Me)^2$]AVP led to a drastic reduction in potency, with the analog possessing only about 1/20,000 times the potency of OT. These results suggest that the conformational relationships of the 1 and 2 positions of AVP are critical to binding at the antidiuretic kidney receptor. It is interesting that many deamino analogs of LVP with large O-alkyl substituents on Tyr^2 apparently can be mixed agonists–antagonists. Finally, the stereochemical requirements of Tyr^2 do not appear to be very important in the antiduretic assay, since [DTyr^2]AVP (Table XVII) has over 40% the potency of AVP and is a full agonist in the renal adenylate cyclase assay as well (Hruby *et al.*, 1979b).

c. Position 3. A change from aromatic to aliphatic amino acids in position 3 of AVP and LVP generally leads to a large decrease in potency in the antidiuretic assay (Berde and Boissonnas, 1968; Walter *et al.*, 1967) (Tables XVII and XXII). One exception is the 1,4-cyclohexadienyl-L-alanine analog of LVP, which retains fairly high antidiuretic assay (Table XXII). Unfortunately, although many substituted analogs of AVP have been prepared (Table XVII), their antidiuretic activities still have not been reported.

d. Position 4. There is a fairly stringent stereochemical requirement for the 4 position of vasopressin for high biological potency, since [DGln^4]AVP (Table XVII) has only about one-thousandth the potency of AVP. On the other hand, the Gln^4 residue can be replaced by a large number of aliphatic side-chain groups with maintenance of high biological potency (Tables XVII and XXII) (Berde and Boissonnas, 1968; Walter *et al.*, 1967). For example, [Abu^4]AVP, [Abu^4]LVP, and [Val^4]AVP all are more potent than the native hormone in the antidiuretic assay (Tables XVII and XXII). On the other hand, [Asn^4]LVP and [Leu^4]LVP have greatly reduced potency.

e. Position 5. When the Asn^5 residue in AVP is replaced by an Asp residue, a quite potent analog in the antidiuretic assay is obtained (Table XVII). However, the dimethylamide analog of LVP (Table XXII) has greatly reduced potency, and an even greater decrease in potency occurs when the Asn^5 residue is replaced by Ala^5 or even more by Ser^5 (Table XXII). Further exploration of the importance and stereostructive requirements of the residue in position 5 with relation to the antidiuretic receptor is needed.

f. Position 6. When the S in the 6 position is replaced by a methylene group as in deamino-[6-carba]AVP (Table XVIII), an analog is obtained that is slightly more potent than [Mpr^1]AVP in the antiduiretic assay. Most of the other substitutions at the Cys^6 (Cys^1) position have already been discussed above. However, it is worth noting again that the 1,6-dicarba analog is less potent than either of the monocarba analogs, suggesting a rather stringent requirement for the correct steric properties in the vicinity of the disulfide bond for maximal potency in the antidiuretic assay.

g. Position 7. Whereas there are very stringent stereostructural requirements for the 7 position of the neurohypophyseal hormones for pressor activity, these requirements are considerably looser for antidiuretic activity. For example, [$\Delta^{3,4}$Pro^7]AVP and the corresponding deamino analog (Table XXVII) are much more potent in the antidiuretic assay than in the pressor assay. In addition, [MeAla^7]AVP, [Sar^7]AVP (Table XVII), and even the azetidine-2-carboxylic acid^7 LVP analog (Table XXII) are quite potent analogs. Furthermore, there is considerable activity even for [DPro^7]LVP (about one-tenth of LVP), suggesting that even the stereochemical requirements for activity in the antidiuretic assay system are not stringent.

h. Position 8. As already pointed out, the 8 position appears to be a critical for both binding and transduction of vasopressin in the antidiuretic assay system. For example, [Ala^8]VP is not able to stimulate renal medullary adenylate cyclase to the same level achieved for vasopressin (Dousa *et al.*, 1971). Nonetheless, both L and D basic amino acids can be accommodated into the 8 position without a very severe loss in potency. In fact, in many but not all cases, the deamino analog with a basic D amino acid in position 8 is more potent than AVP or LVP and as much as 30 times more potent than the corresponding L amino acid residue position-8 analog. Particularly impressive are the very high potencies of [Mpr^1,DArg^8]VP, [Mpr^1,D Dbu^8]VP, [Mpr^1,Dpr^8]VP, and [Mpr^1,NArg^8]VP. These and other results (Tables IV and XVII) suggest that whereas the Arg^8 residue appears to be the best for maximum potency in the antidiuretic assay, sterically smaller amino residues in position 8 are compatible with very high potency, especially in combination with the Mpr^1 substitution. Also noteworthy is that in many cases the D amino acid residue 8-position analog is actually more potent than the all-L analog. Of particular note is [Mpr^1,DArg^8]AVP (DDAVP), which has high antidiuretic potency and very low pressor activity. This compound is now widely used as an antidiuretic in human medicine, including treatment of diabetes insipidus.

i. Position 9. As with other residues found in the acyclic portion of the neurohypophyseal hormones, the C-terminal 9 position appears to be the most flexible in terms of structural requirements for activity in the antidiuretic assay compared to the other neurohypophyseal hormone assay systems examined here. For example, the Ala^9 and DAla^9 analogs of AVP (Tables XVII and XXII) are quite potent. Even the deamido (Table XVII) and des-Gly (Table XXII) analogs have considerable potency. More extensive investigations of analogs structurally modified in this position might prove quite interesting.

4. Multiply Substituted Analogs of the Neurohypophyseal Hormones, Including Oxytocin Analogs

A fairly large number of multiply substituted agonist analogs of the neurohypophyseal hormones have been examined for their antidiuretic activity (Tables XV, XVI, XIX–XXI), including oxytocin analogs (Table IV) (Berde and Boissonnas, 1968; Walter *et al.*, 1967). Particularly impressive are the high potencies of deamino-6-carba-[DArg^8]VP and of the corresponding Orn^8 analogs

(Table XVI), [Mpr[1],$\Delta^{3,4}$Pro[7]]AVP, [Mpr[1],Phe[2],$\Delta^{3,4}$Pro[7]]AVP (Table XIX), and [Asn[4],DArg[8]]VP (Table XIX), all of which are several times more potent than AVP. Equally impressive, though much less potent, are the high activities of deamino-6-carba-OT (Table V), and [DArg[8],Ala[9]]VP (Table XV). In general, however, multiple substitutions that lead to agonist activity pretty much follow what would be expected from the activities of monosubstituted analogs.

5. Summary Comments and Conclusions

Although highly potent and receptor-specific analogs of vasopressin that interact primarily with the antidiuretic receptor are known, in fact, very little is known about the conformational requirements for high potency. Apparently the 1, 3, 5, and 8 positions are critical for binding and transduction, but very little has been done in the way of conformational restructions consistent with agonist activity in the antidiuretic assay system. On the other hand, the very high specificity of DDAVP and a few other similar analogs suggests that we are in a position to derive specific insight into the complementary nature of the antidiuretic receptor structure that will bind to the hormone and transduce the message. It would appear that analogs properly modified at the 4, 7, 9, and possibly the 2 positions could lead to bicyclic analogs with high potency and receptor specificity.

VI. ANTAGONIST ANALOGS OF THE NEUROHYPOPHYSEAL HORMONES THAT COMPETITIVELY BLOCK THE ACTION(S) OF THE NATIVE HORMONES AT RECEPTORS FOR VASOPRESSIN

A. At the Pressor Receptor

Development of antagonists at the pressor receptor paralleled the development of neurohypophyseal hormone antagonists at the uterine receptor, since many of the analogs first found to be antagonists in the uterine assay were also found to be antagonists in the pressor assay. However, there was not a direct correspondence, and just as different structure–activity relationships developed for agonist analogs (*vide supra*), antagonist analogs also were found to possess a unique structure–function profile. The early work has been reviewed (Berde and Boissonnas, 1968; Rudinger and Krejci, 1968), and Hruby and Mosberg (1982) have written a more extensive review of structure–function relationships of pressor antagonist analogs. In this review, we concentrate primarily on providing insight into those aspects of structure important for preventing transduction but still allowing strong interaction with the pressor receptor as measured by high antagonist potency.

1. Fragment Analogs of the Neurohypophyseal Hormones

While removal of the Gly^9 residue from AVP gives an analog with high antidiuretic activity, in the pressor system a weak antagonist is obtained, and this is also the case for the des-Gly-$NH_2{}^9$ deamino-1-carba-DArg^8 analog (Table XII). As expected, $[(C_2H_5)_2Mpr^1]$tocinamide and $[(C_2H_5)_2Mpr^1]$ tocinoic acid (Table II) are weak pressor antagonists.

2. N^α-Extended Analogs of the Neurohypophyseal Hormones

N^α-Acyl-substituted analogs of oxytocin and vasopressin, especially when combined with Tyr^2 O-alkylation, lead to a variety of analogs with inhibitory activities in the pressor assay (Tables III and XIII) (Hruby and Mosberg, 1982). The former modification of neurohypophyseal hormone analogs has not been combined with many other modifications that provide pressor antagonist analogs, and it might be interesting if some of the more potent antagonist analogs (*vide infra*) were N^α-acylated.

3. Monosubstituted Inhibitor Analogs of the Neurohypophyseal Hormones

a. Position 1. Simple acylation of the N^α-amino group of Cys^1 can lead to analogs of vasopressin with weak antagonist activity in the pressor assay (Berde and Boissonnas, 1968; Rudinger and Krejic, 1968; Hruby and Mosberg, 1982). However, substitution of the β-hydrogens of Cys^1 in OT (Table IV) or vasopressin (Tables XVII and XXII) leads invariably to antagonists in the pressor assay. Generally, AVP analogs are more potent antagonists than LVP analogs, which in turn are more potent than oxytocin analogs. Simple structural modifications can provide exceptionally potent pressor antagonists, with $[(C_2H_5)_2Mpr^1]$AVP and $[(CH_2)_5Mpr^1]$AVP being among the more potent pressor antagonists known. It is interesting that whereas $[Hcy^1,Hcy^6]$OT (Table V) does not appear to have antagonist activity in the pressor assay, $[Mpa^1,Hcy^6]$OT (Table V) is a weak pressor antagonist. This illustrates the subtle structural relationships between steric and electronic features of the disulfide bridge, the N^α-amino group of Cys^1, and the Tyr^2 side-chain that determine whether an analog is an agonist or antagonist.

b. Position 2. O-Alkylation of Tyr^2, especially if used in conjunction with acylation of the N^α-amino group, can lead to neurohypophyseal hormones with pressor antagonist activities (Tables XVII and XXII) (Rudinger and Krejci,

1968: Hruby and Mosberg, 1982). However, these analogs often are not very potent and, as will be discussed later, it is the use of O-alkylation of Tyr^2 derivatives in combination with other modification that can give the most potent antagonists.

c. Positions 3, 4, 5, and 6. Single substitutions have not been very successful when made in the 3, 4, 5, or 6 positions of the neurohypophyseal hormones in producing antagonist analogs. It has been reported that [DAla^2]OT (Table IV), [Leu^4]OT (Table IV), [Phe^4]OT (Table IV), and [Igln^5]OT all have some weak antagonist activities, but more potent antagonist analogs have not been developed from those observations.

d. Positions 7, 8, and 9. Again, very few analogs with modifications at single residues in the neurohypophyseal hormones' C-terminal tripeptide moiety have been developed or found. A few exceptions are the 9-deamido analog of AVP (arginine vasopressinoic acid) (Table XVII) and [β-Ala^9]LVP (Table XXII), which have been reported to be weak antagonists on the pressor assay. Interestingly, it also has been reported that [Gly^7,Orn^8]VP (Hruby and Mosberg, 1982) is a partial agonist in the pressor assay system.

4. Multiply Substituted Inhibitor Analogs of the Neurohypophyseal Hormones

From the above discussion it is perhaps not surprising that the most potent antagonists have resulted from substitution of the β-hydrogens of Cys^1 in combination with other residues in the neurohypophyseal hormones. Virtually all analogs with the Pen^1, $(CH_3)_2Mpr^1$, $(C_2H_5)_2Mpr^1$, or $(CH_2)_5Mpr^1$ substitutions are inhibitors (Tables IV–XI, XVIII–XXI, and XXIII). Some of the more potent inhibitors include [$(CH_2)_5Mpr^1$,Tyr(Me)]AVP,[$(CH_2)_5Mpr^1$,DPhe^2]AVP, [$(C_2H_5)_2Mpr^1$,Val^4]AVP, [$(CH_2)_5Mpr^1$,DArg^8]VP, [$(CH_2)_5Mpr^1$,des-Gly^9] AVP, [$(CH_2)_5Mpr^1$,DTyr^2,Val^4]AVP, [$(CH_2)_5$]Mpr^1,DTyr(Me)^2,Val^4]AVP, and [$(CH_2)_5Mpr^1$,Tyr(*i*Pr)^2,Val^4,DArg^8]VP. Although not as potent, the oxytocin analogs [$(CH_3)_2Mpr^1$,Phe^2,Asn^4,Orn^8]OT, [$(CH_3)_2Mpr^1$,Tyr(Me)^2]OT, and [$(C_2H_5)_2Mpr^1$,Tyr(Et)^2,Orn^8]OT are good antagonists. The potencies of these latter analogs are very much in keeping with the 1, 2, 4, and 8(9) positions as being the primary determinants of potency in antagonist analogs. How the 5 position might effect potency in the antagonist series has not been investigated. Finally, it is interesting that [Mpr^1,Ile^4,DArg^8]AVP and [Mpr^1,Leu^4,DArg^8]AVP (Table XIX) are potent antagonists. Very little has been done to explore the structure–function relationships of neurohypophyseal pressor antagonist analogs with respect to positions 5, 6, and 7. In view of the strong evidence that antagonist analogs bind differently to the pressor receptors than do agonists (Hruby, 1981a,b,

1984; Hruby and Mosberg, 1982), such studies could provide new insights into the nature of the hormone antagonist receptor binding interactions, especially if the analogs are further conformationally constrained.

5. Summary Comments and Conclusions

Simple but potent antagonist analogs of vasopressin have been prepared that have high receptor specificity in several cases. A careful investigation of the conformational properties of these inhibitors, coupled with further conformational restrictions, should lead to important insights into those features of structure and conformation that are critical for binding and those important for transduction. There are still some blind spots in the structure–function studies that need to be addressed, but there are clear differences between structure–function relationships of antagonists (and agonists) at the pressor assay (compared to the antidiuretic assay system) that should permit the development of a conformational model for neurohypophyseal hormones specific for the pressor receptor system that can be tested in specific ways. New interests in vasopressin as a pressor agent should speed up this process.

B. Competitive Inhibitors of the Antidiuretic Receptor

Although there are several differences in structure–function relationships at the various receptors for the neurohypophyseal hormones, development of antagonist analogs in the uterus, avian vasodepressor, and pressor assay systems developed more or less simultaneously, especially once the 1-penicillamine and related potent antagonists had been introduced. However, it has been only recently that potent *in vivo* antagonists of AVP at the kidney antidiuretic system have been found. It was reported in 1968 that [Leu^4]OT had weak anti-antidiuretic activity (Chan *et al.*, 1968; Hruby *et al.*, 1969), and this compound also was found to be an inhibitor in the toad urinary bladder assay (Walter *et al.*, 1972). However, efforts to follow up on this lead were unsuccessful because the very weak antagonist potency of the analog made it difficult to obtain reproducible results, and efforts to prepare related analogs with antagonist activities were unsuccessful (Hruby and Mosberg, 1982). Furthermore, even the development of very potent pressor antagonist analogs such as [$(C_2H_5)_2Mpr^1$]AVP (Table XVII) and [$(CH_2)_5Mpr^1$, $Tyr(Me)^2$]AVP did not produce antagonist activities in the antidiuretic assay (Table XVII). These and other results indicated that there was not a straightforward relationship between those structural features that lead to potent antagonism at the pressor receptor and those that will have the same effect at the kidney antidiuretic assay.

In retrospect though, other leads such as the mixed agonist–antagonist analog [Mpr^1,Tyr(Pr)^2]LVP may have provided some insight. It was only when a combination of changes was made at several different amino acids in vasopressin (particularly at positions 1, 2, 4, and 8) that antagonist analogs with reproducible biological antagonist effects (Sawyer *et al.*, 1981) were obtained. Subsequently, other trisubstituted, tetrasubstituted, and even disubstituted analogs were obtained that had good to excellent antagonist activities in the antidiuretic assay. We shall now discuss these results, most of which are found in Tables XVIII, XX, XXI, and XXIII.

1. Multisubstituted Inhibitor Analogs of the Neurohypophyseal Hormones

First, it is interesting to note that disubstituted analogs that have antagonist activities in the antidiuretic assay generally have a D amino acid residue at the 2 position, such as [$(CH_2)_5Mpr^1$,$DPhe^2$]AVP. In this regard it is interesting to note that Larsson *et al.* (1978) reported some time ago that analogs of the type [Mpr^1,Tyr(R)^2]LVP (Table XXII) could have partial antagonist activities. A interesting recent disubstituted analog with inhibitor activities in the antidiuretic assay is [5,8-*cyclo*(Mpr^1,Phe^2,Asp^5,Lys^8)]VP, a bicyclic analog that has weak antagonist activities in the antidiuretic assay (Table XXIII). These studies suggest that it is possible to develop antagonist analogs by appropriate modifications at more than one topologically related part of the peptide hormone. It would appear that both the N-terminal residues and *separately* the C-terminal residues can be modified to give stereostructural changes that lead to antagonist activity. Thus far, to obtain high antagonist potency it has been necessary to modify these further. For example, modification of the C-terminal residues with substitutions at the 1, 2, and 4 positions can provide potent inhibitors in the antidiuretic assay (Table XX). The most powerful antagonists are substituted with a D-aliphatic or a D-aromatic residue in position 2. These include [$(CH_2)_5Mpr^1$,$DIle^2$,Abu^4]AVP, [$(CH_2)_5$-Mpr^1,$DIle^2$,Ile^4]AVP, [$(CH_2)_5Mpr^1$,D-Ile^2,Thr^4]AVP, [$(CH_2)_5Mpr^1$,$DIle^2$,Val^4]AVP, [$(CH_2)_5Mpr^1$,$DPhe^2$,Ile^4]AVP, and [$(CH_2)_5Mpr^1$,$DPhe^2$,Val^4]AVP. Interestingly, when $DIle^2$ is used, there are large differences in the inhibitory potencies in the antidiuretic and pressor assay systems, while when $DPhe^2$ or D Tyr(R)^2 is used the inhibitory potencies in the two assays parallel each other more closely. It is interesting how sensitive antagonist potency can be to small structural changes in these assays. For example, [$(CH_2)_5Mpr^1$,D-Phe^2,Leu^4]AVT is two orders of magnitude less potent than the corresponding Ile^4 analog (Table XX), while the corresponding $DVal^4$ analog is again very potent.

In general, analogs of vasopressin with antagonist activities in the antidiuretic assay system but substituted with a $DArg^8$ instead of an $LArg^8$ residue are less potent antagonists in the antidiuretic assay. For example [$(CH_2)_5Mpr^1$,D-

Phe^2,Val^4,DArg^8]VP has $pA_2 = 7.07$, whereas [$(CH_2)_5$Mpr^1,D-Phe^2,Val^4]AVP has $pA_2 = 8.07$ and [$(CH_2)_5$Mpr^1,DTyr(Me)^2,Val^4]AVP has a $pA_2 = 7.77$, whereas [$(CH_2)_5$Mpr^1,D-Tyr(Me)^2,Val^4,D-Arg^8]VP has $pA_2 = 6.68$. Most interesting is the recent report that des-Gly^9 analogs are essentially equipotent as antagonists as the full analogs (Table XXXI).

2. Discussion and Summary

Examination of the data, especially in Tables XX and XXI, and of the earlier efforts to obtain antagonists for the antidiuretic receptor system suggest that there is a fine structural dividing line between weak agonist analogs and quite potent antagonist analogs. For example, whereas [$(CH_2)_5$Mpr^1,Val^4,D-Arg^8]VP is a weak agonist (Table XX), simple addition of a methyl group to the phenol hydroxyl group of Tyr^2 produces a potent antagonist analog [$CH_2)_5$Mpr^1, Tyr(Me)^2,Val^4,DArg^8]VP, and the corresponding OEt derivative is even more potent. A possible explanation is that the agonist binds to the receptor in a particular way so that appropriate side-chain groups can "move" for transduction, but when just the appropriate lipophilic and/or steric surface presents itself, the molecule when bound to the receptor can no longer interact in such a manner that transduction will occur; thus the analog can competitively inhibit the binding of an agonist by binding in a different manner to the same receptor binding site. site.

Examination of the conformation properties of AVP antagonist analogs for the antidiuretic receptor are needed to obtain critical insights into those conformation structure properties that are critical for inhibiting activity. At this point it appears that the stereostructural properties of the 1, 2, and 4 residues in relation to one another is primarily responsible for not allowing transduction to occur. Although the bicyclic analog of Skala *et al.* (1984) (a very weak antagonist analog) is worthy of further investigation, it is not clear that a disruption of the Asn^5 and Arg^8 side-chain group orientations is responsible for antagonist activities in any other of the much more potent antagonist analogs. Indeed, no changes have been made in the Asn^5 or Arg^8 (Lys^8) side-chain groups for this purpose, although of course orientation of one or both of these residues may be affected by other changes that have been made in the analog. If it is the topological relationships of the Asn^5 and Arg^8 side-chain groups to one another that are critical for transduction at the kidney antidiuretic receptor, as suggested by Walter *et al.* (1977), then the antagonist analogs thus far developed must bind to the receptor in such a manner as to prevent the appropriate relationship of the Asn^5 and either DArg^8 or LArg^8 side-chain moieties. There are other interpretations. For example, it is possible that the Cys^1(Cys^6) side-chain groups and positions 2 and 5 side-chain groups need to related to one another in an appropriate manner for transduction to occur, and apparently this is not possible in the $(CH_2)_5$Mpr^1 analogs. The fact that only the more bulky $(CH_2)_5Mpr^1$ residue gives potent antagonist analogs in the antidiuretic assay argues for this interpretation.

VII. CONCLUDING REMARKS

In this chapter we have examined the known structure–biological activity relationships of the neurohypophyseal hormones and their analogs at the five neurohypophyseal hormone–receptor systems most commonly used to examine these relationships. These studies have shown that each receptor system has different structural and stereochemical requirements for both agonist and antagonist activities. Also, in general, agonist and antagonist analogs for a particular assay system utilize different structural properties for interaction with the receptor. Thus it is possible to design neurohypophyseal hormone analogs with high specificity for a particular receptor. Several approaches have been used to accomplish this kind of specificity. Manning and Sawyer (1984), for example, have emphasized the use of additivity rules for developing selective analogs, especially antagonist analogs. This approach has often worked quite effectively, possibly in part because many of the antagonists that have been developed for the neurohypophyseal hormones have β,β-dialkyl Cys^1 residues, and Hruby and coworkers have shown that this substitution greatly restricts the conformational properties, especially of the 20-membered disulfide-containing ring, including even the side-chain rotamer populations (Hruby 1981a,b; 1984; Hruby and Mosberg, 1982; Hruby *et al.*, 1983b). Often, however, the additivity approach has failed, especially in the agonist series. This is perhaps because insufficient attention has been devoted to considerations regarding the conformational properties of these peptides. Most of the agonist analogs prepared to date are highly flexible, with many conformations easily available to them. Thus they can adopt the conformational features that are needed for strong interaction with a particular receptor. This may explain why there is a strong overlap of activities for the different naturally occurring neurohypophyseal hormones at the several different receptors for these hormones. Such an ability to adapt conformational properties to a particular receptor is undoubtedly one of the reasons why oxytocin, vasopressin, and the other naturally occurring neurohypophyseal hormones have survived so well during evolutionary development.

Another approach that has become increasingly utilized because it offers in principle a rational approach to designing analogs with the specific properties desired is that of specific conformational restriction (Hruby, 1981a,b, 1984; Hruby *et al.*, 1984; Vida and Gordon, 1984). This approach has been applied to neurohypophyseal hormones with considerable success, primarily with inhibitors (Hruby, 1981a, 1985), but, as pointed out in this review, less has been done to restrict the conformation of agonists (Smith and Walter, 1978). We believe this will be possible and will provide critical new insight into the stereostructural requirements for activity at the various receptors for the neurohypophyseal hormones. Since the neurohypophyseal hormones are cyclic compounds, this poses certain problems, but it is likely that careful choice of appropriate modifications will lead to important new insights into the nature of the hormone–receptor interaction and the requirements of the receptor for binding and transducing the message.

This task becomes increasingly important as the number of putative receptors for these hormones continues to grow, especially with regards to effects in the central nervous system (CNS). These effects include (1) learning and behavioral effects, such as in delaying extinction of conditioned avoidance (with OT and AVP apparently having opposite effects in some cases); (2) attenuation of CNS effects related to tolerance and physical dependence of opioids, ethanol, and barbiturates; (3) maternal behavior; (4) acting as a modulator of release of other hormones; (5) stimulation of prostaglandin synthesis and release; (6) natriuretic activity; (7) insulin-like effects at the fat cell; etc. As methods for studying these systems more precisely develop, it is likely that a variety of new insights will be obtained. Indeed, one can confidently predict that although the neurohypophyseal hormones are no longer as widely studied as in the past, those who pursue these objectives will be rewarded with new insights into the structure–function relationships at these "new" receptors for the neurohypophyseal hormones. In some cases, new fundamental knowledge will emerge, and in addition new approaches to the treatment of human and animal disease are likely.

ACKNOWLEDGMENTS

The work done in the laboratory of Victor J. Hruby was supported by grants from the National Science Foundation and the U.S. Public Health Service. Victor J. Hruby was the recipient of a John Simon Guggenheim Fellowship during the time this review was written.

ABBREVIATIONS

Abbreviations follow the IUPAC-IUB Tentative Rules on Biochemical Nomenclature, *J. Biol. Chem.* (1972) **247:**977–983. The following additional abbreviations are used in this chapter:

AT	Aspartocin
AVP	Arginine vasopressin
AVT	Arginine vasotocin

1-Carba 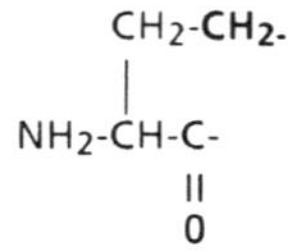 replaces 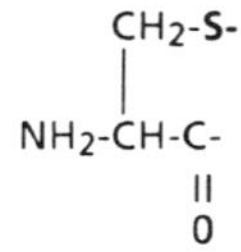 in position 1

6-Carba 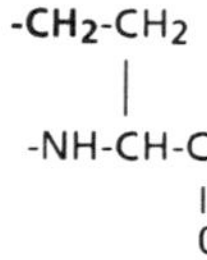 replaces 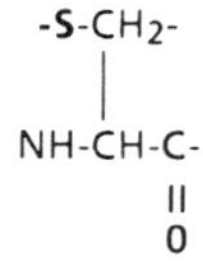 in position 6

Cha — Cyclohexylalanine

Deamino — Deletion of N-terminal amino group, i.e., formal replacement of Cys[1] by 3-mercaptopropionic acid

Deamino-1-carba 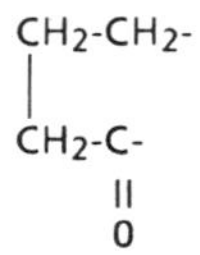 replaces 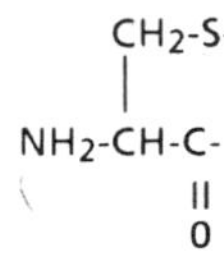 in position 1

Deamino-6-carba 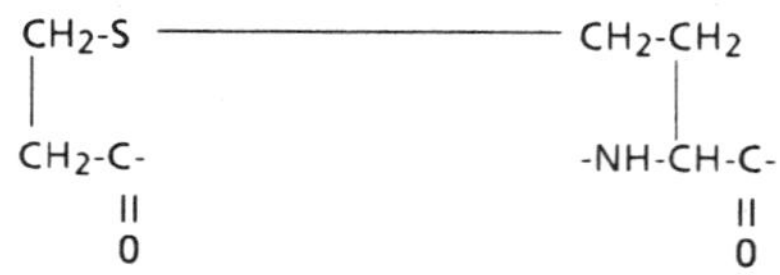 in positions 1 and 6

1,6-Dicarba- 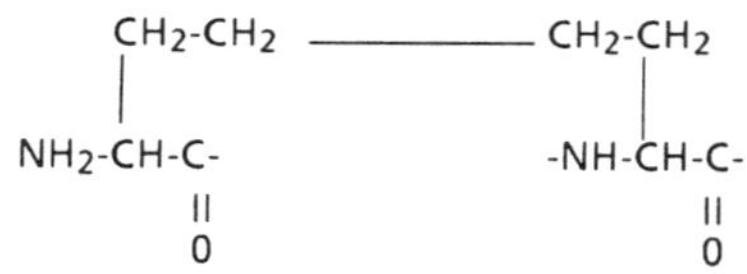 in positions 1 and 6

Deamino-1,6-dicarba 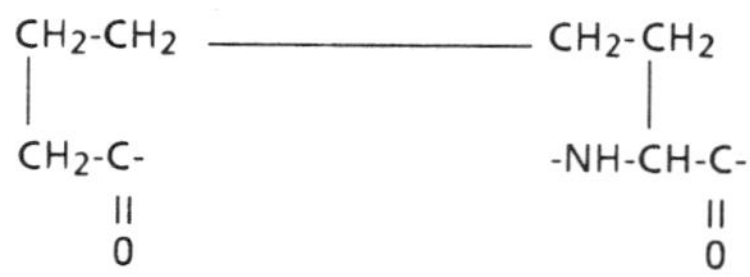 in positions 1 and 6

GT	Glumitocin
HBT	Hydroxybenzotriozole
Hmp	2-Hydroxy-3-mercaptopropionic acid
IT	Isotocin
LVP	Lysine vasopressin
Maa	2-Mercaptoacetic acid
$(CH_3)_2$Maa	2-Mercapto-2,2-dimethylacetic acid
Mba	4-Mercaptobutyric acid
Mpr	3-Mercaptopropionic acid

$(CH_3)_2$Mpr	3-Mercapto-3,3-dimethylpropionic acid
$(C_2H_5)_2$Mpr	3-Mercapto-3,3-diethylpropionic acid
$(CH_2)_5$Mpr	3-Mercapto-3,3-cyclopentamethylene propionic acid
MT	Mesotocin
NArg	Norarginine
OT	Oxytocin
Pen	Penicillamine
PP	Phenylpressin
Pressin	The disulfide-containing ring sequence of vasopressin
Pressinamide	The disulfide-containing ring sequence of vasopressin ending in a carboxamide at the C-terminus
Sec	Selenocysteine
Tocin	The disulfide-containing ring sequence of oxytocin
Tocinamide	The disulfide-containing ring sequence of oxytocin ending in a carboxamide at the C-terminus
VT	Valitocin

REFERENCES

Anagnostaras, P., Cordopatis, P., and Theodoropoulos, D. (1981). *Eur. J. Med. Chem.* **16,** 171–173.

Aoyagi, H., Kondo, M., and Izumiya, N. (1968). *Bull. Chem. Soc. Jpn.* **41,** 2772–2776.

Banerjee, S. N., Diamond, L., Ressler, C., and Sawyer, W. H. (1979). *J. Med. Chem.* **22,** 1487–1492.

Bankowski, K., Manning, M., Haldar, J., and Sawyer, W. H. (1978). *J. Med. Chem.* **21,** 850–853.

Bankowski, K., Manning, M., Seto, J., Haldar, J., and Sawyer, W. H. (1980). *Int. J. Peptide Protein Res.* **16,** 382–391.

Bankowski, K., Misicka, A., and Drabarek, S. (1983). *Polish J. Chem.* **57,** 1045–1046.

Barber, M., Jones, J. H., Stachulski, A. V., Bisset, G. W. Chowdrey, H. S., and Hudson, A. L. (1979). *Int. J. Peptide Protein Res.* **14,** 247–261.

Barth, T., and Kluh, I. (1974). *Collect. Czech. Chem. Commun.* **39,** 506–508.

Barth, T., Krejci, I., Kupkova, B., and Jost, K. (1973). *Eur. J. Pharmacol.* **24,** 183–188.

Barth, T., Slaninova, J., Servitova, L., Lebl, M., Kotsira-Engonopoulos, C., Tzougraki, C., Photaki, I., Krojidlo, M., and Jost, K. (1981). *Collect. Czech. Chem. Commun.* **46,** 2441–2445.

Baxter, J. W. M., Manning, M., and Sawyer, W. H. (1969a). *Biochemistry* **8,** 3592–3597.

Baxter, J. W. M., Wuu, T. C., Manning, M., and Sawyer, W. H. (1969b). *Experientia* **25,** 1127–1128.

Beránková-Ksandrová, Z., Bisset, G. W., Jost, K., Krejci, I., and Pliska, V. (1966). *Br. J. Pharmacol.* **26,** 615–632.

Berde, B., and Boissonnas, R. A. (1968). *In* "Handbook of Experimental Pharmacology," (B. Berde, ed.), Vol. 23, pp. 802–870. Springer-Verlag, New York.

Berde, B., Doepfner, W., and Konzett, H. (1957). *Br. J. Pharmacol.* **12,** 209–214.

Berde, B., Cerletti, A., and Konzett, H. (1961a). *In* "Oxytocin" (R. Caldyro-Barcin and H. Heller, eds.), pp. 247–265. Pergamon Press, Oxford.

Berde, B., Weidmann, H., and Cerletti, A. (1961b). *Helv. Physiol. Pharmacol. Acta* **19,** 285–302.

Berde, B., Huguenin, R. L., and Sturmer, E. (1962). *Experientia* **18,** 444–445.

Bespalova, Z. D., Kaurov, I. A., Martynov, V. F., Natochin, Y. V., Titov, M. I., and Shakhmatova, E. I. (1966). *Vestn. Leningr. Univ.* **4,** 157–159.

Bespalova, Z. D., Martynov, V. F., and Titov, M. I. (1968). *Zh. Obshch. Khim.* **38,** 1684–1687.

Beyerman, H. C., and Maassen van den Brink-Zimmermannova, H. (1968). *Recl. Trav. Chim. Pays-Bas* **87,** 1196–1200.

Bisset, G. W., Clark, B. J., Krejci, I., Polacek, I., and Rudinger, J. R. (1970). *Br. J. Pharmacol.* **40,** 342–360.

Blaha, I., Krchnak, V., and Zaoral, M. (1981). *Collect. Czech. Chem. Commun.* **46,** 2136–2139.

Bodanszky, M., and du Vigneaud, V. (1959). *J. Am. Chem. Soc.* **81,** 1258–1259.

Bodanszky, M., Fagan, D. T., Walter, R., and Smith, C. W. (1978). *J. Med. Chem.* **21,** 306–308.

Bodanszky, M., Chandramouli, N., Martinez, J., and Walter, R. (1979). *J. Med. Chem.* **22,** 270–273.

Bodanszky, M., Tolle, J. C., Seto, J., and Sawyer, W. H. (1980). *J. Med. Chem.* **23,** 1259–1261.

Boissonnas, R. A., Guttmann, St., Huguenin, R. L., Jaquenoud, P.-A., Sandrin, Ed. (1963). *Helv. Chim. Acta,* **46,** 2347–2362.

Botos, C. R., Smith, C. W., Chan, Y.-L., and Walter, R. (1979). *J. Med. Chem.* **22,** 926–931.

Branda, L. A., and du Vigneaud, V. (1966). *J. Med. Chem.* **9,** 169–172.

Branda, L. A., Hruby, V. J., and du Vigneaud, V. (1967). *Mol. Pharmacol.* **3,** 248–253.

Brewster, A. I. R., and Hruby, V. J. (1973). *Proc. Natl. Acad. Sci. USA* **70,** 3806–3809.

Brtnik, F., Barth, T., and Jost, K. (1981). *Collect. Czech. Chem. Commun.* **46,** 278–285.

Brtnik, F., Krejci, I., Kupkova, B., Hrbas, P., Skopkova, J., Barth, T., and Jost, K. (1983). *Collect. Czech. Chem. Commun.* **48,** 2862–2873.

Buku, A., Gazis, D., and Schwartz, I. L. (1984). *Int. J. Peptide Protein Res.* **23,** 551–557.

Burn, H. J., Finney, D. J., and Goodwin, L. G. (1950). *In* ''Biological Standardization'', p. 187. Oxford University Press, New York.

Cash, W. D., and Smith, B. L. (1963). *J. Biol. Chem.* **238,** 994–997.

Chan, W. Y., and du Vigneaud, V. (1962). *Endocrinology* **71,** 977–982.

Chan, W. Y., and du Vigneaud, V. (1970). *J. Pharamacol. Exp. Ther.* **174,** 541–549.

Chan, W. Y., and Hruby, V. J. (1983). *In* ''Peptides: Structure and Function,'' (V. J. Hruby and D. H. Rich, eds.), pp. 857–860. Pierce Chemical Co. Rockford, Il.

Chan, W. Y., and Kelley, N. (1967). *J. Pharmacol. Exp. Ther.* **56,** 150–158.

Chan, W. Y., Hruby, V. J., Flouret, G., and du Vigneaud, V. (1968). *Science* **161,** 280–281.

Chan, W. Y., Chiu, P. J. S., and Sawyer, W. H. (1973). *Proc. Soc. Exp. Biol. Med.* **143,** 954–958.

Chan, W. Y., Powell, A. M., and Hruby, V. J. (1982). *Endocrinology* **111,** 48–54.

Chan, Y. L., and Smith, C. W. (1980). *Biochem. Biophys. Res. Commun.* **96,** 626–631.

Chan, Y. L., Peterson, C. A., Ting, Y. F., and Smith, C. W. (1981). *Life Sci.* **28,** 415–419.

Chauvet, J., Chauvet, M.-T., and Acher, R. (1972). *Experientia* **28,** 1493–1495.

Cheesman, D. W., Schlegel, R., Sagasay, A. M., and Forsham, P. H. (1983). *Endocrinology* **112,** 269–276.

Cort, J. H., Schuck, O., Stribrna, J., Skopkova, J., Jost, K., and Mulder, J. L. (1975). *Kidney Int.* **8,** 292–302.

Cort, J. H., Fric, I., Carlsson, L., Gillessen, D., Bystricky, S., Skopkova, J., Gut, V., Studer, R. O., Mulder, J. L., and Blaha, K. (1976). *Mol. Pharmacol.* **12,** 313–321.

Cort, J. H., Fischman, A. J., Dodds, W. J., Rand, J. H., and Schwartz, I. L. (1981). *Int. J. Peptide Protein Res.* **17,** 14–22.

Cort, J. H., Groginsky, C., and Fischman, A. J. (1982). *Clin. Res.* **30,** 25A.

Datta, H., and Chaudhury, R. R. (1968). *J. Endocrinol.* **42,** 345–346.

Datta, H., and Chaudhury, R. R. (1970). *J. Endocrinol.* **46,** 117–118.

Dousa, T., Hechter, O., Walter, R., and Schwartz, I. L. (1970). *Science* **167,** 1134–1135.

Dousa, T., Hechter, O., Schwartz, I. L., and Walter, R. (1971). *Proc. Natl. Acad. Sci. USA* **68,** 1693–1697.

Drabarek, S., and du Vigneaud, V. (1965). *In* "Peptides, Proceedings of the 6th European Symposium" (L. Zervas, ed.), pp. 229–233. Pergamon, New York.
Dutta, A. S., Anand, N., and Kar, K. (1966a). *J. Med. Chem.* **9,** 497–499.
Dutta, A. S., Anand, N., and Kar, K. (1966b). *Indian J. Chem.* **4,** 488–492.
Dutta, A. S., Anand, N., and Srimal, R. C. (1969). *Indian J. Chem.* **7,** 3–8.
du Vigneaud, V., Ressler, C., Swan, J. M., Roberts, C. W., Katsoyannis, P. G., and Gordon, S. (1953). *J. Am. Chem. Soc.* **75,** 4879–4880.
du Vigneaud, V., Denning, G. S., Drabarek, S., and Chan, W. Y. (1963). *J. Biol. Chem.* **238,** 1560–1561.
du Vigneaud, V., Denning, G. S., Drabarek, S., and Chan, W. Y. (1964). *J. Biol. Chem.* **239,** 472–478/
du Vigneaud, V., Flouret, G., and Walter, R. (1966). *J. Biol. Chem.* **241,** 2093–2096.
Dyckes, D. F., Ferger, M. F., du Vigneaud, V., and Chan, W. Y. (1973). *J. Med. Chem.* **16,** 843–847.
Dyckes, D. F., Nestor, J. J., Jr., Ferger, M. F., and du Vigneaud, V. (1974a). *J. Med. Chem.* **17,** 250–252.
Dyckes, D. F., Nestor, J. J., Jr., Ferger, M. F., du Vigneaud, V., and Chan, W. Y (1974b). *J. Med. Chem.* **17,** 969–972.
Dyckes, D. F., Smith, C. W., Ferger, M. F., and du Vigneaud, V. (1974c). *J. Am. Chem. Soc.* **96,** 7549–7551.
Eisler, K., Rudinger, J., and Sorm, F. (1966). *Collect. Czech. Chem. Commun.* **31,** 4563–4580.
Fahrenholz, F., Thierauch, K.-H., and Crause, P. (1980). *Hoppe-Seyler's Z. Physiol. Chem.* **361,** 153–167.
Fahrenholz, F. Toth, G., Crause, P., Eggena, P., and Schwartz, I. L. (1983). *J. Biol. Chem.* **258,** 14861–14867.
Ferger, M. F., (1974). Personal communication.
Ferger, M. F., Jones, Jr., W. C., Dyckes, D. F., and du Vigneaud, V. (1972). *J. Am Chem. Soc.* **94,** 982–984.
Ferraro, J. J., and du Vigneaud, V. (1966). *J. Am. Chem. Soc.* **88,** 3847–3850.
Ferrier, B. M., and Branda, L. A. (1975). *Can. J. Biochem.* **53,** 21–27.
Ferrier, B. M., and Branda, L. A. (1976a). *Can. J. Biochem.* **54,** 507–511.
Ferrier, B. M., and Branda, L. A. (1976b). *Can. J. Biochem.* **54,** 512–516.
Ferrier, B. M., and du Vigneaud, V. (1966). *J. Med. Chem.* **9,** 55–57.
Ferrier, B. M., Jarvis, D., and du Vigneaud, V. (1965). *J. Biol. Chem.* **240,** 4264–4266.
Fink, M. L., and Bodanszky, M. (1973). *J. Med. Chem.* **16,** 1324–1326.
Flouret, G., and du Vigneaud, V. (1969). *J. Med. Chem.* **12,** 1035–1038.
Fraefel, W., and du Vigneaud, V. (1970a). *J. Am. Chem. Soc.* **92,** 1030–1032.
Fraefel, W., and du Vigneaud, V. (1970b). *J. Am. Chem. Soc.* **92,** 4426–4427.
Fric, I., Kodicek, M., Prochazka, Z., Jost, K., and Blaha, K. (1974). *Collect. Czech. Chem. Commun.* **39,** 1290–1302.
Gazis, D., Roy, U., Schwartz, I. L., and Roy, J. (1983). *Int. J. Peptide Protein Res.* **22,** 450–455.
Gazis, D., Schwartz, I. L., Lammek, B., Grzonka, Z. (1984). *Int. J. Peptide Protein Res.* **23,** 78–83.
Geiger, R., Sturm, K., and Siedel, W. (1968). *Chem. Ber.* **101,** 1223–1231.
Gillessen, D., and du Vigneaud, V. (1967). *J. Biol. Chem.* **242,** 4806–4812.
Gillessen, D., and du Vigneaud, V. (1970). *J. Med. Chem.* **13,** 346–349.
Glass, J. D., and du Vigneaud, V. (1972). *J. Med. Chem.* **15,** 486–488.
Grzonka, Z., Glass, J. D., Schwartz, I. L., and Walter, R. (1974). *J. Med. Chem.* **17,** 1294–1298.
Grzonka, Z., Lammek, B., Gazis, D., and Schwartz, I. L. (1983a). *J. Med. Chem.* **26,** 1786–1787.
Grzonka, Z., Lammek, B., Kasprzykowski, F., Gazis, D., and Schwartz, I. L. (1983b). *J. Med. Chem.* **26,** 555–559.

Guttmann, S., and Boissonnas, R. A. (1963). *Helv. Chim. Acta* **46,** 1626–1636.
Hase, S., Sakakibara, S., Wahrenburg, M., Kirchberger, M., Schwartz, I. L., and Walter, R. (1972a). *J. Am. Chem. Soc.* **94,** 3590–3600.
Hase, S., Sakakibara, S., Wahrenburg, M., and Walter, R. (1972b). *J. Med. Chem.* **15,** 1017–1019.
Hase, S., Schwartz, I. L., and Walter, R. (1972c). *J. Med. Chem.* **15,** 126–128.
Havran, R. T. (1973). *Experientia* **29,** 520–521.
Havran, R. T., and du Vigneaud, V. (1969). *J. Am. Chem. Soc.* **91,** 3626–3628.
Havran, R. T., Schwartz, I. L., and Walter, R. (1969a). *Mol. Pharmacol.* **5,** 83–89.
Havran, R. T., Schwartz, I. L., and Walter, R. (1969b). *J. Am. Chem. Soc.* **91,** 1836–1840.
Havran, R. T., Sapirstein, V. S., Schwartz, I. L., Scott, W. N., and Walter, R. (1973). *Endocrinology* **92,** 226A.
Hechter, O., Terada, S., Spitsberg, V., Nakahara, T., Nakagawaga, S. H., and Flouret, G. (1978). *J. Biol. Chem.* **253,** 3230–3237.
Hermann, P., and Zaoral, M. (1965). *Collect. Czech. Chem. Commun.* **30,** 2817–2826.
Hlavacek, J., Poduska, K., Jost, K., Fric, I., Barth, T., Cort, H. H., Blaha, K., and Sorm, F. (1977). *Collect. Czech. Chem. Commun.* **42,** 1233–1247.
Hlavacek, J., Fric, I., Barth, T., and Jost, K. (1982). *Collect. Czech. Chem. Commun.* **47,** 338–348.
Hope, D. B., and Walti, M. (1973). *Biochem. J.* **135,** 211–213.
Hope, D. B., Murti, V. V. S., and du Vigneaud, V. (1963). *J. Am. Chem. Soc.* **85,** 3686–3688.
Hruby, V. J. (1981a). *In* "Perspectives in Peptide Chemistry" (A. Eberle, R. Geiger, and T. Wieland, eds.), pp. 207–220. Karger, Basel.
Hruby, V. J. (1981b). *In* "Topics in Molecular Pharmacology" (A.S.V. Burgen and G.C.K. Roberts, eds.), pp. 99–126. Elsevier/North Holland, Amsterdam.
Hruby, V. J. (1982). *Life Sci.* **31,** 189–199.
Hruby, V. J. (1984). *In* "Conformationally Directed Drug Design. Peptides and Nucleic Acids as Targets on Templates" (J. A. Vida and M. Gordeon, eds.), pp. 9–27. ACS Symposium Series 251, American Chemical Society, Washington, D.C.
Hruby, V. J. (1985). *In* "Oxytocin: Clinical and Laboratory Studies," (J. A. Amico and A. G. Robinson, eds.), pp. 405–414. Elsevier/North Holland, New York.
Hruby, V. J., and Chan, W. Y. (1971). *J. Med. Chem.* **14,** 1050–1051.
Hruby, V. J., and du Vigneaud, V. (1969a). *J. Med. Chem.* **12,** 731–733.
Hruby, V. J., and du Vigneaud, V. (1969b). *J. Am. Chem. Soc.* **91,** 3624–3626.
Hruby, V. J., and Mosberg, H. I. (1982). *In* "Hormone Antagonists" (M. K. Agarwal, ed.), pp. 433–474. W. de Gruyter, Berlin.
Hruby, V. J., Yamashiro, D., and du Vigneaud, V. (1968). *J. Am. Chem. Soc.* **90,** 7106–7110.
Hruby, V. J., Flouret, G., and du Vigneaud, V. (1969). *J. Biol. Chem.* **244,** 3890–3894.
Hruby, V. J., du Vigneaud, V., and Chan, W. Y. (1970). *J. Med. Chem.* **13,** 185–187.
Hruby, V. J., Ferger, M. F., and du Vigneaud, V. (1971). *J. Am. Chem. Soc.* **93,** 5539–5542.
Hruby, V. J., Smith, C. W., Linn, D. K., Ferger, M. F., and du Vigneaud, V. (1972). *J. Am. Chem. Soc.* **94,** 5478–5480.
Hruby, V. J., Musico, F., Groginsky, C. M., Gitu, P. M., Saba, D., and Chan, W. Y. (1973). *J. Med. Chem.* **16,** 624–629.
Hruby, V. J., Deb, K. K., Yamamoto, D. M., Hadley, M. E., and Chan, W. Y. (1979a). *J. Med. Chem.* **22,** 7–12.
Hruby, V. J., Upson, D. A., Yamamoto, D. M., Smith, C. W., and Walter, R. (1979b). *J. Am. Chem. Soc.* **101,** 2717–2721.
Hruby, V. J., Mosberg, H. I., Hadley, M. E., Chan, W. Y., and Powell, A. M. (1980). *Int. J. Peptide Protein Res.* **16,** 372–381.
Hruby, V. J., Mosberg, H. I., and Viswanatha, V. (1982). *J. Am. Chem. Soc.* **104,** 837–841.

Hruby, V. J., Knittel, J. J., Mosberg, H. I., Rockway, T. W., Wilkes, B. C., and Hadley, M. E. (1983a). *In* "Peptides 1982" (K. Blaha and P. Malon, eds.), pp. 19–30. W. de Gruyter, Berlin.

Hruby, V. J., Rockway, T. W., and Chan, W. Y. (1983b). *Int. J. Peptide Protein Res.* **21,** 24–34.

Hruby, V. J., Krstenansky, J. L., and Cody, W. L. (1984). *Ann. Rev. Med. Chem.* **19,** 303–312.

Huffman, W. F., and Moore, M. L. (1984). U.S. Patent No. 4,469,679.

Huffman, W. F., Ali, F. E., Bryan, W. M., Callahan, J. F., Moore, M. L., Silvestri, J. S., Yim, N. C. F., Kinter, L. B., McDonald, J. E., Ashton-Shue, D., Stassen, F. L., Heckman, G. D., Schmidt, D. B., and Sulat, L. (1985). *J. Med. Chem.* **28,** 1759–1760.

Huguenin, R. L. (1964). *Helv. Chim. Acta* **47,** 1934–1941.

Huguenin, R. L. (1966). *Helv. Chim. Acta* **49,** 711–725.

Huguenin, R. L. and Boissonnas, R. A. (1961). *Helv. Chim. Acta* **44,** 213–231.

Huguenin, R. L., and Boissonnas, R. A. (1962). *Helv. Chim. Acta* **45,** 1629–1643.

Huguenin, R. L., and Boissonnas, R. A. (1963). *Helv. Chim. Acta* **46,** 1669–1676.

Huguenin, R. L., and Boissonnas, R. A. (1966). *Helv. Chim. Acta* **49,** 695–705.

Huguenin, R. L., and Guttmann, S. (1965). *Helv. Chim. Acta* **48,** 1885–1898.

Huguenin, R. L., Sturmer, E., Boissonnas, R. A., and Berde, B. (1965). *Experientia* **21,** 68–69.

Jager, G., and Geiger, R. (1973). *Liebigs Ann. Chem.* 1928–1933.

Jaquenoud, P.-A. (1965). *Helv. Chim. Acta* **48,** 1899–1911.

Jaquenoud, P.-A., and Boissonnas, R. A. (1959). *Helv. Chim. Acta* **42,** 788–793.

Jaquenoud, P.-A., and Boissonnas, R. A. (1961). *Helv. Chim. Acta* **44,** 113–122.

Jarvis, D., and du Vigneaud, V. (1967). *J. Biol. Chem.* **242,** 1768–1771.

Jarvis, D., Bodanszky, M., and du Vigneaud, V. (1961). *J. Am. Chem. Soc.* **83,** 4780–4784.

Jones, D. A., and Sawyer, W. H. (1980). *J. Med. Chem.* **23,** 696–698.

Jones, W. C., Nestor, J. J., Jr., and du Vigneaud, V. (1973). *J. Am. Chem. Soc.* **95,** 5677–5679.

Jost, K., and Sorm, F. (1971a). *Collect. Czech. Chem. Commun.* **36,** 234–245.

Jost, K., and Sorm, F. (1971b). *Collect. Czech. Chem. Commun.* **36,** 297–298.

Jost, K., and Sorm, F. (1971c). *Collect. Czech. Chem. Commun.* **36,** 2795–2808.

Jost, K., Rudinger, J., and Sorm, F. (1963). *Collect. Czech. Chem. Commun.* **28,** 2021–2030.

Jost, K., Debabov, V. G., Nesvadba, H., and Rudinger, J. (1964). *Collect. Czech. Chem. Commun.* **29,** 419–432.

Jost, K., Barth, T., Krejci, I., Fruhaufova, L., Prochazka, Z., and Sorm, F. (1973). *Collect. Czech. Chem. Commun.* **38,** 1073–1083.

Jost, K., Prochazka, Z., Cort, J. H., Barth, T., Skopkova, J., Prusik, Z., and Sorm, F. (1974). *Collect. Czech. Chem. Commun.* **39,** 2835–2856.

Kasafirek, E., Rabek, V., Rudinger, J., and Sorm, F. (1966). *Collect. Czech. Chem. Commun.* **31,** 4581–4591.

Katsoyannis, P. G., and du Vigneaud, V. (1958). *J. Biol. Chem.* **233,** 1352–1354.

Kaurov, O. A., Martynov, V. F., and Mikhailov, Y. D. (1971). *Zh. Obshek. Khim.* **41,** 1413.

Kaurov, O. A., Martynov, V. F., Mikhailov, Y. D., and Auna, Z. P. (1972). *Zh. Obshch. Khim.* **42,** 1654.

Kaurov, O. A., Martynov, V. F., and Popernatskii, O. A. (1973a). *Zh. Obshch. Khim.* **43,** 909–917.

Kaurov, O. A., Martynov, V. F., and Smirnova, M. P. (1973b). *Zh. Obshch. Khim.* **43,** 217–218.

Kaurov, O. A., Natochin, Yu. V., Popernatskii, O. A., and Shakhmatova, E. I. (1973c). *Zh. Obshch. Khim.* **43,** 2521–2528.

Keller, O., and Rudinger, J. R. (1974). *Helv. Chim. Acta* **57,** 1253–1259.

Kimbrough, R. D., Cash, W. D., Branda, L. A., Chan, W. Y., and du Vigneaud, V. (1963). *J. Biol. Chem.* **238,** 1411–1414.

Kolc, J., Zaoral, M., and Sorm, F. (1967). *Collect. Czech. Chem. Commun.* **32,** 2667–2671.

Krchnak, V., Zaoral, M., and Machova, A. (1979). *Collect. Czech. Chem. Commun.* **44,** 2161–2164.

Krojidlo, M., Barth, T., Servitova, L., Dobrovsky, K., Jost, K., and Sorm, F. (1975). *Collect. Czech. Chem. Commun.* **40,** 2708–2717.

Krojidlo, M., Barth, T., Blaha, K., and Jost, K. (1976). *Collect. Czech. Chem. Commun.* **41,** 1954–1958.

Kruszynski, M., Lammek, B., Manning, M., Seto, J., Haldar, J., and Sawyer, W. H. (1980). *J. Med. Chem.* **23,** 364–368.

Larsson, L.-E., Lindeberg, E. G. G., and Melin, P. (1977). *Int. J. Pept. Protein Res.* **9,** 182–186.

Larsson, L.-E., Lindeberg, G., Melin, P., and Pliska, V. (1978). *J. Med. Chem.* **21,** 352–356.

Lebl, M., and Barth, T. (1983). *In* "Peptides: Structure and Function" (V. J. Hruby and D. H. Rich, eds.), pp. 357–360. Pierce Chemical Company, Rockford, Il.

Lebl, M., and Jost, K. (1978). *Collect. Czech. Chem. Commun.* **43,** 523–534.

Lebl, M., Barth, T., and Jost, K. (1978a). *Collect. Czech. Chem. Commun.* **43,** 1300–1305.

Lebl, M., Barth, T., and Jost, K. (1978b). *Collect. Czech. Chem. Commun.* **43,** 1538–1556.

Lebl, M., Bojanovska, V., Barth, T., and Jost, K. (1978c). *Experientia* **34,** 1543–1545.

Lebl, M., Barth, T., and Jost, K. (1979a). *Collect. Czech. Chem. Commun.* **44,** 2563–2572.

Lebl, M., Bojanovska, V., Barth, T., and Jost, K. (1979b). *Collect. Czech. Chem. Commun.* **44,** 2573–2582.

Lebl, M., Dimeli, A., Bojanovska, V., Slaninova, J., Barth, T., and Jost, K. (1979c). *Collect. Czech. Chem. Commun.* **44,** 2556–2562.

Lebl, M., Barth, T., and Jost, K. (1980a). *Collect. Czech. Chem. Commun.* **45,** 2855–2864.

Lebl, M. Barth, T., Skopkova, J., and Jost, K. (1980b). *Collect. Czech. Chem. Commun.* **45,** 2865–2871.

Lebl, M., Machova, A., Hrbas, P., Barth, T., and Jost, K. (1980c). *Collect. Czech. Chem. Commun.* **45,** 2714–2723.

Lebl, M., Hrbas, P., Skopkova, J., Slaninova, J., Machova, A., Barth, T., and Jost, K. (1982a). *Collect. Czech. Chem. Commun.* **47,** 2540–2560.

Lebl, M., Pospisek, J., Hlavacek, J., Barth, T., Malon, P., Servitova, L., Hauzer, K., and Jost, K. (1982b). *Collect. Czech. Chem. Commun.* **47,** 689–701.

Lebl, M., Barth, T., Servitova, L., Slaninova, J., and Jost, K. (1985a). *Collect. Czech. Chem. Commun.* **50,** 132–145.

Lebl, M., Hruby, V. J., Slaninova, J., and Barth, T. (1985b). *Collect. Czech. Chem. Commun.* **50,** 418–427.

Lindeberg, G. E. (1975). *Int. J. Peptide Protein Res.* **7,** 395–401.

Lindeberg, G. E., Bodanszky, M., Acosta, M., and Sawyer, W. H. (1974). *J. Med. Chem.* **17,** 781–783.

Lindeberg, G. E., Melin, P., and Larsson, L.-E. (1976). *Int. J. Peptide Protein Res.* **8,** 193–198.

Lindeberg, G. E., Karlsson, S. M., and Melin, P. (1977). *Int. J. Peptide Protein Res.* **10,** 240–244.

Lowbridge, J., Manning, M., Haldar, J., and Sawyer, W. H. (1977a). *J. Med. Chem.* **20,** 120–123.

Lowbridge, J., Manning, M., Haldar, J., and Sawyer, W. H. (1977b). *J. Med. Chem.* **20,** 1173–1176.

Lowbridge, J., Manning, M., Seto, J., Haldar, J., and Sawyer, W. H. (1979). *J. Med. Chem.* **22,** 565–569.

Lozhkina, A., Samavtsen, V., and Martynov, M. (1973). *Khim. Phir. Svedin.* **89.**

Lundell, E. O., and Ferger, M. F. (1975a). *Bioorg. Chem.* **4,** 377–384.

Lundell, E. O., and Ferger, M. F. (1975b). *J. Med. Chem.* **18,** 1045–1046.

Lundell, E. O., Smith, C. W., and Ferger, M. R. (1975). *J. Med. Chem.* **18,** 1262–1264.

Manning, M., and du Vigneaud, V. (1965). *J. Am. Chem. Soc.* **87,** 3978–3982.

Manning, M., and Sawyer, W. H. (1984). *Trends Neurosci.* **7,** 6–9.

Manning, M., Coy, E. J., and Sawyer, W. H. (1970). *Biochemistry* **9,** 3925–3929.

Manning, M., Baxter, J. W. M., Wuu, T. C., Smart-Abbey, V., Morton, K., Coy, E. J., and Sawyer, W. H. (1971a). *J. Med. Chem.* **14,** 1143–1144.

Manning, M., Coy, E. J., and Sawyer, W. H. (1971b). *Experientia* **27,** 1372–1374.

Manning, M., Balaspiri, L., Acosta, M., and Sawyer, W. H. (1973a). *J. Med. Chem.* **16,** 975–978.

Manning, M., Coy, E. J., Acosta, M., and Sawyer, W. H. (1973b). *J. Med. Chem.* **16,** 836–839.

Manning, M., Coy, E. J., Sawyer, W. H., and Acosta, M. (1973c). *J. Med. Chem.* **16,** 463–466.

Manning, M., Lowbridge, J., Haldar, J., and Sawyer, W. H. (1976). *J. Med. Chem.* **19,** 376–380.

Manning, M., Lowbridge, J., Stier, Jr., C. T. Haldar, J., and Sawyer, W. H. (1977). *J. Med. Chem.* **20,** 1228–1230.

Manning, M., Lowbridge, J., Seto, J., Haldar, J., and Sawyer, W. H. (1978). *J. Med. Chem.* **21,** 179–182.

Manning, M., Lammek, B., Kolodziejczyk, M., Seto, J., and Sawyer, W. H. (1981). *J. Med. Chem.* **24,** 701–706.

Manning, M., Klis, W. A., Olma, A., Seto, J., and Sawyer, W. H. (1982a). *J. Med. Chem.* **25,** 414–419.

Manning, M., Lammek, B., Kruszynski, M., Seto, J., and Sawyer, W. H. (1982b). *J. Med. Chem.* **25,** 408–414.

Manning, M., Olma, A., Klis, W. A., Kolodziejczyk, M., Seto, J., and Sawyer, W. H. (1982c). *J. Med. Chem.* **25,** 45–50.

Manning, M., Olma, A., Klis, W. A., Seto, J., and Sawyer, W. H. (1983). *J. Med. Chem.* **26,** 1607–1613.

Manning, M., Nawrocka, E., Misicka, A., Olma, A., Klis, W. A., Seto, J., Sawyer, W. H. (1984a). *J. Med. Chem.* **27,** 423–429.

Manning, M., Olma, A., Klis, W., Kolodziejczyk, A., Nawrocka, E., Misicka, A., Seto, J., and Sawyer, W. H. (1984b). *Nature (London)* **308,** 652–653.

Marbach, P., and Rudinger, J. (1974a). *Experientia* **30,** 696.

Marbach, P., and Rudinger, J. (1974b). *Helv. Chim. Acta* **57,** 403–414.

Martynov, V. F., Bespalova, Z. D., Titov, M. I., and Shurukhina, S. I. (1967). *Zh. Obsch. Khim.* **37,** 583–587.

Meienhofer, J., and du Vigneaud, V. (1961). *J. Am. Chem. Soc.* **83,** 142–145.

Meienhofer, J., Trzeciak, A., Havran, R. T., and Walter, R. (1970). *J. Am. Chem. Soc.* **92,** 7199–7202.

Meienhofer, J., Trzeciak, A., Dousa, T., Hechter, O., Havran, R. T., Schwartz, I. L., and Walter, R. (1971). *In* "The Peptides 1969," p. 157–160. American Elsevier, New York.

Melin, P., Trojnar, J., Vilhardt, H., and Aakerlund, M. (1983). *In* "Peptides: Structure and Function" (V. J. Hruby and D. H. Rich, eds.), pp. 361–364. Pierce Chemical Company, Rockford, Il.

Meraldi, J.-P., Hruby, V. J., and Brewster, A. I. R. (1977). *Proc. Natl. Acad. Sci. USA* **74,** 1373–1377.

Midura-Nowaczek, K., and Drabarek, S. (1978). *Polish J. Chem.* **52,** 341–345.

Moore, G. (1978). *Biochem. J.* **173,** 403–409.

Moore, M. L., Huffman, W. F., Roberts, G. D., Rottschaefer, S., Sulat, L., Stefankiewics, J. S., and Strassen, F. (1985). *Biochem. Biophys. Res. Commun.* **121,** 878–883.

Moore, S., Felix, A. M., Meienhofer, J., Smith, C. W., and Walter, R. (1977). *J. Med. Chem.* **20,** 495–500.

Morikawa, T., and Sakakibara, S. (1979). *In* "Proc. 8th Symp. on Peptide Chemistry" (T. Kaneko, ed.), pp. 106–109. Peptide Research Foundation, Osaka, Japan.

Nestor, J. J., Jr., Ferger, M. F., and du Vigneaud, V. (1974). Unpublished results.

Nestor, J. J., Jr., Ferger, M. F., and Chan, W. Y. (1975a). *J. Med. Chem.* **18,** 1022–1024.

Nestor, J. J., Jr., Ferger, M. F., and du Vigneaud, V. (1975b). *J. Med. Chem.* **18,** 284–287.

Nesvadba, H., Honzl, J., and Rudinger, J. (1963). *Collect. Czech. Chem. Commun.* **28,** 1691–1705.

Niedrich, H. J. (1972). *J. Prakt. Chem.* **314,** 769–779.

Papsuevich, G. S., and Chipens, G. (1970). *Zh. Obshch. Khim.* **40,** 709–710.

Papsuevich, G. S., Krikis, A. Y., and Chipens, G. I. (1972). *Zh. Obshch. Khim.* **42,** 224–229.

Photaki, I., and du Vigneaud, V. (1965). *J. Am. Chem. Soc.* **87,** 908–913.

Photaki, I., Tzougraki, C., and Kotsira-Engonopoulos, C. (1979). *Int. J. Peptide Protein Res.* **13,** 426–433.

Pliska, V., and Krejci, I. (1966). *Arch. Int. Pharmacodyn.* **161,** 289–297.

Pliska, V., and Marbach, P. (1978). *Eur. J. Pharmacol.* **49,** 213–222.

Pliska, V., Vasak, J., Rufer, M., and Rudinger, J. R. (1973). *Experientia* **29,** 171–173.

Pliska, V., Marbach, P., Vasak, J., and Rudinger, J. (1977). *Experientia* **33,** 367–369.

Polacek, I., Krejci, I., Nesvadba, H., and Rudinger, J. R. (1970). *Eur. J. Pharmacol.* **9,** 239–245.

Prochazka, Z., Barth, T., Cort, J. H., Jost, K., and Sorm, F. (1978a). *Collect. Czech. Chem. Commun.* **43,** 655–663.

Prochazka, Z., Krejci, I., Kupkova, B., Slaninova, J., Bojanovska, V., Prusik, Z., Vosekalna, I. A., Malon, P., Barth, T., Fric, I., Blaha, K., and Jost, K. (1978b). *Collect. Czech. Chem. Commun.* **43,** 1285–1299.

Prochazka, Z., Lebl, M., Servitova, L., Barth, T., and Jost, K. (1981). *Collect. Czech. Chem. Commun.* **46,** 947–956.

Prochazka, Z., Lebl, M., Barth, T., Hlavacek, J., Trka, A., Budesinsky, M., and Jost, K. (1984). *Collect. Czech. Chem. Commun.* **49,** 642–652.

Rekowski, P., Lammek, B., Melin, P., Ragnarsson, U., and Kupryszewski, G. (1985). *Acta. Chem. Scand. B* **39,** 453–457.

Rosamond, J. D., and Ferger, M. F. (1976). *J. Med. Chem.* **19,** 873–876.

Roy, C., Barth, T., and Jard, S. (1975). *J. Biol. Chem.* **250,** 3149–3156.

Roy, J., Havran, R. T., Schwartz, I. L., and Walter, R. (1975). *Int. J. Peptide Protein Res.* **7,** 171–178.

Roy, J., Gazis, D., and Schwartz, I. L. (1980a). *Int. J. Peptide Protein Res.* **16,** 106–110.

Roy, J., Johnson, M., Dubin, S., Gazis, D. C., and Schwartz, I. L. (1980b). *Int. J. Peptide Protein Res.* **15,** 279–284.

Roy, J., Johnson, M., Gazis, D., and Schwartz, I. L. (1980c). *Int. J. Peptide Protein Res.* **16,** 55–60.

Roy, J., Gazis, D., Shakman, R., and Schwartz, I. L. (1982). *Int. J. Peptide Protein Res.* **20,** 35–42.

Roy, U., Gazis, D., Dal Pan, G., Schwartz, I. L., and Roy, J. (1983a). *Int. J. Peptide Protein Res.* **22,** 525–538.

Roy, U., Gazis, D., Schwartz, I. L., and Roy, J. (1983b). *Int. J. Peptide Protein Res.* **22,** 398–403.

Rudinger, J. (1971). *In* "Drug Design" (J. Ariens, ed.), Vol. 2, pp. 319–419. Academic, New York.

Rudinger, J., and Krejci, I. (1968). *In* "Handbook of Experimental Pharmakologie" (B. Berde, ed.), pp. 748–801. Springer Verlag, New York.

Rudinger, J., Kesarev, O. V., Poduska, K., Pickering, B. T., Dyball, R. E. J., Ferguson, D. R., and Ward, W. R. (1969). *Experientia* **25,** 680–682.

Rudinger, J., Pliska, V., and Krejci, I. (1972). *Recent Progr. Horm. Res.* **28,** 131–172.

Sakakibara, S., and Hase, S. (1968). *Bull. Chem. Soc. Japan* **41,** 2816.

Sawyer, W. H., and Manning, M. (1971). *J. Endocrinol.* **49,** 151–165.

Sawyer, W. H., and Manning, M. (1973). *Ann. Rev. Pharmacol.* **13,** 5–17.

Sawyer, W. H., Wuu, T. C., Baxter, J. W. M., and Manning, M. (1969). *Endocrinology* **85,** 385–388.

Sawyer, W. H., Baxter, J. W. M., Manning, M., Heinicke, E. A., and Perks, A. M. (1970). *Gen. Comp. Endocrinol.* **15,** 52–58.

Sawyer, W. H., Acosta, M., Balaspiri, L., Judd, J., and Manning, M. (1974). *Endocrinology* **94,** 1106–1115.

Sawyer, W. H., Haldar, J., Gazis, D., Seto, J., Bankowski, K., Lowbridge, J., Turan, A., and Manning, M. (1980). *Endocrinology* **106,** 81–91.

Sawyer, W. H., Pang, P. K., Seto, J., McEnroe, M., Lammeck, B., and Manning, M. (1981). *Science* **212,** 49–51.

Schild, H. O. (1949). *Br. J. Pharmacol.* **4,** 277–280.

Schillinger, E., Loge, O., Schrolder, E., Klieger, E., and Luebke, K. (1972). *Eur. J. Biochem.* **27,** 473–481.

Schneider, C. H., and du Vigneaud, V. (1962). *J. Am. Chem. Soc.* **84,** 3005–3008.

Schulz, H., and du Vigneaud, V. (1966a). *J. Med. Chem.* **9,** 647–650.

Schulz, H., and du Vigneaud, V. (1966b). *J. Am. Chem. Soc.* **88,** 5015–5018.

Schulz, H., and du Vigneaud, V. (1967). *J. Med. Chem.* **10,** 1037–1039.

Schwyzer, R. (1977). *Ann. NY Acad. Sci.* **297,** 3–26.

Skala, G., Smith, C. W., Taylor, C. J., and Ludens, J. H. (1984). *Science* **226,** 443–445.

Smith, C. W., and Ferger, M. F. (1976). *J. Med. Chem.* **19,** 250–254.

Smith, C. W., and Walter, R. (1978). *Science* **199,** 297–299.

Smith, C. W., Hruby, V. J., and Ferger, M. F. (1974). *J. Med. Chem.* **17,** 873–876.

Smith, C. W., Ferger, M. F., and Chan, W. Y. (1975). *J. Med. Chem.* **18,** 822–825.

Smith, C. W., Botos, C. R., Skala, G., and Walter, R. (1977). *J. Med. Chem.* **20,** 1457–1460.

Smith, C. W., Skala, G., and Walter, R. (1978a). *J. Med. Chem.* **21,** 115–117.

Smith, C. W., Stahl, G. L., Walter, R. (1978b). Unpublished results.

Smith, C. W., Walter, R., Moore, S., Makofske, R. C., Meienhofer, J. (1978c). *J. Med. Chem.* **21,** 117–120.

Smith, C. W., Skala, G., and Walter, R. (1980a). *Int. J. Peptide Protein Res.* **16,** 365–371.

Smith, C. W., Walter, R., Stavropoulos, G., and Theodoropoulos, D. (1980b). *J. Med. Chem.* **23,** 217–219.

Smyth, D. G. (1967). *J. Biol. Chem.* **242,** 1592–1598.

Smyth, D. G. (1970). *Biochim. Biophys. Acta* **200,** 395–403.

Snell, C. R., and Smyth, D. G. (1977). *Biochem. J.* **165,** 43–47.

Stadel, J. M., Goodman, D. B. P., Galardy, R. E., and Rasmussen, H. (1978). *Biochemistry* **17,** 1403–1408.

Strub, K. M., Cort, J. H., Daum, A., Gillessen, D., Trzeciak, A., Studer, R. O., and Rudinger, J. (1973). *Experientia* **29,** 76.

Takashima, H., and du Vigneaud, V. (1970). *J. Am. Chem. Soc.* **92,** 2501–2504.

Takashima, H., Fraefel, W., and du Vigneaud, V. (1969). *J. Am. Chem. Soc.* **91,** 6182–6185.

Takashima, H., Hruby, V. J., and du Vigneaud, V. (1970). *J. Am. Chem. Soc.* **92,** 677–680.

Theodoropoulos, D., Liakopoulou, M., and Smith, C. W. (1976). *In* "Peptides 1976" (A. Loffet, ed.), pp. 505–509. University of Brussels, Belgium.

Ting, Y. F., Smith, C. W., Stahl, G. L., Walter, R., Cordopatis, P., and Theodoropoulos, D. (1980). *J. Med. Chem.* **23,** 693–695.

Turan, A., Manning, M., Haldar, J., and Sawyer, W. H. (1977). *J. Med. Chem.* **20,** 1169–1172.

van Nispen, J. W., Hannink, J. A. J., Schoffelmeer, M. S., Janssen, W. P. A., Polderdijk, J. P., and Greven, H. M. (1984). *Recl. J. R. Neth. Chem. Soc.* **103,** 68–74.

Vavrek, R. J., Ferger, M. F., Allen, G. A., Rich, D. H., Blomquist, A. T., and du Vigneaud, V. (1972). *J. Med. Chem.* **15,** 123–126.

Vida, J. A., and Gordon, M., eds. (1984). "Conformationally Directed Drug Design. Peptides and Nucleic Acids as Templates or Targets," pp. 1–260. ACS Symposium Series 251, American Chemical Society, Washington, D.C.

Walter, R. (1977). *Fed. Proc.* **36,** 1872–1878.

Walter, R., and Chan, W. Y. (1967). *J. Am. Chem. Soc.* **89,** 3892–3898.
Walter, R., and Schwartz, I. L. (1966). *J. Biol. Chem.* **241,** 5500–5503.
Walter, R., and Schwartz, I. L. (1968). *Life Sci.* **7,** 545–548.
Walter, R., and du Vigneaud, V. (1966). *Biochemistry* **5,** 3720–3727.
Walter, R., Schwartz, I. L., and Rudinger, J. (1967). *Am. J. Med.* **42,** 653–677.
Walter, R., Schwartz, I. L., Darnell, J. H., and Urry, D. W. (1971). *Proc. Natl. Acad. Sci. U.S.A.* **68,** 1355–1359.
Walter, R., Kirchberger, M. L., and Hruby, V. J. (1972). *Experientia* **28,** 959–960.
Walter, R., Schwartz, I. L., Hechter, O., Dousa, T., and Hoffman, P. L. (1973). *Endocrinology* **91,** 39–48.
Walter, R., Yamanaka, T., and Sakakibara, S. (1974). *Proc. Natl. Acad. Sci. USA* **71,** 1901–1905.
Walter, R., Smith, C. W., and Roy, J. (1976a). *Proc. Natl. Acad. Sci. USA* **73,** 3054–3058.
Walter, R., Smith, C. W., Roy, J., and Formento, A. (1976b). *J. Med. Chem.* **19,** 822–825.
Walter, R., Smith, C. W., Mehta, P. K., Boonjarern, S., Arruda, J. A. L., and Kurtzman, N. A. (1977). *In* "Disturbances in Body Fluid Osmolality" (T. E. Andreoli, J. Gratham, and F. C. Rector, eds.), pp. 1–36. American Physiological Society, Bethesda, Md.
Walter, R., Skala, G., and Smith, C. W. (1978). *J. Am. Chem. Soc.* **100,** 972–973.
Walti, M., and Hope, D. B. (1974). *J. Chem. Soc. Perkins Trans.* 202–205.
Walti, M., and Hope, D. B. (1975). *J. Chem. Soc. Perkins Trans.* 1691–1694.
Wille, M. A., du Vigneaud, V., and Chan, W. Y. (1972). *J. Med. Chem.* **15,** 11–12.
Wyssbrod, H. R., Ballardin, A., Gibbons, W. A., Roy, J., Schwartz, I. L., and Walter, R. (1975). *In* "Peptides: Chemistry, Structure and Biology" (R. Walter, and J. Meienhofer, eds.), pp. 815–822. Ann Arbor Science, Ann Arbor, Mich.
Yamanaka, T., and Sakakibara, S. (1974). *Bull. Chem. Soc. Jpn.* **47,** 1228–1232.
Yamanaka, T., Hase, S., Sakakibara, S., Schwartz, I. L., Dubois, B. M., and Walter, R. (1970). *Mol. Pharmacol.* **6,** 474–480.
Yamashiro, D., and du Vigneaud, V. (1968). *J. Am. Chem. Soc.* **90,** 487–490.
Yamashiro, D., Aanning, H. L., and du Vigneaud, V. (1965). *Proc. Natl. Acad. Sci. USA* **54,** 166–171.
Yamashiro, D., Gillessen, D., and du Vigneaud, V. (1966). *Biochemistry* **5,** 3711–3720.
Yamashiro, D., Havran, R. T., Aanning, H. L., and du Vigneaud, V. (1967). *Proc. Natl. Acad. Sci. USA* **57,** 1058–1061.
Zaoral, M. (1965). *Collect. Czech. Chem. Commun.* **30,** 1853–1868.
Zaoral, M., and Blaha, I. (1977). *Collect. Czech. Chem. Commun.* **42,** 3654–3657.
Zaoral, M., and Brtnik, F. (1975). *Collect. Czech. Chem. Commun.* **40,** 905–912.
Zaoral, M., and Flegel, M. (1972a). *Collect. Czech. Chem. Commun.* **37,** 1539–1545.
Zaoral, M., and Flegel, M. (1972b). *Collect. Czech. Chem. Commun.* **37,** 2639–2643.
Zaoral, M., and Krchniak, V. (1977). *Collect. Czech. Chem. Commun.* **42,** 3500–3509.
Zaoral, M., and Sorm, F. (1965). *Collect. Czech. Chem. Commun.* **30,** 2813–2816.
Zaoral, M., and Sorm, F. (1966a). *Collect. Czech. Chem. Commun.* **31,** 90–97.
Zaoral, M., and Sorm, F. (1966b). *Collect. Czech. Chem. Commun.* **31,** 310–314.
Zaoral, M., Kasafirek, E., Rudinger, J., and Sorm, F. (1965). *Collect. Czech. Chem. Commun.* **30,** 1869–1873.
Zaoral, M., Kolc, J., and Sorm, F. (1967a). *Collect. Czech. Chem. Commun.* **32,** 1242–1249.
Zaoral, M., Kolc, J., and Sorm, F. (1967b). *Collect. Czech. Chem. Commun.* **32,** 1250–1257.
Zaoral, M., Kolc, J., Korenczki, F., Cerneckij, V. P., and Sorm, F. (1967c). *Collect. Czech. Chem. Commun.* **32,** 843–853.
Zaoral, M., Kolc, J., and Sorm, F. (1970). *Collect. Czech. Chem. Commun.* **35,** 1716–1726.
Zaoral, M., Brtnik, F., Barth, T., and Machova, A. (1976a). *Endocrinol. Exp.* **10,** 183–191.

Zaoral, M., Brtnik, F., Barth, T., and Machova, A. (1976b). *Collect. Czech. Chem. Commun.* **41,** 2088–2095.

Zaoral, M., Flegel, M., Barth, T., and Machova, A. (1978). *Collect. Czech. Chem. Commun.* **43,** 511–522.

Zaoral, M., Brtnik, F., Flegel, M., Barth, T., and Machova, A. (1979a) *Collect. Czech. Chem. Commun.* **44,** 1179–1186.

Zaoral, M., Krchnak, V., Brtnik, F., Machova, A., and Skopkova, J. (1979b). *Collect. Czech. Chem. Commun.* **44,** 2447–2450.

Zaoral, M., Krchnak, V., Skopkova, J., and Machova, A. (1979c). *Collect. Czech. Chem. Commun.* **44,** 3133–3136.

Zhuze, A. L., Jost, K., Kasafirek, E., and Rudinger, J. (1964). *Collect. Czech. Chem. Commun.* **29,** 2648–2662.

Chapter 5

The Conformation of Neurohypophyseal Hormones

JUDITH C. HEMPEL
Smith Kline and French Laboratories
Philadelphia, Pennsylvania 19101

I. INTRODUCTION

Oxytocin was the first peptide hormone to be isolated, to be sequenced, and to have its identity confirmed through synthesis (Du Vigneaud, 1966). Oxytocin induces milk ejection and uterine contraction in mammals. Vasopressin is the antidiuretic hormone (V2 receptor) and has potent pressor activity (V1 receptor). The antidiuretic hormone is arginine vasopressin (AVP) in most mammals, but is lysine vasopressin (LVP) in pig and a few related species.

The hormones are nonapeptides having a 20-membered "hexapeptide" ring (created by a disulfide bond) and a tripeptide tail. The primary structure for oxytocin is

$$\begin{array}{l}\text{Cys}^1\text{-Tyr}^2\text{-Ile}^3\text{-Gln}^4\text{-Asn}^5\text{-Cys}^6\text{-Pro}^7\text{-Leu}^8\text{-Gly}^9\text{-NH2}\\ \text{S}\text{———————————}\text{S}\end{array}$$

and for arginine vasopressin (AVP) it is

$$\begin{array}{l}\text{Cys}^1\text{-Tyr}^2\text{-Phe}^3\text{-Gln}^4\text{-Asn}^5\text{-Cys}^6\text{-Pro}^7\text{-Arg}^8\text{-Gly}^9\text{-NH2}\\ \text{S}\text{————————————}\text{S}\end{array}$$

LVP is identical to AVP, except that lysine is substituted for arginine in the 8 position. Oxytocin's structure differs from the vasopressins only at residues 3 and 8.

A major hypothesis in the peptide literature is that the biological specificity and activity of these peptide hormones is directly related to their molecular conformation. A significant effort has gone into the study of conformational

THE PEPTIDES, VOLUME 8

preferences for oxytocin, vasopressin, and various peptide analogs. This literature has been reviewed by a number of authors (Glickson, 1975; Hruby, 1981, 1982, 1984; Hruby and Mosberg, 1981, 1982; Hruby *et al.*, 1983a; Smith, 1981; Smith *et al.*, 1977; Urry *et al.*, 1981; Walter, 1972, 1977; Walter *et al.*, 1971, 1977a). These summaries provide a variety of perspectives of the literature.

In this chapter we present hypotheses in the literature related to the conformation of the hormones at their receptors; we review physical studies that contribute to our understanding of the conformational preferences of the hormones; and we examine very briefly the influence of these concepts on the design of agonists and antagonists at the neurohypophyseal hormone receptors. The neurohypophyseal hormones are, as we shall see, flexible molecules that can assume a variety of conformations.

II. THEORIES RELATING CONFORMATION TO ACTIVITY

Fisher's "lock-and-key" model for hormone–receptor interactions assumes that binding of the hormone to the receptor is based on "molecular recognition." As illustrated in Fig. 1 (Kessler, 1982), binding of the hormone is assumed to lead to conformational changes in the receptor that cause "transduction" of biological signal. Agonists are compounds that bind to the receptor and transduce signal, while substrates that bind but do not induce the signal necessary for biological activity are competitive inhibitors or antagonists.

Oxytocin and vasopressin are full agonists at each other's receptors but show distinct specificity (high potency) for their own receptors (Roy *et al.*, 1975). The structural characteristics of the side-chains at residues 3 and 8 (which differ in the two hormones) therefore affect binding (i.e., potency) more than transduction of message (i.e., intrinsic activity). Structural features of oxytocin related to its binding and activity at the uterine receptor have been extensively probed through the synthesis and testing of peptide analogs, as have those of vasopressin at the

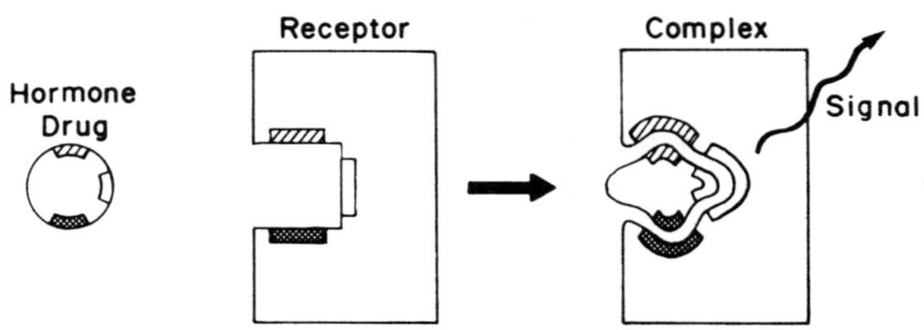

Figure 1. Model of hormone– or drug–receptor interaction [From Kessler (1982), with permission.]

antidiuretic receptor. Structure–activity relationships for these hormones are discussed in detail in Chapter 4 (this volume).

A cooperative model for the active conformation of oxytocin at the uterine receptor was proposed by Walter and co-workers in 1971 and for the active conformation of lysine vasopressin at the antidiuretic receptor in 1977 (Walter *et al.*, 1977a). The cooperative models incorporate two levels of inductive reasoning: first, the derivation of a solution conformation model for the hormone in dimethyl sulfoxide (DMSO) using information from proton nuclear magnetic resonance (nmr) studies; and second, the extrapolation of the solution model to a model for the hormone at its receptor. The cooperative models for hormone activity include the observation (Walter, 1972) that hydrophilic side-chains are implicated in the expression of activity characteristic for the individual hormone.

Features of the molecular topology, or three-dimensional shape, of the cooperative models proposed for oxytocin and vasopressin are summarized here. These models propose active and binding elements and the three-dimensional topology of the hormones when bound to the receptor. These hypotheses have stimulated a significant number of studies to address questions related to hormone conformation in solution and at the receptor, but, most importantly, they have stimulated synthetic studies in which molecules with a desired activity have been designed (i.e., deduced) from the active conformation hypotheses.

A. Oxytocin at the Uterine Receptor

The biologically active conformation model proposed for oxytocin at the uterotonic receptor is illustrated in Fig. 2. This molecular conformation is charac-

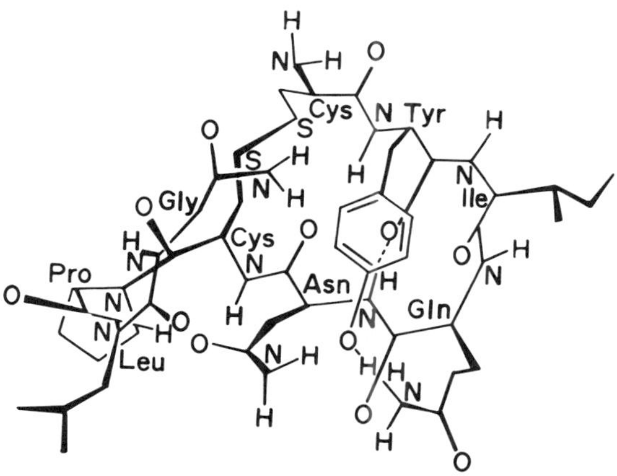

Figure 2. Conformation proposed for oxytocin while bound to the uterine receptor. The conformation is viewed from the top. The top face is characterized by a hydrophilic cluster containing the Tyr[2] hydroxyl, Gln[4], Asn[5], and glycinamide[9] carboxamide groups. [From Smith (1981), with permission.]

terized by two peptide-backbone beta-turn regions, one at the Ile^3 to Gln^4 region of the 20-membered hexapeptide ring and the other in the tripeptide tail. The side-chains and the tail are oriented up and over the hexapeptide ring. The bottom face of the model is nearly featureless and extends conceptually through the hydrophobic "edge" created by the side-chains of Ile^3, Gln^4, Pro^7, and Leu^8 as they bend over the hexapeptide ring. These four residues fall at the corner positions of beta-turn regions and are proposed to be binding elements. The opposite face of the molecule is characterized by the hydrophilic elements of the hormone. The hydroxyl group of the Tyr^2 side-chain, along with the Asn^5 carboxamide, appear to be important active elements (Walter *et al.*, 1971; Smith, 1981).

B. Vasopressin at the Antidiuretic Receptor

The biologically active conformation model proposed for vasopressin at the antidiuretic receptor is similar to that proposed for oxytocin (Walter *et al.*, 1977a). However, as illustrated in Fig. 3, the side-chains of Tyr^2 and Phe^3 point out and away from the hexapeptide ring. The other side-chains of the hexapeptide ring extend up and over the ring. The hydrophobic face of this model extends through a hydrophobic "edge" created by Gln^4, Asn^5, Cys^6, and Pro^7 and includes the hydrocarbon portion of Lys^8. The hydrophilic face of the molecule is characterized by a cluster of carboxamide groups of Gln^4, Asn^5, and the glycinamide terminus, and the basic moiety of residue 8 (Walter *et al.*, 1977a; Smith, 1981).

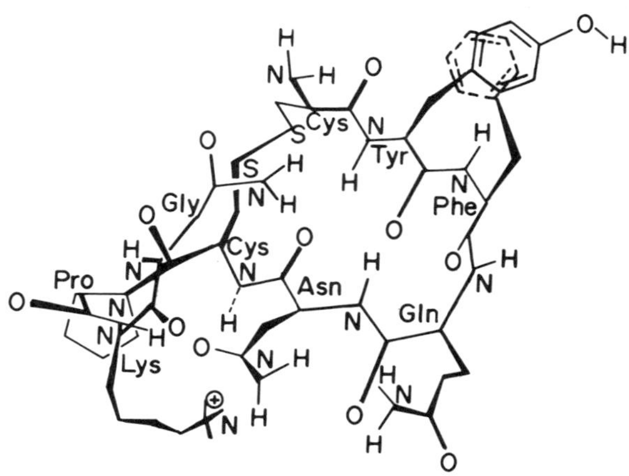

Figure 3. Conformation proposed for vasopressin while bound to the antidiuretic receptor. The conformation is viewed from the top. The top face is characterized by a hydrophilic cluster containing the Gln^4, Asn^5, and glycinamide^9 carbonamide groups, as well as the basic moiety of residue 8. [From Smith (1981), with permission.]

III. CONFORMATIONAL ASSIGNMENTS

Our knowledge of the three-dimensional character of the neurohypophyseal hormones comes from a variety of physical techniques—X-ray crystallography, nuclear magnetic resonance, circular dichroism, infrared and Raman spectroscopy—and, to a lesser degree, through energy-based calculations of molecular conformational preference. Experimental studies of the neurohypophyseal hormones and related analogs to determine conformation have been confined for the most part to solution studies. The interpretation of such studies must always take into consideration the possibility of multiple peptide conformations in solution and the influence of the solvent on peptide conformation (Glickson, 1975; Urry *et al.*, 1981; Kessler, 1982).

All physical studies reported in the literature for oxytocin and vasopressin are not reviewed in this chapter. Rather, there is an arbitrary focus on studies that provide conformational interpretations. In practice, these studies deal mainly with peptide backbone conformation. Figures 4–7, which illustrate conformational features proposed in the literature for the neurohypophyseal hormones, were generated for this chapter using the Smith Kline & French (SK&F) Molecular Modelling System, a modified version of SYBYL (registered trademark of Tripos Associates in Clayton, Mo.).

A. X-Ray Crystallography

The experimental technique that provides the most structural information is X-ray crystallography, and the crystal structures of a significant number of peptides have been determined. However, because of the intrinsic flexibility of most small peptides and the influence of crystal-packing forces on molecular conformation, the relationship of crystal to solution conformation and, more importantly, to "active" conformation hypotheses remains unresolved. Nevertheless, a crystal structure provides a three-dimensional picture at atomic resolution of the peptide in a conformation that is both thermodynamically and kinetically accessible to that peptide. Oxytocin, vasopressin, and closely related analogs have proven difficult to crystallize. However, crystal structure determinations have recently been reported for deamino-oxytocin (Blundell and Wood, 1982; Wood *et al.*, 1986) and for pressinoic acid, which is the cyclic moiety of vasopressin (Langs *et al.*, 1986). These crystal structure determinations represent an exciting new development in the elucidation of conformational preference for this class of molecules.

There are at least two well-defined conformers in the crystal structure of deamino-oxytocin (Wood *et al.*, 1986). The conformers are characterized by two

Table I. Dihedral Angles of Beta-Turn Types[a]

Type	Phi (i + 1)	Psi (i + 1)	Phi (i + 2)	Psi (i + 2)
Type I	−60	−30	−90	0
Type I′	60	30	90	0
Type II	−60	120	80	0
Type II′	60	−120	−80	0
Type III	−60	−30	−60	−30
Type III′	60	30	60	30

[a]Data from Smith and Pease (1980).

beta turns in the peptide backbone, a type II beta turn at the Ile^3 to Gln^4 region of the hexapeptide ring and a type III beta turn at the Pro^7 to Leu^8 region of the tail. Beta turns are characterized by a hydrogen bond between the carbonyl oxygen of residue i (i.e., the first residue of the turn) and the amide hydrogen of residue $i + 3$ (i.e., the residue located three residues toward the carboxy terminus of the peptide). The dihedral angles of standard beta-turn types are given in Table I (Smith and Pease, 1980). The center residues of the beta-turn sequence identify the turn region of the peptide backbone.

The hexapeptide rings of the conformers of deamino-oxytocin in this crystal structure are characterized by two transannular hydrogen bonds like the idealized beta-cross structure illustrated in model A of Fig. 4. However, the conformation of the ring in the crystal is not planar (like model A of Fig. 4) but is bowed. The curved shape of the molecule is completed by the tail, oriented away from the ring. All peptide bonds are trans. There are at least two possible conformations for the disulfide bridge in the crystal structure of deamino-oxytocin that differ in the chirality of the disulfide torsion (Wood *et al.*, 1986).

By contrast, the crystal structure of pressinoic acid has only one conformer (Langs *et al.*, 1986). The backbone conformation of the hexapeptide ring has two beta bends: a distorted type II′ beta bend at the Phe^3 to Gln^4 region of the hexapeptide ring and a type I beta turn at the Gln^4 to Asn^5 region of the ring. Although the type II′ beta bend is not "closed" by a hydrogen bond, the backbone conformation can be characterized as having two interlocking beta turns (like model C in Fig. 4). The observation of the type II′ beta bend is of particular interest. As implied by the beta-turn nomenclature (Table I), Phe^3 is characterized by a positive phi value. However, positive phi values are normally associated with D rather than L amino acids. The pressinoic acid crystal as a whole has induced a conformation in an individual residue (Phe^3), which is not a favored conformation for that residue as an isolated residue. Low energy conformations of this type (with positive phi values for Phe^3) were previously reported in molecular dynamics modelling studies of LVP (Hager *et al.*, 1985).

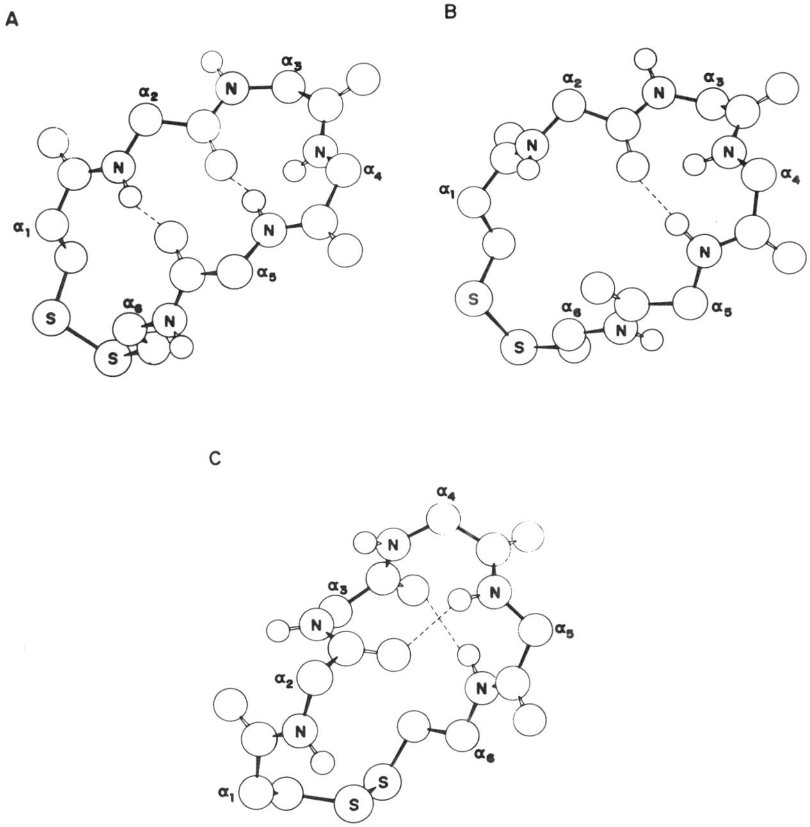

Figure 4. Conformations illustrating transannular hydrogen-bonding patterns proposed for the hexapeptide ring of (A) deamino-oxytocin and deaminovasopressin in DMSO; (B) oxytocin and vasopressin in DMSO; and (C) ocytocin and vasopressin in water.

B. Nuclear Magnetic Resonance Spectroscopy

Nuclear magnetic resonance (nmr) spectroscopy provides precisely determined experimental parameters that contain conformational information. Nevertheless, the technique has its difficulties. The assignment of nmr spectral features to structural features of the molecule is often a formidable task, and once the nmr parameters are assigned they must then be interpreted to yield conformational information. Correlations (Karplus curves) for vicinal coupling constants between protons attached to two bonded atoms yield multiple values of dihedral torsion angles for a given nmr coupling constant. This multiplicity introduces ambiguity into conformational assignments.

Problems associated with conformational averaging in the interpretation of peptide spectra are well known (Glickson, 1975; Urry *et al.*, 1981; Kessler, 1982). Conformational averaging introduces uncertainty into peptide-backbone conformational assignments based on analysis of J coupling constants using a Karplus-type correlation such as the Bystrov relation (Bystrov, 1976) for the protons attached to the alpha carbon of an amino acid residue and the preceeding amide nitrogen. The nmr studies summarized in the sections that follow suggest that oxytocin and vasopressin can undergo conformational averaging in the peptide backbone of the hexapeptide ring in water (but presumably not in DMSO). A cyclic peptide is of necessity more restricted than a linear peptide, and nmr relaxation studies indicate that the tripeptide tail is much more flexible than the hexapeptide ring in both DMSO and water, as expected. Physical studies of antagonists of oxytocin that incorporate Pen (rather than Cys) at position 1 strongly suggest that the hexapeptide ring of these molecules is more conformationally constrained than that of oxytocin.

The nmr studies in general do not yield enough parameters to fully define all the degrees of freedom of a peptide. Hence, peptide models play an important role in the interpretation of nmr results. Solution conformation hypotheses are derived in practice by building molecular models consistent with conformational constraints derived from experiment, using hand held molecular models or, more recently, computer-based molecular modelling. A solution conformation model was proposed by Urry and Walter in 1971 for oxytocin and deamino-oxytocin in DMSO using proton nmr studies. This model generated a variety of questions, which were subsequently addressed in physical studies, for oxytocin, deamino-ocytocin, vasopressin, and related analogs. Studies in DMSO, for the most part, are in agreement with the model. However, the model does not hold in water. A full solution model has not been defined for water, although features of a solution conformation model in water have been proposed. Due to conformational averaging, these features may not simultaneously characterize a single conformation.

It seems clear from solution studies reported below that the neurohypophyseal hormones and related agonist analogs are characterized by a beta turn at the residue 3–4 region of the hexapeptide ring in DMSO, as proposed in the Urry and Walter solution model for oxytocin. (This beta turn was assigned for oxytocin as a type II beta turn on the basis of nmr studies of constrained oxytocin analogs.) A second beta turn at the residue 7–8 region of the tripeptide tail in the solution model for oxytocin and deamino-oxytocin in DMSO probably does not arise for vasopressin. It is interesting to note that the beta-turn regions of conformers of deamino-oxytocin in the recent crystal structure correspond to the beta-turn regions proposed in the model for oxytocin and deamino-oxytocin in DMSO. There is a type II beta turn at the residue 3–4 region of these conformers (Wood *et al.*, 1986).

1. Oxytocin

a. Peptide Backbone Conformation. Proton nmr studies as a function of temperature suggest that oxytocin and deamino-oxytocin are characterized by a hydrogen bond between the Tyr^2 CO and the Asn^5 NH in DMSO (Urry *et al.*, 1970). The Asn NH resonance is observed to shift very little with temperature but is shifted to higher fields, which is indicative of a beta turn or a beta fold involving that amide. A beta-turn assignment at the Ile^3 to Gln^4 region of these molecules is further supported by a small J for residue $i + 1$ (Ile^3) and a large J for residue $i + 2$ (Gln^4) of the proposed beta turn. Idealized type I and type II beta turns (Smith and Pease, 1980) are illustrated as stereo pairs in Figs. 5 and 6, along with J coupling constants for the two central residues of the turn region estimated using the Bystrov relation (Bystrov, 1976).

Deamino-oxytocin is, in addition, assigned a second transannular hydrogen bond between the Tyr NH and the Asn carbonyl using similar arguments (Urry *et al.*, 1970). Hydrogen bonding patterns assigned to the 20-membered hexapeptide rings of deamino-oxytocin and oxytocin in DMSO are illustrated schematically as models A and B in Fig. 4.

Urry and Walter (1971) proposed a model for the solution conformation for oxytocin in DMSO that is consistent both with hydrogen bonding assignments and with conformational information implied by J coupling constants for the NH-CH dihedral angles. The model incorporates two beta turns: (1) Tyr-Ile-Gln-Asn in the peptide ring, and (2) Cys-Pro-Leu-Gly in the peptide tail.

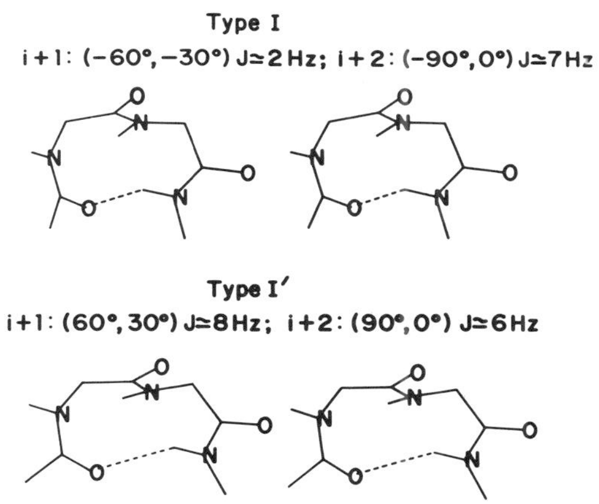

Figure 5. Type I and type I′ beta-turn models. The (phi, psi) values for the second ($i+1$) and third ($i+2$) residues of the beta turn (Smith and Pease, 1980) are indicated along with a J coupling constant estimated (Bystrov, 1976) for an L residue at that position. Stereo figures illustrate the conformation of the 10-membered hydrogen-bonded ring characteristic of this class of turn.

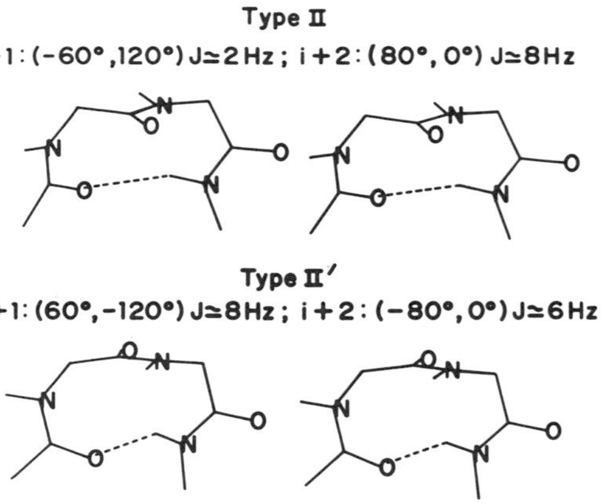

Figure 6. Type II and type II′ beta-turn models. The (phi, psi) values for the second ($i+1$) and third ($i+2$) residues of the beta turn (Smith and Pease, 1981) are indicated along with a J coupling constant estimated (Bystrov, 1976) for an L residue at that position. Stereo figures illustrate the conformation of the 10-membered hydrogen-bonded ring characteristic of this class of turn.

A hydrogen-bonding interaction was proposed between the amino group at the Gly terminus and the carbonyl of the Asn side-chain in this model, along with side-chain orientations consistent with backbone conformation. This model for the solution conformation of oxytocin can be visualized using Fig. 2, which illustrates the active conformation model proposed for oxytocin at the uterine receptor. The two models differ only in that the Tyr^2 side-chain is hypothesized to point out and away from the hexapeptide ring in DMSO and to fold over the hexapeptide ring in the active conformation at the uterine receptor. The placement of the Tyr side-chain has received considerable discussion in the literature. This solution model, derived inductively from proton nmr studies of oxytocin in DMSO, led directly to the proposal of the cooperative model for the activity of oxytocin at the uterine receptor (Walter *et al.*, 1971).

In water, diminished rates of peptide proton exchange for Cys^6 and Asn^5 NH groups of oxytocin and [Glu^4] oxytocin suggest that these amides are hydrogen bonded (Sarathy *et al.*, 1981a; Walter *et al.*, 1981). The small coupling constant observed for residue 4 of oxytocin and [Glu^4] oxytocin are consistent with a conformation for oxytocin in water that has a beta turn with a hydrogen bond between the Ile^3 carbonyl and the Cys^6 NH. It has been proposed on the basis of these observations that a conformation with a hydrogen bond between the Ile^3 CO and the Cys^6 NH is in equilibrium with a conformation with a beta turn (like that postulated for oxytocin in DMSO) with a hydrogen bond between the Tyr^2 carbonyl and the Asn^5 NH (Sarathy *et al.*, 1981a). These hydrogen-bonding patterns are illustrated schematically in model C of Fig. 4. This assignment is in agreement with the observation of a high barrier to hydrogen exchange for the

Asn^5 and Cys^6 NH groups of oxytocin in water (Krauss and Cowburn, 1981). The high barrier to hydrogen exchange was interpreted by Krauss and Cowburn (1981) as due to hydrogen bonding for Cys^6 NH but to steric hindrance rather than hydrogen bonding in the case of Asn^5.

b. Beta-Turn Assignment. As Walter and co-workers (1977b) point out, L residues at positions $i + 1$ and $i + 2$ of the proposed beta turn of oxytocin, which prefer negative phi values, are consistent with either a type I or a type II beta turn. Idealized type I and type II beta turns are illustrated as stereo pairs in Figs. 5 and 6. Substitution of a D residue at position 3 with the retention of the L residue at position 4 allows only a type II′ beta turn, since D residues prefer positive phi values. Substitution of a D residue at position 4 with the retention of the L residue at position 3 "allows" only a type II beta turn. The similarity of the proton nmr spectrum of [$DAla^4$] oxytocin in DMSO to that of oxytocin and differences in the spectrum of [$DAla^3$] oxytocin relative to that of oxytocin (Walter *et al.*, 1977b) suggest that the backbone conformation of [$DAla^3$]oxytocin may be quite different from that of oxytocin while that of [$DAla^4$]oxytocin may be quite similar. These authors tentatively suggest that oxytocin is characterized by a type II beta turn.

Proton nmr studies of [Pro^3,Gly^4]oxytocin in DMSO are also consistent with a beta turn at the residue 3–4 level for this molecule. Conformational restrictions on the allowed phi values of the proline at position 3 are consistent only with a type II beta turn. [Phi for proline is constrained to be between $-60°$ and $-80°$.] The similarity of the nmr spectrum of this molecule to that of oxytocin suggests a type II beta turn for oxytocin (Wyssbrod *et al.*, 1975; Ballardin *et al.*, 1978).

Carbon-13 nmr studies of [Pro^3]oxytocin, [Pro^3,Gly4]oxytocin, and [Leu^2,Pro^3,Gly^4]oxytocin in water demonstrate that the introduction of a Pro residue at position 3 of oxytocin causes conformational heterogeneity in the 20-membered ring due to cis–trans interconversion in the peptide bond linking Tyr^2 to Pro^3 (Deslauriers *et al.*, 1978). In addition, [Pro^3,Gly^4]oxytocin exhibits an increased rate of motion for the alpha carbon of Gly^4 relative to the other alpha carbons of the cyclic moiety, rather than the expected decrease of mobility in this region due to a beta turn favored by the Pro-Gly substitution (Deslauriers *et al.*, 1977).

These observations for [Pro^3,Gly^4]oxytocin in water suggest conformational flexibility in the beta-turn region. This observation may be solvent-dependent.

c. Conformational Averaging. Deuterium nmr studies in water of oxytocin (selectively deuterated in Cys^6, in Cys^1, and in Tyr^2) suggest that reorientation of oxytocin about an intramolecular axis is more frequent than is the reorientation of the molecule as a whole. Hence the conformation of oxytocin in water can only be a time-averaged conformation (Glasel *et al.*, 1973).

There is no evidence for any cis–trans isomerization of peptide bonds in oxytocin based on carbon-13 nmr studies in water and in DMSO (Deslauriers *et al.*, 1972). Carbon-13 studies support the interpretation that the tail of oxytocin and of LVP is more mobile than the ring in both water and DMSO (Deslauriers *et al.*, 1973; Walter *et al.*, 1974a).

d. Side-Chain Orientation. Alpha-hydrogen to beta-hydrogen coupling constants for oxytocin have been determined from proton nmr studies in water and are interpreted to indicate significant populations of the staggered rotomers about the C-alpha to C-beta bond of the Tyr^2 residue and to suggest significant population of the staggered rotomers of Gln^4 and Asn^5 as well. However, it is proposed that the half cystyl-1 and half cystyl-6 residues are not averaging but are characterized by one of two sets of nearly eclipsed conformations for the chi^1 values of Cys^1 and Cys^6 ($-122°$ for Cys^1 and $+116°$for Cys^6 or $-118°$ for Cys^1 and $+4°$ for Cys^6) (Wyssbrod *et al.*, 1979). Further studies (Fischman *et al.*, 1980) support the assignment of chi^1 for Cys^1 as $-120°$ and chi^1 for Cys^6 as $+120°$. This disulfide bridge model (with right-handed disulfide) is illustrated as a stereo pair in Fig. 7.

Deslauriers and co-workers (1974a) have compared proline ring mobility in a series of proline-containing peptide peptides. The T^1 relaxation times in solution show that rapid interconversion of proline ring conformers occurs in peptides. In luteinizing hormone–releasing hormone (LHRH) the beta and gamma carbons of the proline ring are equally mobile, suggesting rapid interconversion of half-chair forms of the pyrrolidine ring puckered at the beta and gamma carbons. In oxytocin the mobility of the beta carbon of the proline ring is greatest. This observation is consistent with an *endo–exo* flip of the beta carbon. In LVP the

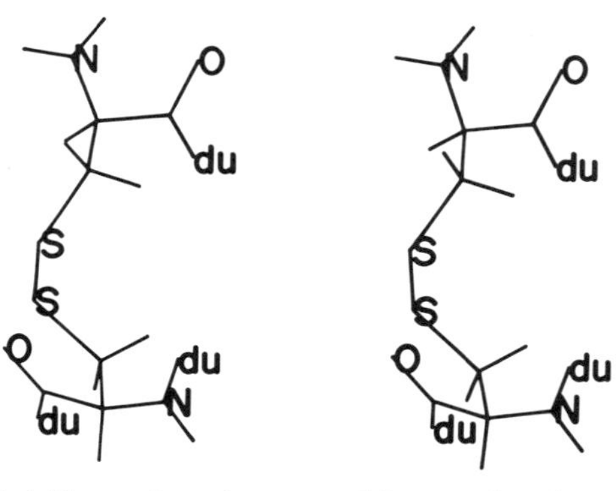

Figure 7. Disulfide bridge conformation proposed for oxytocin and vasopressin in solution: chi^1 for Cys^1 is $-120°$; chi^1 for Cys^6 is $+120°$; disulfide torsion is $+93°$. Dummy (du) atoms indicate points of attachement of the disulfide bridge to the hexapeptide ring and to the tripeptide tail.

mobility of the gamma carbon is greatest. This observation is consistent with an *endo–exo* flip of the gamma carbon.

e. Amino Terminus Effects. Enhanced hydrogen exchange rates were observed for the NH of Tyr[2] of oxytocin in water, presumably due to the charged nature of the amino terminus of Cys[1] (Krauss and Cowburn, 1981). The carbon-13 resonances of the amino acids of the ring of oxytocin in water shift as the pH is changed from 3.0 to 8.4, but no such shift is observed for the amino acids of the tail (Deslauriers *et al.*, 1974b). It is proposed that the shift in carbon-13 resonances is due to a conformational change in the ring introduced by protonation of the amino terminus of Cys[1], which is transmitted via perturbation of the dihedral angle of the disulfide (Deslauriers *et al.*, 1974b).

f. Conformational Influence of Tail. Studies to investigate the influence of the tail on the conformation of the peptide ring have also been reported. Proton nmr spectra of tocinamide (the ring of oxytocin) and of deaminotocinamide (the ring of deamino-oxytocin) suggest that the conformations of these molecules differ from each other and from the conformations of analogs with tails (Brewster *et al.*, 1972). It is suggested that the C-terminus carboxamide protons of tocinamide and of deaminotocinamide are hydrogen bonded to the Asn side-chain carbonyl. The diminished rate of peptide hydrogen exchange for the Cys[6] NH of tocinoic acid suggests that this NH is hydrogen-bonded in water (Sarathy *et al.*, 1981b).

Proton nmr studies of [DPro[7]]oxytocin and [Mpr[1],DPro[7]]oxytocin suggest that the tail has a different orientation with respect to the ring in these compounds than it does in oxytocin (Brewster *et al.*, 1973a,b). A preliminary study of oxytocin that monitors through-space interactions of protons suggests that the tail of oxytocin does not fold back over the hexapeptide ring in DMSO (Ford and Gibbons, 1979). This study represents the first application of nuclear Overhauser enhancement (NOE) techniques to the neurohypophyseal hormones, and the orientation of the tail of oxytocin is deduced from the observation of an NOE between delta hydrogens of Pro[7] and the alpha hydrogen of Cys[6].

2. Vasopressin

a. Backbone Conformation. Walter and co-workers (1972) compare the proton spectrum of LVP in DMSO to that of oxytocin and conclude that both molecules are characterized by a beta turn with a hydrogen bond between the Tyr carbonyl and the Asn NH. They point out that the J coupling constants of Ile[3] and Gln[4] of oxytocin are 4.4 Hz and 6.6 Hz, respectively, in agreement with a type II beta-turn hypothesis. The J coupling constants of Phe[3] and Gln[4] of LVP are 8 Hz and 6 Hz, respectively, and are not consistent with the same hypothesis. Nevertheless, on the basis of temperature-dependence studies, the NH of Asn^5 for both oxytocin and LVP is proposed to be involved in hydrogen bonding. It is postulated that a beta turn is characteristic of both molecules, and the deviation in the expected coupling constants for LVP may be due to an interaction between the aromatic rings of Tyr and Phe (Walter *et al.*, 1972). Conformational differences in the turn region are also possible.

Walter *et al.* (1974b) report that evidence for a beta turn in the tail of AVP and arginine vasotocin is no more than "suggestive." The active conformation model for vasopressin at the antidiuretic receptor (Fig. 3) incorporates solution conformation features of vasopressin in DMSO. DMSO solution and cooperative models for vasopressin differ only in the orientation proposed for the side-chains of Tyr^2 and Phe^3. These rings are stacked in the cooperative model but not in the DMSO solution model. The authors suggested that they would be stacked in water. Considerable effort has been directed toward demonstrating stacking of the aromatic rings of Tyr^2 and Phe^3 in solution. The weight of the evidence does not support stacking in either water or DMSO, although a proposal of stacking for deamino-LVP in water has not been refuted.

The carbon-13 spectra of oxytocin, LVP, and AVT in DMSO indicate that the peptides are characterized by all trans peptide bonds; that the Pro residue is unaffected by the structural substitutions in the ring or tail; and that substitutions in the ring represented by this set of compounds do not perturb the resonances of the tail (Deslauriers *et al.*, 1973). The tail is more mobile than the ring in both water and DMSO (Deslauriers *et al.*, 1973; Walter *et al.*, 1974a).

Glickson *et al.* (1972) conclude from proton nmr studies in DMSO that deamino-LVP and deamino-8-tosyllysine vasopressin—molecules in which the amino terminus is removed and in the case of the tosyllysine analog the cationic side-chain of the lysine residue is blocked—that the peptide ring is characterized by two transannular hydrogen bonds between the Tyr carbonyl and Asn NH and between the Tyr NH and the Asn carbonyl.

A diminished rate of peptide hydrogen exchange for the Cys^6 NH of LVP suggests that this NH is hydrogen-bonded in water (Sarathy *et al.*, 1981a). Nitrogen-15 nmr studies of oxytocin and AVP in water suggest that the backbone conformations of the two hormones are quite similar (Live, 1979a,b).

Transannular hydrogen-bonding patterns proposed for LVP in DMSO, deamino-LVP in DMSO, and LVP in water conform to those proposed for oxytocin in DMSO, deamino-oxytocin in DMSO, and oxytocin in water, as illustrated in Fig. 4.

b. Side-Chain Orientation. Proton nmr studies of AVP in water suggest that the chi^1 values of Cys^1 and Cys^6 of AVP are similar to those of oxytocin in water and that there is no specific stacking between Tyr^2 and Phe^3 rings for AVP in aqueous solution (Cowburn *et al.*, 1981). Proton nmr studies in water suggest that the side-chain conformations of oxytocin and AVP are largely unaffected by structural differences in these two molecules (Cowburn *et al.*, 1983; Wyssbrod *et al.*, 1981a). However, proton nmr studies of deamino LVP in water have been interpreted to suggest that the Tyr^2 and Phe^3 rings are stacked for deamino LVP in water (Von Dreele *et al.*, 1972).

Carbon-13 studies indicate that the proline ring undergoes conformational transitions (Deslauriers *et al.*, 1974a) in LVP and that the mobility of the gamma carbon is greatest. This observation is consistent with an *endo–exo* flip of the gamma carbon. The carbon-13 relaxation times observed for the proline ring of LVP in these studies have also been analyzed in terms of a single ring conformation (Shekar *et al.*, 1984).

c. Amino Terminus Effects. Walter *et al.* (1973) find that carbon-13 spectra of AVP, LVP, oxytocin, and AVT are sensitive to pH, which is not observed with the corresponding deamino analogs. This is interpreted to suggest a perturbation in the disulfide bridge due to protonation of the terminal amino of the Cys^1.

3. Analog Studies

a. Agonists. Wyssbrod and co-workers (1981a) conclude on the basis of proton nmr studies of [7(3,4-dehydroproline)]AVP and AVP in water that differences in the biological activities of these compounds are not due to differences in conformation. The proton nmr spectra of these two compounds in water are very similar. Similar comparative studies of AVP with D- and L-arginine at position 8 show that the proton nmr spectra of these compounds in water is also very similar, and this similarity is interpreted to suggest that there is extensive conformational averaging in the tail, since the change in chirality of a residue might otherwise be reasonably expected to introduce a major perturbation in conformation (Wyssbrod *et al.*, 1981b).

b. Antagonists. Proton nmr studies of [Pen^1]oxytocin and a series of oxytocin antagonists with penicillamine at position 1 are characterized by both very large (>8 Hz) and very small (<3 Hz) *J* coupling constants for the hexapeptide ring. The observation of large (or small) *J* coupling constants is not consistent with extensive conformational averaging, because of the functional form of the correlation function (Kessler, 1982). Near-zero temperature dependence for the Ile^3 and Asn^5 NH groups suggests that these NH groups are hydrogen bonded, and hydrogen bonds from the Asn NH to the Ile CO and from the Ile NH to the Pen CO, are proposed for this class of molecules (Hruby *et al.*, 1983a; Hruby and Mosberg, 1981). Carbon-13 nmr studies of [Pen^1,Cle^2]oxytocin and of [Pen^1]oxytocin in DMSO indicate that these molecular conformations are similar to each other and different from that of oxytocin. The study implies that the antagonists have similar dynamic properties. It is suggested that this class of antagonists is characterized by a less flexible solution conformation than are oxytocin agonist molecules (Hruby *et al.*, 1983b). An early hypothesis by this group was that the Tyr^2 side-chain was constrained to point away from the peptide ring in this class of antagonists and for this reason could not assume the "active" conformation postulated in the cooperative model for the activity of oxytocin at the uteronic receptor. However, this hypothesis would appear not to hold on the basis of further physical studies and structure activity relationship (SAR) studies (Hruby *et al.*, 1983b).

Hruby and Mosberg (1981, 1982) suggest that the intrinsic conformational flexibility of oxytocin (and by implication that of vasopressin and other peptide hormones) is an important component of agonist activity. They note that since antagonists lack the ability to transduce biological messages, properties of the agonist hormone that they lack may represent properties that are necessary for the transduction of biological messages (Hruby, 1981). They hypothesize that an agonist of a peptide hormone has necessary *dynamic* properties that lead the receptor to transduce its message.

4. Peptide–Protein Interactions

It would be extremely interesting to study the conformation of the hormone bound to the receptor of interest. The recognition of a peptide by a protein of necessity takes into account the drug surface features, that is to say, the topology of the molecule, not just the dihedral angles of the backbone. The electrostatic shape of the molecules (not just the steric shape) is critical (Walter, 1972; Robson, 1984).

There is no conformational information for oxytocin or vasopressin bound to their receptors. However, binding of oxytocin, lysine vasopressin, and a variety of peptide analogs to purified neurophysins from oxytocin-containing neurosecretory granules has been investigated structurally, primarily through nmr studies. These studies were reviewed in 1979 and provide a conceptual model for peptide–protein interactions like those of the hormone and its receptor (Breslow, 1979). The amino terminus is essentially required for binding of the hormones to neurophysin (this is not true for binding to the receptor), and residues 1–3 appear to interact directly with the protein upon binding. Carbon-13 nmr studies indicate that Tyr^2 of oxytocin undergoes hindered rotation when bound to neurophysin (Blumenstein *et al.*, 1980) and that Ile^3 is involved in interactions between oxytocin and neurophysin but Gly^9 is not (Cohen *et al.*, 1977).

C. Circular Dichroism Spectroscopy

Circular dichroism (CD) is a descriptor of the conformation of the molecule as a whole and provides a qualitative measure of peptide conformation within a series of structurally related analogs, i.e., similar versus different. The optical activity of peptides and proteins results from the electronic interactions of independent chromophores oriented by a chiral secondary structure as well as from the chirality of individual amino acids. There is currently no fully satisfactory interpretive procedure for chirooptical phenomena (Blaha, 1981).

The CD spectrum of oxytocin is a superposition of two or three peptide-backbone bands, three bands due to the tyosine aromatic side-chain, and two bands due to the disulfide group. The CD spectrum of vasopressin is similar (Tu *et al.*, 1979). The π to π^* peptide band at 200 nm is a sensitive indicator of changes in peptide-backbone conformation. A complex positive band at 225 nm contains the B_{1u} band of the aromatic side-chain of tyrosine and the peptide n to π^* band. The disulfide chromophore contributes a relatively strong positive band at around 250 nm and a weak negative band at 275 nm. The positive band at 250 nm can be assigned to conformers with a disulfide torsion angle of around 90° and the negative band to conformers with a significantly different disulfide torsion angle (differing by 30° or more) (Blaha, 1981).

1. Disulfide Chirality

Urry and co-workers (1968) assigned the predominant disulfide torsion angle of oxytocin to be right-handed based on comparisons of the CD spectrum with that of model compounds. CD studies of two model oxytocin "hexapeptides" in which residues 3, 4, and 5, and residues 4 and 5 have been substituted by glycine suggest that a right-handed disulfide torsion angle of approximately 90° is characteristic of the 20-membered ring of this class of peptides (Fric *et al.*, 1980).

CD studies of analogs of oxytocin in which one or both sulfur atoms are replaced by methylene groups have been compared and are interpreted to suggest that there is a conformation-stabilizing interaction between the aromatic side-chain of Tyr and the sulfur atom of Cys^6 (Blaha, 1981).

2. Amino Terminus Effects

Protonation of the amino terminus of oxytocin, [Ile^2]oxytocin, and [Gly^2]oxytocin diminishes the intensity of the CD band near 250 nm (Beychok and Breslow, 1968). Protonation of the amino terminus is presumed to result in an interaction of the amino terminus with the aromatic ring of Tyr^2. Similar pH effects are observed for analogs of oxytocin in which one or both of the sulfur atoms of the disulfide bridge are substituted by a methylene group (Fric *et al.*, 1974a,b). The 250-nm band is only barely discernible in CD studies of deamino-oxytocin analogs, and this observation is not pH-dependent (Beychok and Breslow, 1968; Fric *et al.*, 1975).

D. Infrared and Laser Raman Spectroscopy

Infrared (ir) and Raman spectroscopy probe the vibrational patterns of peptides. Normal mode analysis can distinguish beta turns in peptide chains from alpha-helical and beta-sheet structure. Specifically, the amide III vibrational band of a beta turn is predicted to lie at higher energies than the amide III bands of alpha-helix or beta-sheet features (Bandekar and Krimm, 1979). Disulfide bands yield estimates of the absolute magnitude of the disulfide torsion (Maxwell and Scheraga, 1977; Sugeta *et al.*, 1973).

1. Oxytocin and Vasopressin

The amide III band of oxytocin suggests that this molecule is characterized by a backbone conformation incorporating a beta turn (as a solid and in water and DMSO solution) (Tu *et al.*, 1978). Laser Raman studies of oxytocin, [Ile^8]oxytocin (mesotocin), [Arg^8]oxytocin (vasotocin), and AVP suggest that the peptide backbone conformations for these peptides are similar. All show a disulfide band at 512 cm^{-1}, implying a *gauche–gauche–gauche* conformer for the C—S—S—C bridge of the molecule with a satellite band characteristic of some conformer with disulfide torsion significantly different from 90° (probably 120°) (Tu *et al.*, 1979). This disulfide bridge conformational assignment is consistent with the nmr assignment for oxytocin and vasopressin (Fig. 7). Raman spectra for oxytocin and vasopressin in water (pH 4) show similar disulfide bands (Maxwell and Scheraga, 1977).

2. Oxytocin Agonists and Antagonists

CD and laser Raman studies of oxytocin agonists and antagonists indicate conformational differences in the two classes of molecules (Hruby and Mosberg, 1982; Hruby *et al.*, 1982). The CD spectra of oxytocin antagonists [Pen^1, Thr^4]oxytocin and [Pen^1,Phe^2,Thr^4]oxytocin are similar to each other and different from the CD spectra of oxytocin, [DCys^1]oxytocin, [DTyr^2]oxytocin, and [D Asn^5]oxytocin. Laser Raman studies show differences in the amide III band for the two classes of molecules and suggest that the disulfide region of the agonists is either *G–G–G* about the C—S—S—C torsions, according to the interpretive scheme of Sugeta and co-workers (1973), or is characterized by torsions of approximately 20–30° about the C—S bonds and a torsion of approximately 85° about the S—S bond according to the interpretative scheme of Van Wart and Scheraga (1976). By contrast, the disulfide region of the antagonists is interpreted to be either *G–G–T* or *T–G–T*, according to the interpretive scheme of Sugeta and co-workers. These studies represent the only studies to date for analogs in which Pen is substituted for Cys. The relevance of the interpretative scheme of Sugeta and co-workers to this class of molecules is not clearly established; however, the implications of these studies are intriguing, and further physical studies of antagonists of the neurohypophyseal hormones are clearly indicated.

IV. ENERGY-BASED MOLECULAR MODELLING

Force fields to calculate the energy of a peptide are similar in concept to those used in the interpretation of vibrational specta using normal mode analysis, except that they include nonbonding interactions, both electrostatic and hydrophobic. The calculation is an empirical procedure, and the validity of the calculation derives from the force-field parameters that relate the structural properties of model compounds to the compound of interest. Several different force fields have been used in modelling studies of the neurohypophyseal hormones. The reader is referred to the original papers for the specifics of each study (Gibson and Scheraga, 1967; DeCoen and Ralston, 1973; Honig *et al.*, 1973; Kotelchuck *et al.*, 1972a,b; Nikiforovich *et al.*, 1979; Hagler *et al.*, 1985).

Two types of energy-based modelling studies of the neurohypophyseal hormones have been undertaken: (1) modelling studies to examine specifics of the solution conformation model proposed by Urry and Walter (1971) for oxytocin on the basis of nmr studies in DMSO, and (2) studies undertaken in the hope of identifying either the global minimum for these molecules and/or obtaining an overview of the conformation space of these molecules. The global minimum of a molecule refers to its lowest minimum-energy conformation. If the global minimum is significantly lower in energy than all others, then it is the most favored conformation. It seems unlikely that oxytocin and vasopressin can be characterized by a single, highly favored conformation, on the basis of physical studies that suggest conformational averaging and changes in conformation with solvent and pH. As outlined below, modelling studies support this interpretation.

A. Solution Conformation Models

Kotelchuck and co-workers (1972a,b) modelled the conformation proposed for oxytocin in DMSO (Urry and Walter, 1971) and investigated the energetics of variants of the ''beta sheet'' class of ring conformations. This conformational class has a roughly planar 20-membered ring with the disulfide bridge ''below'' the plane of the ring and a beta turn at the Ile^3 to Gln^4 region of the ring. Multiple minimum-energy conformations within this class of ring conformations were identified, but none of these conformations had two transannular hydrogen bonds like the idealized beta-sheet structure (conformation A in Fig. 4). Some of the low-energy conformations found in this study had no transannular hydrogen bonds; others had only the hydrogen bond associated with the beta turn itself. No energetic preference was established for type II versus type I beta turns. None of the conformations modelled had a beta turn at the Pro^7 to Leu^8 level of the tripeptide tail like that proposed for the solution conformation of oxytocin in DMSO.

B. Conformational Searching

1. Stereoalphabet Search

DeCoen and Ralston (1973) searched for low-energy conformations of the 20-membered ring of oxytocin that have each of the amino acids in a minimum-energy conformation for the amino acid as an isolated residue. This search of the conformation space of the oxytocin ring spanned the conformations defined by all possible ''words'' derived from the letter codes of the stereoalphabet defined in column 1 of Table II. Only 224 ''words'' in this search are compatible with disulfide ring closure of the hexapeptide unit. (These 224 conformations are characterized by a distance of 3.9 ± 0.5 Å between the beta carbon of Cys^1 and the beta carbon of Cys^6.) Twelve low-energy conformations for oxytocin and for desamino-oxytocin were identified in this study (badda*b, bcbaac, bcbacb, bcbacc, bca*beb, bca*bec, bdaacc, bebacb, cea*bec, a*a*bacc, badaed, a*ea*bec). Reported structures do not include the disulfide bridge. The conformations are numbered 1–12 according to the scheme set by the authors. This numbering scheme corresponds to their placement in the list above.

The conformation bdaacc (number 7) corresponds approximately to a beta-cross structure with two hydrogen bonds between the NH and CO groups of tyrosine and asparagine residues. This hexapeptide ring is not planar. This hydrogen-bonding feature was proposed for desamino-oxytocin from nmr studies (Urry *et al.*, 1970) and characterized the crystal structure of this molecule, which is also nonplanar (Wood *et al.*, 1986). The word badaed (number 11) is the backbone conformation that permits the best spatial arrangement of the side-chains. It is a conformation in which the turn regions of the 20-membered ring no longer lies in the residue 3–4 region like bdaacc and the conformations investigated by Kotelchuck and co-workers above. Rather, the disulfide bridge and the residue 3–4 region lie along the ''long'' axis of the 20-membered ring. Beta

Table II. Stereoalphabet Code[a]

	DeCoen and Ralston			Nikiforovich *et al.*	
	phi	psi		phi	psi
a	−60	−60	R	−60	−60
b	−160	+160	B	−140	+140
c	−140	+80		—	—
d	−80	+160		—	—
e	−80	+80		—	—
a*	+60	+60	L	+60	+60

[a]From DeCoen and Ralston (1973) and Nikiforovich *et al.* (1979).

turns in this nomenclature are sequences of two letters, ac and ae (type I), da* and ea* (type II), and aa (type III) or a*a* (type III′) (DeCoen and Ralston, 1977).

Nikiforovich and co-workers (1979) used similar conformation-building and energy-testing procedures to search the conformation space of oxytocin. Individual amino acid conformations used in this search are defined in column 2 of Table II. Three ring classes (BBRRBB, BBRRRB, and BBRBLB) are compatible with disulfide ring closure. (These conformations are characterized by a distance of less than 7 Å between the alpha carbon of Cys^1 and the alpha carbon of Cys^6, the constraint chosen in this study to insure that disulfide ring closure is possible.) The two lowest-energy conformations of the oxytocin ring belong to the ring class BBAABB, and these authors suggest that the oxytocin ring is experimentally constrained to this class of conformations. The energy-minimization studies suggest that the tail of oxytocin is much more flexible than the oxytocin ring. The two lowest-energy ring conformations are labeled I and II by the authors. Features of the beta turn in the oxytocin ring proposed by Urry and Walter (1971) are practically identical with those of conformation I. Conformation II lacks the beta turn characteristic of conformation I.

2. Molecular Dynamics

Recently, a molecular-dynamics simulation of lysine vasopressin has provided a new perspective on the conformation space of this class of molecules and a view of their dynamic nature (Hagler *et al.*, 1985). In molecular dynamics, as in the more common energy-minimization procedure or normal mode analysis, the potential energy of the molecule is expressed in terms of an analytical force field. Having specified the potential energy of the molecule and the initial coordinates of the molecule, Newton's equations are integrated forward in time to obtain the trajectories of the atoms and the associated conformational fluctuation and transitions of the peptide hormone. The dynamics simulation can be viewed as a ''movie'' that illustrates vibrational and rotational motion of the molecule as it undergoes conformational transitions. Instantaneous snapshots from the dynamics trajectory of LVP are illustrated as stereo pairs in Fig. 8.

This trajectory began from a beta-sheet ring conformation (A in Fig. 4). As the dynamics simulation progresses, the hexapeptide ring opens up. This is illustrated in Fig. 8 by the snapshots after 0.2 psec and 1.4 psec of simulation. The tripeptide tail exhibited the range of conformational flexibility in the molecular-dynamics simulation that has been inferred from physical studies and previous energy-based modelling studies. This is illustrated by a comparison of the orientation of the tripeptide tail in the snapshot after 3.9 psec of simulation with that of the earlier snapshots in Fig. 8.

That the molecule traversed various conformational states rather than undergo-

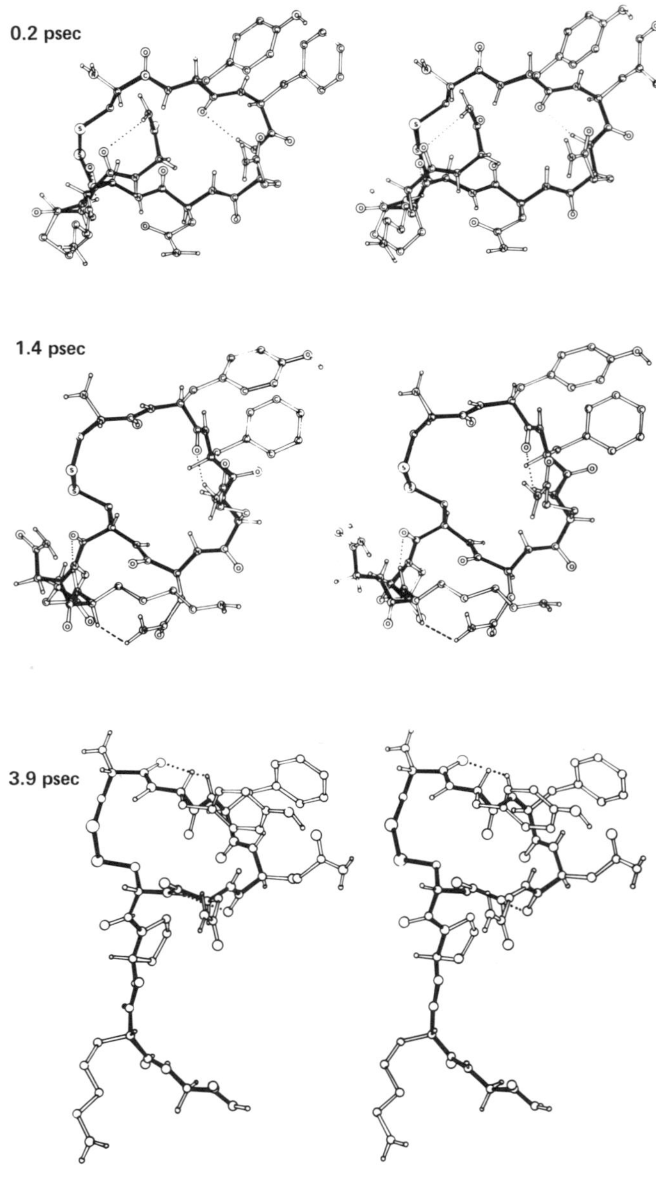

Figure 8. Instantaneous snapshots of conformations from a molecular-dynamics simulation of LVP. Top: After 0.2 psec of simulation. Middle: After 1.4 psec of simulation. Bottom: After 3.9 psec of simulation. [Figure after Hagler *et al.* (1985), with permission.]

ing fluctuations about a single conformation has been confirmed by minimizing the energy of numerous instantaneous conformations at different points along the trajectory. These results represent only a short trajectory, and the results of a much longer simulation are under analysis.

C. Ring Conformations

Flexible molecules like peptides are characterized by many minimum-energy conformations. For relatively small molecules (like cyclohexane), minimum-energy conformations (like the boat, chair, and twisted-boat conformations of cyclohexane) are easily visualized. Peptides generally have too many minimum-energy conformations to easily visualize them individually and collectively. However, the minimum-energy conformations of the cyclic peptides oxytocin and vasopressin generated in modelling studies can be categorized by the nature of the 20 membered ring of the molecules. Ring conformations for which torsion angles were reported in the literature were built using the SK&F Molecular Modelling System, and the three-dimensional topology of the rings was analyzed (Hempel, 1985).

Twenty-membered ring conformations generated in the modelling studies include ring conformations with a turn at the residue 3–4 region. The more "planar" of this class of ring conformations do not prefer to form two transannular hydrogen bonds between the NH and CO of residue 2 and the CO and NH of residue 5 of the hexapeptide ring (Kotelchuck *et al.*, 1972a,b). Ring conformations of this class that deviate from planarity can pick up the two transannular hydrogen bonds between residues 2 and 5 [conformation I of Nikiforovich *et al.* (1979); conformation 7 of DeCoen and Ralston (1973)].

A multiplicity of conformations with a turn in the peptide backbone of the hexapeptide ring at the residue 3–4 region characterize the short molecular-dynamics trajectory reported for LVP by Hagler and co-workers in 1985. This report represents the first application of molecular-dynamics simulation to peptides. The simulation was "focused" in this region of conformation space by the choice of initial starting structure. This study identified low-energy and minimum-energy conformations of LVP in which individual residues are not characterized by minimum-energy conformations of the amino acid as an isolated residue. This result suggests that even though stereoalphabet searches yield low-energy conformations of a cyclic peptide, their view of the conformation space of the molecule may be more restricted than has been previously appreciated.

The topology of the 20-membered ring can change significantly through a reorientation of the turn regions within the ring. Geometry changes of this type are exemplified by the DeCoen and Ralston conformations (all except conformation 7) in which residues 3 and 4 lie along a "long" axis of a hexapeptide ring.

Early computational studies by Gibson and Scheraga (1967) generated minimum-energy conformations for oxytocin and LVP of this type. It is interesting to note that the conformations of Gibson and Scheraga (1967) incorporate a left-handed disulfide bond. Low-energy conformations of the 20-membered hexapeptide ring can incorporate left- and right-handed disulfide torsion angles.

Honig and co-workers (1973) have proposed a minimum-energy conformation for neurohypophyseal hormones derived from a model conformation built using an empirical statistical analysis of amino acid conformations in protein structures (model I) and a model (model II) constructed to be consistent with tetrahedral binding geometry for Cu(II) by the amide NH groups of residues 1–4 of the hexapeptide ring. The rings in these molecular conformations are more "round" than "cigar-shaped," and in this sense are different from either of the two general classes of peptide rings discussed above.

V. SUMMARY AND CONCLUSIONS

A. Solution Conformation

Physical studies have demonstrated that oxytocin, vasopressin, and agonist analogs prefer a similar conformation in DMSO. The conformation of these molecules is solvent-dependent (water versus DMSO), and protonation of the amino terminus of the neurohypophyseal hormones has a significant effect on conformation. These observations are consistent with conformational flexibility in the peptide backbone of the hexapeptide ring, and evidence to date suggests significant conformational averaging in the ring in water (but not DMSO). That is to say, the 20-membered hexapeptide ring is not self-constrained to a narrowly defined ring conformation (i.e., a global minimum). Modelling studies support this interpretation.

The solution conformation model proposed by Urry and Walter in 1971 for oxytocin and for deamino-oxytocin in DMSO provides our current working hypothesis for the DMSO solution conformation of neurohypophyseal hormones and closely related agonist analogs. The model proposes a beta turn at the residue 3–4 region of the hexapeptide ring, with a transannular hydrogen bond from the Tyr^2 carbonyl to the Asn^5 amide, and includes a second transannular hydrogen bond between the Tyr^2 amide and the Asn^5 carbonyl for deamino-oxytocin. This hypothesis is supported by a host of solution studies for the neurohypophyseal hormones and agonist analogs. The recent crystal structure determination for deamino-oxytocin (Wood *et al.*, 1986) unequivocally establishes the experimental relevance of this class of ring conformation for deamino-oxytocin.

The Urry and Walter model for oxytocin in DMSO incorporates hypotheses related to side-chain and tail orientation. However, physical studies suggest conformational averaging of the side-chains and tail in solution, and for the most part the physical studies do not define preferred orientations in solution. The preliminary report of a nuclear Overhauser enhancement study for oxytocin in DMSO (Ford and Gibbons, 1979) is interpreted to suggest that the orientation of the tail in oxytocin is away from the hexapeptide ring, rather than over the ring as proposed in solution models. This report suggests further NOE investigations for this class of molecules, although there are difficulties in interpretation of results (Smith and Pease, 1980).

The hypothesis that the antagonist activity of oxytocin antagonists incorporating Pen at position 1 is linked to conformational constraint is extremely interesting. Integration of conformational properties of both oxytocin and vasopressin antagonists into binding conformation hypotheses is clearly an area for further investigation. Perceived differences in the conformation of the disulfide bridge region of the hexapeptide ring for antagonists incorporating Pen at position 1 relative to that of oxytocin strongly suggest further conformational studies of constrained analogs. Interpretations of physical studies of agonists have suggested that there is conformational flexibility in the disulfide region of these molecules. Although a relatively "fixed" disulfide bridge conformation is proposed for oxytocin and vasopressin in DMSO, it is postulated that conformational changes observed with pH are linked to perturbation of the disulfide bridge by the protonation of the amino terminus. Conformers that differ in the chirality of the disulfide torsion have been identified in the crystal structure of deamino-oxytocin (Wood *et al.*, 1986).

Three-dimensional features of the solution model for oxytocin in DMSO were transferred almost directly into cooperative models for oxytocin at the uterine receptor and for vasopressin at the antidiuretic receptor. The continuing viability of both solution and cooperative models suggests that the conformation of the neurohypophyseal hormones in DMSO and that at the receptor are similar.

B. Analog Design

Analogs have been designed to probe the active conformation hypotheses proposed for oxytocin and vasopressin. For example, hydrogen-bonding patterns incorporated into the active-conformation hypothesis for oxytocin at the uterine receptor have been investigated by Roy and co-workers (1982, 1983). Their work demonstrates that the biologically active conformation of oxytocin does not have an absolute requirement for a hydrogen-bonding interaction between the carboxamide carbonyl of Asn^5 and the peptide NH of Leu^8, since an analog of oxytocin incorporating *N*-methylleucine in the 8 position and an analog with an ester linkage replacing the amide linkage between residues 7 and 8 retain the biological activity of oxytocin (although with lower potency). Numerous SAR studies in the literature probe aspects of the active and binding elements proposed in the active-conformation hypotheses for oxytocin and vasopressin. Specifics of active-conformation hypotheses, such as hydrogen bonds, appear much less important to activity than molecular topology.

Conformation–activity studies for oxytocin and vasopressin were reviewed by Smith in 1981. Of particular interest is the report of vasopressin agonists, [7-(3,4)dehydroproline]AVP and the deamino analog of this molecule, that show enhanced *in vivo* agonist potency over that of AVP. The design rationale for introducing double bonds in Pro^7 (identified as a binding element in the biologically active conformation hypothesis for vasopressin at the antidiuretic receptor) was the hope that deformable electron clouds with their ability to undergo $\pi-\pi$ or other nonbonded interactions might enhance receptor binding of the analog, provided the steric fit at the receptor were correct. It was reasoned that the side-chains of the residues at the corners of the beta turns of the receptor model, like Pro^7, are the most exposed binding elements (Smith *et al.*, 1977; Smith and Walters, 1978). The enhanced potency observed for these analogs supports the design rationale, although there are ambiguities in dealing with *in vivo* data.

A new class of vasopressin antagonist at the antidiuretic receptor has been reported that was designed using the active-conformation hypothesis for vasopressin at the antidiuretic receptor (Skala *et al.*, 1984). The active conformation hypothesis for vasopressin proposes that the basic moiety of residue 8 (Lys or Arg) is in close proximity to the Asn^5 side-chain carboxamide. It was reasoned that structural modifications that perturb the orientation and cooperativity of these active elements but do not unduly reduce binding might be expected to lead to antidiuretic inhibitors. The antagonist analog reported in this study is a bicyclic analog of vasopressin in which a side-chain representing Lys^8 is joined to that of Asp^5 through a carboxamide bond. The remainder of the vasopressin molecule is deaminovasopressin with Phe (rather than Tyr) at the 2 position. This analog is reported to antagonize antidiuretic response (Skala *et al.*, 1984), in support of the design hypothesis.

In conclusion, basic understanding of the conformation and topology of neurohypophyseal hormones has been achieved through a variety of physical studies. Conformational understanding, along with active-conformation hypotheses and SAR studies, has stimulated a conformational approach to the design of peptide analogs of the neurohypophyseal hormones.

REFERENCES

Ballardin, A., Fischman, A. J., Gibbons, W. A., Roy, J., Schwartz, I. L., Smith, C. W., Walter, R. and Wyssbrod, H. R. (1978). *Biochemistry* **17,** 4443–4454.

Bandekar, J., and Krimm, S. (1979). *Proc. Natl. Acad. Sci. USA* **76,** 774–777.

Beychok, S., and Breslow, E. (1968). *J. Biol. Chem.* **243,** 151–154.

Blaha, K. (1981). *In* "Perspectives in Peptide Chemistry and Biochemistry" (A. Eberle, R. Geiger, and T. Wieland, eds.), pp. 272–282. Karger, Basel.

Blundell, T., and Wood, S. (1982). *Annu. Rev. Biochem.* **51,** 123–154.

Blumenstein, M., Hruby, V. J., and Viswanatha, V. (1980). *Biochem. Biophys, Res. Commun.* **94,** 431–437.

Breslow, Ester. (1979). *Annu. Rev. Biochem.* **48,** 251–274.

Brewster, A. I., Glasel, J. A., and Hruby, V. J. (1972). *Proc. Natl. Acad. Sci. USA* **69,** 1470–1474.

Brewster, A. I., Richard., and Hruby, V. J. (1973a). *Proc. Natl. Acad. Sci. USA* **70,** 3806–3809.

Brewster, A. I., Hruby, V. J., Glasel, J. A., and Tonelli, A. E. (1973b). *Biochemistry* **12,** 5294–5304.

Bystrov, V. F. (1976). *Prog. NMR Spectroscopy* **10,** 41–81.

Cohen, P., Convert, O., Griffin, J. H., Nicolas, P., and DiBello, C. (1977). *In* "Peptides" (Proc. 5th Am. Pept. Symp.) (M. Goodman and J. Meienhofer, eds.), pp. 176–178. Wiley, New York.

Cowburn, D., Live, D. H., and Agosta, W. C. (1981). *In "Biomolecular Stereodynamics"*. (Proc. Symp.) (S. Ramaswamy, ed.), Vol. 2, pp. 345–351. Adenine Press. New York.

Cowburn, D., Live, D. H., Fischman, A. J., and Agosta, W. C. (1983). *J. Am. Chem. Soc.* **105,** 7435–7442.

DeCoen, J. L., and Ralston, E. (1973). In "Peptides" (H. Nesvadba, ed)., (Proc. Eur. Peptide Symp.) pp. 335–342. North-Holland, Amsterdam.

DeCoen, J. L., and Ralston, E. (1977). *Biopolymers* **16,** 1929–1943.

Deslauriers, R., Walter, R., and Smith, I. C. P. (1972). *Biochem. Biophys. Res. Commun.* **48,** 854–859.

Deslauriers, R., Walter, R., and Smith, I. C. P. (1973). *In* "Proceedings of the International Conference on Stable Isotopes in Chemistry, Biology and Medicine. (P. Klein, ed). pp. 64–71. National Technical Information Service, Springfield, Va.

Deslauriers, R., Smith, I. C. P., and Walter, R. (1974a). *J. Biol. Chem.* **249,** 7006–7010.

Deslauriers, R., Walter, R., and Smith, I. C. P. (1974b). *Proc. Natl. Acad. Sci. USA* **71,** 265–268.

Deslauriers, R., Smith, I. C. P., Levy, G. C., Orlowski, R., and Walter, R. (1977). In "Peptides" (*Proc. Am. Pept. Symp.*) (M. Goodman and J. Meienhofer, eds.), pp. 168–170. Wiley, New York.

Deslauriers, R., Walter, R., Smith, I. C. P., Levy, G. C., Orlowski, R., and Walter, R. (1978). *J. Am. Chem. Soc.* **100,** 3912–3918.

Du Vigneaud, V. (1966). "A Trail of Sulfur Research from Insulin to Oxytocin. Nobel Lectures in Chemistry, 1942–1963," Vol. 3, pp. 446–463. Elsevier, Amsterdam.

Fischman, A. J., Live, D., Wyssbrod, H. R., Agosta, W. C., and Cowburn, D. (1980). *J. Am. Chem. Soc.* **102,** 2533–2539.

Ford, J. J., and Gibbons, W. A. (1979). *In* "Peptides" (Proc. 6th Am. Pept. Symp.) (E. Gross and J. Meienhofer, eds.), pp. 857–879. Pierce Chemical Co., Rockford, Ill.

Fric, I., Kodicek, M., Jost, K., and Blaha, K. (1974a). *Collect. Czech. Chem. Commun.* **39,** 1271–1289.

Fric, I., Kodicek, M., Prochaszka, Z., Jost, K., and Blaha, K. (1974b). *Collect. Czech. Chem. Commun.* **39,** 1290–1302.

Fric, I., Kodicek, M., Flegel, M., and Zaoral, M. (1975). *Eur. J. Biochem.* **56,** 493–502.

Fric, I., Leonteva, L. I., Malon, P., Jost, K., and Blaha, K. (1980). *Collect. Czech. Chem. Commun.* **45,** 1109–1131.

Gibson, K. D., and Scheraga, H. A. (1967). *Proc. Natl. Acad. Sci. USA* **58,** 1317–23.

Glasel, J. A., McKelvy, J. F., Hruby, V. J., and Spatola, A. F. (1973). *Ann. NY Acad. Sci.* **222,** 778–788.

Glickson, J. D. (1975). "Peptides: Chemistry, Structure and Biology" (R. Walter and J. Meienhofer, eds.), pp. 782–802. Ann Arbor Scientific, Ann Arbor, Mich.

Glickson, J. D., Urry, D. W., Havran, R. T., and Walter, R. (1972). *Proc. Natl. Acad. Sci. USA* **69,** 2136–2140.

Hagler, A. T., Osguthorpe, D. J., Dauber-Osgutharpe, P., and Hempel, J. C. (1985). *Science* **227,** 1309–1315.

Hempel, J. C. (1985). "Peptides, Structure, and Function" (Proc. 9th Am. Peptide Symp.) (C. M. Deber, V. J. Hruby, and K. D. Kopple, eds.) pp. 141–144. Pierce Chemical Co., Rockford, Illinois.

Honig, B., Kabat, E. A., Katz, C., Levinthal, C., and Wu, T. T. (1973). *J. Mol. Biol.* **80,** 277–295.

Hruby, V. J. (1981). *Top. Mol. Pharmacol.* **1,** 99–126.

Hruby, V. J. (1982). *Life Sci.* **31,** 189–199.

Hruby, V. J. (1984). *ACS Symp. Ser.* **251,** 9–27.

Hruby, V. J., and Mosberg, H. I. (1981). *Dev. Endocrinol.* **13,** 227–237.

Hruby, V. J., and Mosberg, H. I. (1982). *In* "Hormone Antagonists" (M. K. Agarwal, ed). pp. 433–473. de Gruyter, Berlin.

Hruby, V. J., Mosberg, H. I., Fox, J. W., and Tu, A. T. (1982). *J. Biol. Chem.* **257,** 4916–4924.

Hruby, V. J., Mosberg, H. I., Sawyer, T. K., Knittel, J. J., Rockway, T. W., Ormberg, J., Darman, P., Chan, W. Y., and Hadley, M. E. (1983a). *Biopolymers* **22,** 517–530.

Hruby, V. J., Rockway, T. W., Viswanatha, V., and Chan, W. Y. (1983b). *Int. J. Pept. Protein Res.* **21,** 24–34.

Kessler, H. (1982). *Angew. Chem. Int. Ed. Engl.* **21,** 512–523.

Kotelchuck, D., Scheraga, H. A., and Walter, R. (1972a). *In* "Chemistry and Biology of Peptides" (Proc. Am. Pept. Symp.) (J. Meienhofer, ed.), pp. 35–43. Ann Arbor, Mich.

Kotelchuck, D., Scheraga, H. A., and Walter, R. (1972b). *Proc. Natl. Acad. Sci. USA* **69,** 3629–3633.

Krauss, E. M., and Cowburn, D. (1981). *Biochemistry* **20,** 671–679.

Langs, D. A., Smith, G. D., Stezowski, J. J., and Hughes, R. E. (1986). *Science* (in press).

Live, D. (1979a). *In* "Peptide Structure, Biology, and Function". (Proc. Am. Pept. Symp.) (E. Gross, J. Meienhofer, eds.), pp. 221–224. Pierce Chemical Co., Rockford, Ill.

Live, D. H., Wyssbrod, H. R., Fischman, A. J., Agosta, W. C., Bradley, C. H., and Cowburn, D. (1979b). *J. Am. Chem. Soc.* **101,** 474–479.

Maxwell, F. R., and Scheraga, H. A. (1977). *Biochemistry* **16,** 4443.

Nikiforovich, G. V., Leonova, V. I., Galaktionov, S. G., and Chipens, G. I. (1979). *Int. J. Pept. Protein Res.* **13,** 363–373.

Robson, B. (1984). *CRC Crit. Rev. Biochem.* **14,** 273–296.

Roy, C., Barth, T., and Jard, S. (1975). *J. Biol. Chem.* **250,** 3157–3168.

Roy, J., Gazis, D., Shakman, R., and Schwartz, I. L. (1982). *Int. J. Pept. Protein Res.* **20,** 35–42.

Roy, U., Gazis, D., Schwartz, I. L., and Roy, J. (1983). *Int. J. Pept. Protein Res.* **22,** 398–403.

Sarathy, K. P., Krishna, N. R., Lenkinski, R. E., Glickson, J. D., Hruby, V. J., and Walter, R. (1981a). *In* "Peptide Synthesis, Structure and Function" (Proc. Am. Pept. Symp.) (D. Rich, E. Gross, eds.), pp. 359–362. Pierce Chemical Co., Rockford, Ill.

Sarathy, K. P., Krishna, N. R., Lenkinski, R. E., Glickson, J. D., Rockway, T. W., Hruby, V. J., and Walter, R. (1981b). *Dev. Endocrinol.* **13,** 211–226.

Skala, G., Smith, C. W., Taylor, C. J., and Ludens, J. H. (1984) *Science* **226,** 443.

Shekar, S. C., Sankaram, M. B., and Easwaran, K. R. K. (1984). *Int. J. Pept. Protein Res.* **23,** 166–173.

Smith, C. W. (1981). *Dev. Endocrinol.* **13,** 23–35.

Smith, C. W., and Walter, R. (1978). *Science* **199,** 297–9.

Smith, C. W., Botos, C. R., and Walter R. (1977). *In* "Peptides" (Proc. 5th Am. Pept. Symp.) (M. Goodman and J. Meienhofer, eds.) pp. 161–164.

Smith, J. A., and Pease, L. G. (1980). *CRC Crit. Rev. Biochem.* **8,** 315–399.

Sugeta, H., Go, A., and Miyazawa, T. (1973). *Bull. Chem. Soc. Jpn.* **46,** 3407–3411.

Tu, A. T., Bjarnason, J. B., and Hruby, V. J. (1978). *Biochem. Biophys. Acta* **533,** 530–533.

Tu, A. T., Lee, J., Deb, K. K., and Hruby, V. J. (1979). *J. Biol. Chem.* **254,** 3272–3278.

Urry, D. W., and Walter, R. (1971). *Proc. Natl. Acad. Sci. USA* **68,** 956–958.

Urry, D. W., Quadrifoglio, F., and Walter, R. (1968). *Proc. Natl. Acad. Sci. USA* **60,** 967–974.

Urry, D. W., Ohnishi, M., and Walter, R. (1970). *Proc. Natl. Acad. Sci. USA* **66,** 111–116.

Urry, D. W., Venkatachalam, C. M., Khaled, M. A., and Prasad, K. U. (1981). *Dev. Endocrinol.* **13,** 169–193.

Van Wart, H. E., and Scheraga, H. A. (1976). *J. Phys. Chem.* **80,** 1812.

Von Dreele, P. H., Brewster, A. I., Scheraga, H. A., Ferger, M. F., and Du Vigneaud, V. (1971). *Proc. Natl. Acad. Sci. USA* **68,** 1028–1031.

Walter, R. (1972). *In* "Structure–Activity Relationships of Proteins, Polypeptides and Hormones" (Proc. Int. Symp.) (M. Margoulies and F. Greenwood, eds.), pp. 181–193, 456–457. Excerpta Medica, Amsterdam.

Walter, R. (1977). *Fed. Proc. Am. Soc. Exp. Biol.* **36,** 1872–1878.

Walter, R., Schwartz, I. L., Darnell, J. H., and Urry, D. W. (1971). *Proc. Natl. Acad. Sci. USA* **68,** 1355–1359.

Walter, R., Glickson, J. D., Schwartz, I. L., Havran, R. T., Meienhofer, J., and Urry, D. W. (1972). *Proc. Natl. Acad. Sci. USA* **69,** 1920–1924.

Walter, R., Prasad, K. U. M., Deslauriers, R., and Smith, I. C. P. (1973). *Proc. Natl. Acad. Sci. USA* **70,** 2086–2090.

Walter, R., Smith, I. C. P., and Deslauriers, R. (1974a). *Biochem. Biophys. Res. Commun.* **58,** 216–221.

Walter, R., Ballardin, A., Schwartz, I. L., Gibbons, W. A., and Wyssbrod, H. R. (1974b). *Proc. Natl. Acad. Sci. USA* **71,** 4528–4532.

Walter, R., Smith, C. W., Mehta, P. K., Boonjarerh, S., Arruda, J. A L., Kurtzman, N. A. (1977a). *In* "Disturbances in Body Fluid Osmolality" (T. Andreoli, J. Grantham, and C. Floyd, eds.), pp. 1–36. American Physiological Society, Bethesda, Md.

Walter, R., Wyssbrod, H. R., and Glickson, J. D. (1977b). *J. Am. Chem. Soc.* **99,** 7326–7332.

Walter, R., Smith, C. W., Sarathy, K. P., Pillai, R. P., Krishna, N. R., Lenkinski, R. E., Glickson, J. D., and Hruby, V. J. (1981). *Int. J. Pept. Protein Res.* **17,** 56–64.

Wood, S. P., Tickle,, I. J., Treharne, A. M., Pitts, J. E., Mascarenhas, Y , Li, J. Y., Husain, J., Cooper, S., Blundell, T. L., Hruby, V. J., Buku, A., Fischman, A. J., and Wyssbrod, H. R. (1986). *Science* **232,** 633–636.

Wyssbrod, H. R., Ballardin, A., Gibbons, W. A., Roy, J., Schwartz, I. L., and Walter, R. (1975). *In* "Peptides: Chemistry, Structure and Biology" (Proc. Am. Pept. Symp.) (R. Walter, J. Meienhofer, eds.), pp. 815–822. Ann Arbor Science, Ann Arbor, Mich.

Wyssbrod, H. R., Fischman, A. J., Live, D. H., Hruby, V. J., Agarwal, N. S., and Upson, D. A. (1979). *J. Am. Chem. Soc.* **101,** 4037–4043.

Wyssbrod, H. R., Fischman, A. J., Wittbold, W. M., Smith, C. W., Walter, R., and Schwartz, I. L. (1981a). *Int. J. Pept. Protein Res.* **17,** 48–55.

Wyssbrod, H. R., Fischman, A. J., Wittbold, W. M., Kondor, P., Cort, J. H., and Schwartz, I. L. (1981b). *Adv. Physiol. Sci. Proc. Int. Congr.* **13,** 93–99.

Chapter 6

Central Nervous System Effects of Neurohypophyseal Peptides*

PAULA L. HOFFMAN

National Institute on Alcohol Abuse and Alcoholism
Division of Intramural Clinical and Biological Research
Bethesda, Maryland 20892

> I missed her. Missing her reminded me of my one night in the House of Blue Lights, because I'd gone there out of missing someone else. I'd gotten drunk to begin with, then I'd started hitting vasopressin inhalers. If your main squeeze has just decided to walk out on you, booze and vasopressin are the ultimate in masochistic pharmacology; the juice makes you maudlin and the vasopressin makes you remember, I mean really remember. Clinically they use the stuff to counter senile amnesia, but the street finds its own uses for things. So I'd bought myself an ultra-intense replay of a bad affair; trouble is, you get the bad with the good. Go gunning for transports of animal ecstasy and you get what you said, too, and what she said to that, how she walked away and never looked back.
>
> *From "Burning Chrome"*
> *by William Gibson*
> Copyright Omni Publications International, 1982

I. INTRODUCTION

Our understanding of the mechanisms by which human memory operates derives to a large degree from studies of amnesic subjects—those in whom

*Abbreviations: icv, intracerebroventricular; sc, subcutaneous; AVP, [8-arginine]vasopressin; LVP, [8-lysine]vasopressin; DGAVP, [des-9-glycinamide,8-arginine]vasopressin; DGLVP, [des-9-glycinamide,8-lysine]vasopressin; DDAVP, [1-β-mercaptopropionic acid, 8-D-arginine]vasopressin; dPTyr(Me)AVP, [1-deaminopenicillamine,2-(*O*-methyl)tyrosine,8-arginine]vasopressin; $d(CH_2)_5$-Tyr(OEt)VAVP, [1-(β-mercapto-β,β-cyclopentamethylenepropionic acid),2-(*O*-ethyl)tyrosine,4-valine,8-arginine]vasopressin; $ACTH_{4-10}$, peptide comprising amino acid residues 4–10 of adrenocorticotropic hormone.

THE PEPTIDES, VOLUME 8

memory is lost (Squire, 1982). It is clear from such studies, and is, in fact, self-evident, that severe loss of memory is incompatible with a "normal" human life. Considering the increasing loss of memory that occurs during aging, or in conditions such as Alzheimer's disease (Squire and Davis, 1981), it is not surprising that the idea of a drug that can enhance or reinstate memory is quite appealing. The hormone of the posterior pituitary gland, arginine vasopressin (AVP), has been studied for nearly two decades with respect to its ability to modify memory and learning. As the investigations have proceeded, controversies have arisen regarding experimental results themselves, as well as the interpretation of these results (see, for example, Sahgal, 1984; Messing and Sparber, 1984; Strupp and Weingartner, 1984). In part, these difficulties occur because of the complexities inherent in defining and measuring "memory," particularly in animals. At a very basic level, most investigators recognize short-term and long-term memory. Material that is learned is first stored in short-term memory, and this memory can be converted to long-term memory by a process called consolidation (Squire, 1982). Another aspect of memory that is often studied is the "retrieval" of information which has been stored in memory. In addition to the basic divisions of long- and short-term memory, consolidation, and recall, or retrieval, it appears (for example) that verbal memory may have different characteristics from nonverbal memory (see Sahgal, 1984), which can be a major problem in comparing animals and humans. Additionally, it is very difficult to dissociate the confounding influence of attentional or motivational factors in learning and memory processes from the measurement of those processes *per se* (see Strupp and Weingartner, 1984). Another complicating factor in studies of a drug's effect on memory is the possibility that the drug will affect *performance* of a behavioral task (e.g., by causing sedation or activation), rather than learning or memory of the task. The question of which of these factors is most affected has arisen as vasopressin has been tested in different behavioral paradigms, and it has yet to be completely resolved.

Nevertheless, there is overwhelming evidence that vasopressin and related peptides, in addition to possessing peripheral, endocrinological activities, also influence memory and behavior, by acting at either central or peripheral sites. Some indirect evidence for a central site of action has been the localization of vasopressin and oxytocin in extrahypothalamic areas of brain (see Swaab, 1982). Cell bodies in the supraoptic and paraventricular nuclei of the hypothalamus have been found to have projections to limbic and other brain areas, in addition to the long-recognized projections to the posterior pituitary (Swaab, 1982). Although the anatomical localization of memory is not clear, the hippocampus and other limbic structures have been strongly implicated in particular aspects of acquisition of information and memory storage (Squire, 1982; Moore and Solomon, 1980). Vasopressin-containing neuronal cell bodies have also been found in the suprachiasmatic nucleus (Swaab, 1982). Those neurons project to the hypothalamus, as well as to the organum vasculosum of the lamina terminalis (OVLT), an area outside the blood–brain barrier. It is possible that such vasopressin-containing neurons could be a source of vasopressin in areas of the brain other than the hypothalamus, and that this extrahypothalamic vasopressin is responsible for the behavioral effects of the hormone. Alternatively, it has been suggested that plasma vasopressin may be carried to its sites of action in the brain via the cerebrospinal fluid (CSF) (de Wied, 1977). However, as will be discussed later, available evidence does not support this hypothesis.

In addition to conceptual problems—i.e., the exact aspect(s) of learning or memory that vasopressin modulates, and whether that modulation is a result of central or peripheral effects—there are also a number of technical problems that complicate the literature. Thus, different investigators use peptides of varying degrees of purity, different doses, and different routes of administration, as well as different experimental conditions, even in similar paradigms. Because, in animal studies, behavioral responses are used as a measure of hormonal effects on memory, consistent handling of the animals and precise control of all experimental variables are essential. The goal of this review is to analyze the results obtained in animal and human studies, taking into account, where possible, the factors listed above, and to try to reach conclusions about the ability of vasopressin to modulate memory, as well as its site(s) and possible mechanism of action.

II. ANIMAL STUDIES

A. Neurohypophyseal Peptides and Avoidance Responses

One of the most widely used paradigms for testing the role of vasopressin in memory has been the measurement of peptide effects on active and passive avoidance behaviors (see van Wimersma Greidanus *et al.*, 1983). When active avoidance behavior is studied, the animal (usually a rat) is exposed to a conditioned stimulus (e.g., a light or a sound), followed, at some time interval, by an unconditioned stimulus, usually footshock. The animal can escape the footshock by (for example) entering another chamber, or by jumping onto a pole (unconditioned response). During the training period, consisting of a set number of trials, the animal learns to avoid the shock by making the avoidance response as soon as the conditioned stimulus occurs (this avoidance is termed the conditioned response). Memory is then tested by measuring the conditioned avoidance response during retention tests held at some interval after training.

When passive avoidance responding is measured, the animal is usually placed in a lighted box, with access to a dark box, which is preferred by rodents. When it enters the dark box, the animal is given a footshock. Thereafter, during testing, the time until the dark box is entered (latency) is measured; in this instance, avoidance of the shock (i.e., long latency) requires only that the animal remain where it is in the lighted box, and thus is passive, rather than active. In each of these paradigms, several factors can be measured: rate of *acquisition* of the response is a measure of learning ability, and *retention* of the response over time is taken as a measure of memory. The response can also be *extinguished,* if the footshock is no longer paired with the conditioned stimulus or with entrance into the dark box. The rate of extinction is often used as a measure of "forgetting," although it can also be viewed as the ability to learn that the situation has changed (see Strupp and Weingartner, 1984; Messing and Sparber, 1984). The time at which vasopressin is administered in these paradigms has been used to assess the peptide's action on various aspects of learning or memory. Thus, if peptide is administered prior to training, its effects on acquisition of the response, or learning, can be determined, with certain caveats as mentioned above, regarding effects on performance. When the drug is administered immediately after training and retention is tested some time later, drug effects on "consolidation" of memory are thought to be measured. If the drug is given immediately prior to retention testing, then effects on retrieval of information may be assessed. Finally, in many of the studies of deWied's group (see below), the rate of extinction of the avoidance response has been tested, with peptide given immediately after training or either prior to or immediately following the first extinction session (see van Wimersma Greidanus *et al.,* 1975a, 1983).

The studies of deWied and his colleagues form a major portion of the literature dealing with the effects of vasopressin on active and passive avoidance behavior. In 1965, de Wied showed that the increase in rate of extinction of an active avoidance response that occurred in hypophysectomized animals could be prevented if the rats were given subcutaneous injections of pitressin tannate, an extract of the posterior pituitary (de Wied, 1965). Later experiments showed that the active ingredient in pitressin was lysine vasopressin (LVP), and that this peptide could also *inhibit* (i.e., reduce the rate of) extinction of an avoidance response in intact rats, if the peptide was administered subcutaneously (sc) either during acquisition of the response or during its extinction (de Wied and Bohus, 1966; de Wied, 1971). Some of the most dramatic aspects of these findings were the long-term nature of the effect (extinction had still not occurred several days after cessation of hormone treatment) and the relatively low doses that were effective (1 μg of LVP in a single injection) (de Wied, 1971; de Wied *et al.,* 1976). These data led de Wied to postulate that vasopressin might be important in the formation of "long-term" memory (de Wied, 1971, 1977; van Wimersma Greidanus *et al.,* 1983). That is, it was felt that vasopressin treatment strengthened the memory of training, making it more difficult for the animals to "forget." In some other studies by de Wied and his colleagues, vasopressin was also found to enhance the *acquisition* of a conditioned avoidance response, which was regarded as an effect on learning (Bohus *et al.*, 1973; Lande *et al.*, 1971).

The work of this group of investigators has addressed a number of important issues. It was shown in an early study that LVP also maintained (inhibited extinction of) a passive avoidance response in rats (Ader and de Wied, 1972), and that the behavioral effects of the peptide depended on the time of treatment, relative to the time of testing. That is, LVP appeared to be effective if given within 1 hr of behavioral training or testing, in either active or passive avoidance paradigms, but not if given 6 hr prior to training or testing (Bohus *et al.*, 1972; de Wied, 1971). This time course of peptide action suggested that it is necessary for vasopressin to be associated with specific environmental cues in order for the peptide to "maintain new behavior patterns" (Bohus *et al.*, 1972); these findings were also consistent with the idea that vasopressin, administered at a critical time, could enhance the normally occurring consolidation of memory (see van Wimersma Greidanus *et al.*, 1983).

A major question that has been raised by these studies, and that is still not totally resolved, is whether vasopressin acts at a central or peripheral site to produce its effects. De Wied and his colleagues, using the pradigm of peptide-induced inhibition of extinction of active and passive avoidance responses, have tried various experimental approaches in order to answer this question and, on the whole, their results are in line with a central site of action for vasopressin. In an early study, it was shown that an analog of LVP, [des-9-glycinamide]LVP (DGLVP), which had been reported to have very little peripheral endocrine activity (Dekanski, 1952; see Walter *et al.*, 1975), could still increase resistance to extinction of active avoidance behavior (Lande *et al.*, 1971; de Wied *et al.*, 1972). In contrast, angiotensin, which is a more powerful pressor agent than vasopressin, did not affect this measure of "memory" (de Wied, 1971). Furthermore, having found that arginine vasopressin (AVP), the naturally occurring hormone of the rat, could maintain avoidance responding as well as LVP (de Wied, 1976), de Wied and his colleagues performed two sets of experiments to ascertain more directly whether the hormones maintain avoidance behavior by actions in the central nervous system (CNS). First, it was shown that a single, intracerebroventricular (icv) injection of AVP (0.025 ng), given immediately after training, could induce resistance to extinction of an active avoidance response (de Wied, 1976). This dose was much lower than that necessary when AVP was administered sc. In the second set of experiments, AVP was effectively removed from cerebrospinal fluid by icv administration of antiserum to AVP (van Wimersma Greidanus *et al.*, 1975b; van Wimersma Greidanus and de Wied, 1976b). One microliter of antiserum, with a binding capacity of about 2.5 ng of AVP, given immediately after or $\frac{1}{2}$ hr before the learning trial, induced a *deficit* in passive avoidance behavior when retention was tested 24 or 48 hr later. Thus, antiserum to vasopressin, administered centrally, had an effect on behavior that was opposite to that of the hormone but did not alter the endocrine response to vasopressin. However, systemic administration of a large amount of antiserum, which *did* affect the endocrine response to AVP, *did not* affect passive avoidance behavior, supporting the idea of a central site of action for the hormone in maintaining such behavior (van Wimersma Greidanus *et al.*, 1975a,b). Another interesting aspect of this study was the finding that antiserum to AVP, administered icv, did not affect passive avoidance performance when this behavior was measured either 2 min or 1 hr after training—i.e., antiserum did not affect learning or short-term memory, but did appear to specifically inhibit memory consolidation.

While all of these results support the hypothesis that there is a central site of action for vasopressin, they are in some ways inconsistent with de Wied's earlier work. Thus, hypophysectomy should theoretically affect *only* plasma levels of vasopressin (since the synthesis of vasopressin in hypothalamus should not be eliminated by this procedure). However, hypophysectomy was shown to increase the rate of extinction of active avoidance behavior in rats (de Wied, 1965). If this behavioral change were due to the drop in circulating levels of vasopressin, then peripheral administration of antibody might have also been expected to increase the rate of extinction of avoidance behavior. On the other hand, it is possible that during hypophysectomy, hypothalamic vasopressin-containing cell bodies were inadvertently damaged or destroyed, lowering brain levels of AVP as well as plasma levels.

In any case, the finding that antibodies to vasopressin, injected into CSF, can facilitate extinction of avoidance behaviors is compatible with a role for *endogenous* vasopressin in the normal maintenance of avoidance behavior. To further test this postulate, de Wied and others have made use of the Brattleboro rat, which is genetically deficient in vasopressin (Valtin and Schroeder, 1954). Homozygous animals of this strain display diabetes insipidus (DI), and no vasopressin is detectable in their brains (Morris, 1982). Recent reports indicate that the genetic defect in these animals resides in inefficient translation of the RNA coding for vasopressin (Schmale *et al.*, 1984). In studies by de Wied and his colleagues, homozygous DI rats, in contrast to heterozygous littermates that produce vasopressin, did not retain a passive avoidance response (de Wied *et al.*, 1975). If these animals were tested immediately after the learning trial, they showed complete passive avoidance behavior, but this behavior was no longer present at 24 hr after the trial. Subcutaneous injection of 1 μg of AVP or DGLVP allowed these animals to maintain passive avoidance behavior (de Wied *et al.*, 1975). These studies supported the hypothesis that endogenous vasopressin was not required for learning but was necessary for consolidation of memory (de Wied *et al.*, 1975).

In general, the work of de Wied's group is internally consistent and supports a role for centrally acting vasopressin in the formation of long-term memory. These workers have contributed enormously to the acceptance of the concept that peptides may act as neuromodulators that can influence behavior or memory. Because of the importance of this concept, a number of other groups have attempted to confirm the effects of vasopressin on active and passive avoidance behaviors. Koob *et al.* (1981) reported that systemic injection of 1 μg of AVP significantly inhibited extinction of a pole-jump active avoidance response in rats, the test most often used by de Wied's group (see van Wimersma Greidanus *et al.*, 1983). Wang (1972) also found that synthetic DGLVP inhibited extinction of active avoidance behavior in rats, and Kovács *et al.* (1978, 1979a,b,c,) found that AVP or LVP, given after training or before testing, increased passive avoidance responding, as compared to that of saline-treated animals. Krejčí *et al.* (1979) found that LVP, and analogs of vasopressin that are resistant to metabolic degradation, enhanced retention of passive avoidance behavior in rats when these peptides were administered systemically. The analogs had a somewhat longer duration of action than the parent hormone. These authors also noted a sedative effect of some of the hormone analogs, although the short-lived sedation did not affect the results of the passive avoidance studies (Krejčí *et al.*, 1979).

Studies of other types of avoidance behavior have also been performed. Vawter and Green (1980) tested the effect of DGLVP on resistance to extinction of a conditioned taste aversion. Rats that were given 1.0 μg of DGLVP subcutaneously, either during conditioning or extinction or both, showed increased resistance to extinction of this conditioned avoidance response. Similarly, treatment of young or aged rats with 1 μg/kg of LVP during the extinction period prolonged a conditioned taste aversion in both groups (Cooper *et al.*, 1980). Thus, vasopressin appeared to inhibit extinction when the aversive stimulus was an internal cue, as well as the usual external cue. In a passive avoidance paradigm using mice, in which the aversive stimulus was attack by a trained fighter mouse, subcutaneous administration of 0.8 μg of LVP, either just after acquisition of the response or just prior to retention testing, maintained the avoidance response (Leshner and Roche, 1977; Roche and Leshner, 1979). The effect of LVP was not apparent until 48 hr after acquisition, and it lasted for 240 hr. LVP had no effect on the *acquisition* of the response in this paradigm (Leshner and Roche, 1977; Roche and Leshner, 1979). Davis *et al.* (1982a) reported that AVP—but not [1-deamino,8-D-arginine]AVP (DDAVP) (Davis, 1984)—maintained conditioned avoidance responding in chicks. The animals were trained in a one-trial passive avoidance task to suppress their spontaneous pecking behavior. That is, while chicks normally peck at a visually attractive target, this target is made aversive when it is precoated with ethanol. Chicks that received AVP intraventricularly (0.25 μg) or subcutaneously (2.5 μg), at 1 or 9 (but not 59) min after training, displayed reduced pecking behavior, compared to controls, when they were tested at 24 hr after peptide administration

On the other hand, in some experiments, no effect of vasopressin on avoidance behavior could be observed. In a rather extensive study, Hostetter *et al.* (1980) examined the effect of several doses of LVP on passive avoidance behavior in rats. In these studies, the intensity of the footshock was varied, since the ''memory'' of training is known to depend on shock intensity (e.g., see Koob *et al.*, 1981). Hostetter *et al.* (1980) found that, while latencies in the retention tests did depend on the level of the shock, LVP had no effect on these latencies. Similarly, Gold and van Buskirk (1976) found that a subcutaneous injection of LVP immediately after training of rats did not affect passive avoidance behavior in later retention sessions. In both of these studies, doses of hormone were reported as international units (IU), with no indication of IU/mg, so that it is difficult to compare the doses with those used by other workers. It is of interest that Gold and van Buskirk did find that posttraining systemic injections of epinephrine, norepinephrine, or adrenocorticotropic hormone (ACTH) enhanced avoidance latencies in the same paradigm in which LVP was ineffective. There are a number of possible explanations for an inability to find an effect of vasopressin on avoidance behavior. Perhaps the simplest is differences in the purity or potency of the peptide(s) used. Commercial preparations often contain impurities that may have behavioral effects, or that simply reduce the amount of peptide administered. Often, chromatographic procedures do not distinguish such impurities and only biological testing procedures are adequate; such bioassays are rarely performed. Rigter (1982) has described the importance of *behavioral* variables in evaluating the effect of vasopressin in the passive avoidance paradigm. His experiments showed that pretraining experience—i.e., allowing the rats to practice the procedure and familiarizing them with the apparatus by allowing them to explore it—was necessary in order for the effects of vasopressin to be observed. Whether this phenomenon was due to differences in the strength of the memory trace, habituation, or some other factor was not clear; however, the lack of pretraining in studies such as that by Hostetter *et al.* (1980) may have contributed to their inability to observe any effect of vasopressin on avoidance behavior. Hagan (1982) also reported that the response to vasopressin could be altered by manipulation of behavioral parameters prior to peptide treatment. In his experiments, LVP, administered at 60 min after avoidance training, *increased* the rate of extinction of the avoidance response. However, in these studies, another group of rats was subjected to response prevention, i.e., the animals were prevented from making an avoidance response upon presentation of the conditioned stimulus but were then not shocked. In this group of rats, avoidance responding during the extinction session was already reduced, compared to controls (i.e., extinction of the avoidance behavior had already begun). Under these conditions, LVP now *inhibited* extinction, as found in many other studies (see above). This result emphasizes the importance of the behavioral state of the animals, and of their baseline levels of avoidance responding, when peptide effects are being assessed.

These types of experiments provide possible explanations for difficulties that some investigators have reported in replicating the work of de Wied and his colleagues. It seems clear that if the experimental conditions are carefully monitored, one can observe that vasopressin and related peptides will maintain avoidance responding (e.g., Rigter, 1982). However, the *interpretation* of such results still raises a number of questions. According to de Wied and co-workers, the results indicate (1) that vasopressin acts centrally to (2) consolidate the formation of long-term memory, and (3) that endogenous hormone is important in this process under normal circumstances. Each of these points has been challenged by other investigators.

1. Does Vasopressin Act Centrally to Affect Avoidance Behavior?

The strongest evidence *against* a central site of action for AVP derives from the work of Koob and colleagues, who agree that vasopressin inhibits extinction of active avoidance behavior (Koob *et al.*, 1981), but who have proposed that the peripheral effects of vasopressin may be responsible for this action (LeMoal *et al.*, 1981; Ettenberg *et al.*, 1983a,b). Thus, doses of AVP that, administered subcutaneously, inhibited extinction of an active avoidance response, also induced an early pressor response (LeMoal *et al.*, 1981). Furthermore, an antagonist peptide, dPTyr(Me)AVP, that prevented the pressor response, also prevented the behavioral response (LeMoal *et al.*, 1981). A similar result was obtained when an appetitive task (see below) was used (Ettenberg *et al.*, 1983b). While the behavioral and pressor responses could be unrelated, these data could also be interpreted to indicate that the vascular response to AVP, or other systemic effects of the peptide, is an aversive stimulus that alerts or arouses the animal, and thereby causes persistence of escape behaviors (LeMoal *et al.*, 1981). In a later study, these authors more extensively investigated the aversive properties of AVP (Ettenberg *et al.*, 1983a). In these experiments, the effect of AVP on an appetitively motivated task (see below), rather than on an aversively motivated task, was determined. AVP reduced the time necessary for water-deprived rats to locate a water tube in an environment to which they had been exposed once before. Lithium chloride, which causes illness in the animals and is often used as an aversive stimulus, also reduced the time necessary to find the water. This suggested that AVP, rather than affecting "memory," might be arousing the animals due to its aversive properties. In separate experiments, AVP was found to be aversive when tested in conditioned taste and place aversion paradigms (Ettenberg *et al.*, 1983a). Interestingly, in these experiments, [des-9-glycinamide]AVP (DGAVP), an analog reported to have greatly reduced peripheral activity, did not have aversive properties and also did not improve performance in the water-finding test. These results are provocative and challenge both the interpretation of the behavioral effects of AVP as direct effects on memory and the central site of action of AVP, particularly in light of the inactivity of DGAVP.

However, there are a number of questions that can be raised about this study.

It is surprising, for example, that the appetitively motivated task that was chosen was one that involved water deprivation, since AVP has profound effects on water balance and since its secretion is stimulated during water deprivation. Such effects may well confound the results. Furthermore, the suggestion that the pressor effect of AVP is aversive and is responsible for the behavioral effects of the peptide is not consistent with previous experiments showing that angiotensin II, a more powerful endogenous pressor substance, does not affect avoidance responding (de Wied, 1971). Finally, although DGAVP was not behaviorally active in the study of Ettenberg *et al.* (1983a), both DGAVP and DGLVP have been reported by several other investigators to alter behavior (e.g., Vawter and Green, 1980; Wang, 1972; de Wied *et al.*, 1972; Ramaekers *et al.*, 1977). It should be noted that the purity and potency of different batches of DGAVP are particularly difficult to assess, due to the reduced endocrinological activities of this peptide. Therefore, negative results obtained with this analog should be cautiously interpreted.

There is also a relatively large body of evidence, based on several experimental approaches (some of which are described above), that supports a central site of action for AVP. For example, small quantities of vasopressin or DGAVP, when administered into CSF (de Wied, 1976, Koob *et al.*, 1981) or directly into specific brain areas (Kovács *et al.*, 1979c), have potent effects on avoidance behavior, and the effects of these peptides can be eliminated by discrete lesions of certain brain areas (van Wimersma Greidanus *et al.*, 1974, 1979b; van Wimersma Greidanus and de Wied, 1976a; and see below). Some investigators have reported that AVP can increase blood pressure by acting at central sites (Matsuguchi *et al.*, 1982; Berecek *et al.*, 1984), although others report no effect on blood pressure when AVP is administered intracerebroventricularly (de Wied *et al.*, 1984). Thus, it is not inconceivable that AVP, when administered directly into the brain or CSF, may still induce behavioral effects by acting centrally to produce a peripheral response (i.e., increased blood pressure). However, it is difficult to estimate whether doses of AVP used in behavioral experiments would be expected to cause centrally mediated changes in blood pressure. Furthermore, a later paper by de Wied and co-workers provided convincing evidence against such an interpretation (de Wied *et al.*, 1984). These authors found that AVP, as well as a behaviorally active AVP fragment, administered icv, did not increase blood pressure but did maintain passive avoidance behavior. They also showed that *central* administration of a peptide antagonist of AVP blocked the behavioral, but not the pressor, effects of systemically administered AVP, and that the AVP fragment, given systemically, produced behavioral but not pressor responses (de Wied *et al.*, 1984). These results strongly support the hypothesis that the effects of vasopressin on avoidance behavior are mediated directly by central mechanisms and do not result from the aversive stimulus of a pressor response. On the other hand, Ettenberg (1984) reported that not only the behavioral effect of subcutaneously administered AVP (enhanced memory in an appetitive task), but also its aversive properties (conditioned taste aversion and place preference), were attenuated by icv administration of the peptide antagonist [1-deaminopenicillamine,2-(*O*-methyl)tyrosine]AVP, dPTyr(Me)AVP. In this study, only a very high dose of antagonist (10 μg) was effective, suggesting to the author that the site of action of the antagonist in brain was far removed from the lateral ventricle. This result differs from that of de Wied *et al.* (1984), who used a different antagonist—[1-(β-mercapto-β,β-cyclopentamethylenepropionicacid), 2-(*O*-methyl)tyrosine]AVP—and found that 3 ng of antagonist, given icv, blocked the behavioral but not the pressor effect of AVP. Ettenberg (1984) suggested that the peptide antagonist blocked the effect of AVP on memory as a result of its blockade of AVP-induced arousal, but suggested, based on the potency of AVP and related peptides to enhance memory when given icv, that AVP may alter behavior via independent mechanisms when administered centrally or peripherally (Ettenberg, 1984). A similar conclusion—i.e., that AVP is aversive, but that the aversive properties and effects on memory or behavior may be dissociable—was reached by Tam *et al.* (1985). These authors found that the aversive effect of AVP (in place preference and taste aversion paradigms) was not blocked by lesions of the dorsal noradrenergic bundle in brain, a procedure that does block the behavioral effect of AVP (see Section VII). Thus, it is possible that both central *and* peripheral effects of AVP may contribute to its effects on learning and memory.

One of the most persistent arguments opposing the hypothesis that AVP acts centrally has been that it is not clear how the hormone, when administered peripherally, reaches its site of action in the brain. The half-life of circulating vasopressin is on the order of minutes (Lauson, 1974; Gaziz, 1978; Janáky *et al.*, 1982), while its behavioral actions may persist for days or weeks (de Wied, 1971). Furthermore, the hormone apparently does not easily cross the blood–brain barrier or the blood–CSF barrier (Vorherr *et al.*, 1968; Zaidi and Heller, 1974; Ang and Jenkins, 1982; Jones and Robinson, 1982). Possible explanations for the activity of systemically administered vasopressin include a site of action at areas outside the blood–brain barrier (e.g., the median eminence or circumventricular organs) or a role for longer-acting and more hydrophobic fragments of vasopressin that are produced in plasma and can reach the brain. This latter possibility is discussed in detail in Section VII.

2. *Are the Effects of Vasopressin on Avoidance Behavior Indicative of Effects on Memory?*

While it seems clear that vasopressin can modify avoidance behavior, there is still some controversy as to the exact nature of the effect. De Wied has maintained that vasopressin enhances consolidation of memory, or formation of long-term memory, even though the hormone may not always modify short-term memory (see van Wimersma Greidanus *et al.*, 1983; King and de Wied, 1974). Other investigators, particularly Sahgal (1984), suggest that vasopressin does not directly affect any aspect of memory or learning *per se*, but instead increases the arousal or alertness of the animals and thereby alters avoidance behavior. Such an increase in arousal could be due to the aversive properties of the hormone, as described above, but could also arise from direct effects of AVP or an active hormone fragment on the neuronal systems in brain that mediate arousal. Some of the most convincing evidence for this hypothesis is derived from dose–response studies of the effects of vasopressin on avoidance behavior. Vasopressin, like many other neuropeptides, has been reported to exhibit a bell-shaped, or "inverted U-shaped," dose–response pattern. This pattern was shown by Hagan *et al.* (1982), who used doses of LVP ranging from 0.036 to 2.97 μg per rat in an active avoidance paradigm. In their experiments, lower doses of vasopressin enhanced avoidance responding, with the maximal response at 0.11 μg. Doses of 0.33 and 0.99 μg were less effective, and 2.97 μg of vasopressin reduced avoidance responding. Crine (1982) also reported that when retention was tested at 384 hr after training in a passive avoidance response, a high dose of LVP (20 IU/kg, which had been given immediately after training) *reduced* avoidance responding. A bell-shaped dose–response pattern is characteristic for drugs that increase arousal, since a small increase in arousal enhances motivation and, therefore, performance, while a larger increase in arousal distracts the animal and disrupts performance (see Sahgal, 1984). The results of Sahgal's studies also support the idea that vasopressin affects behavior by increasing alertness or arousal. He found that there was a bimodal pattern of passive avoidance responding in animals tested for retention at 24 hr after training (Sahgal, 1984; Sahgal *et al.*, 1982; Sahgal and Wright, 1983). That is, there was a group of animals with long latencies to enter the dark box ("good memory") and a group with short latencies. This distribution was more pronounced in animals that had been treated with AVP after training than in those that had been treated with saline. As discussed above, both low and high levels of arousal produce poor performance in behavioral tasks (see Saghal, 1984), and it was therefore hypothesized that AVP improves performance in animals with low levels of arousal (causing increased latencies in these animals) and disrupts learning in those that are already at or above the optimal level of arousal for passive avoidance behavior. This hypothesis was supported by the study of Saghal and Wright (1983), who found that substances known to affect arousal (i.e., amphetamine and chlordiazepoxide) had effects that were similar to vasopressin on passive avoidance retention. Crine (1984) has also provided evidence for an interaction between vasopressin and the state of "arousal" of the animal, which influences passive avoidance performance. He reported that passive avoidance behavior was impaired in rats that were housed in isolated conditions, and that this impairment was reduced in rats treated with a low dose of AVP. It was speculated that AVP increased the low arousal level in the isolated animals. It should be emphasized that in these studies, as well as others discussed so far, the interval between treatment with peptide (or other drugs) and retention testing is long enough (when peptide is administered before testing) to ensure that direct effects of the drugs on activity (e.g., sedation) do not influence the results.

It appears that only one study has attempted to directly address the question of whether AVP affects memory *per se,* or acts on related processes. $ACTH_{4-10}$ has been suggested to affect motivation or attention (e.g., van Riezen *et al.*, 1977), and this peptide, like AVP, inhibits extinction of active avoidance behavior (de Wied and Bohus, 1966). It was found that transection of the fornix and stria terminalis completely eliminated the ability of $ACTH_{4-10}$ to inhibit extinction but did not alter this effect of DGLVP (van Wimersma Greidanus *et al.*, 1979a). This result was interpreted to mean that the peptides inhibited extinction via different mechanisms: i.e., $ACTH_{4-10}$ by affecting attention, and DGLVP by affecting memory (van Wimersma Greidanus *et al.*, 1979a). However, in spite of such an interpretation, it is still quite difficult to satisfactorily distinguish between effects on arousal or motivation and effects on memory.

3. Is Endogenous Vasopressin Involved in Memory under Normal Circumstances?

The evidence presented by de Wied and co-workers in favor of a role for endogenous hormone in memory processes consists primarily of experiments showing that antibodies to vasopressin produce opposite behavioral effects to the hormone, and that Brattleboro rats, which are deficient in vasopressin, also have memory deficiencies. In a recent study, Koob and co-workers (1981) found that a peptide antagonist of AVP, dPTyr(Me)AVP, when administered subcutaneously (100 μg/rat), produced behavioral effects opposite to those of AVP—i.e., it *facilitated* extinction of an active avoidance response. The antagonist also facilitated extinction when it was administered intracerebroventricularly, but only at a very high dose (10 μg). The necessity for the high dose was suggested to result from the increased hydrophobicity of the analog, as compared to AVP, which would mean that the antagonist would more easily pass the blood–CSF barrier, thus lowering its concentration in the CSF (Koob *et al.*, 1981). In any case, the behavioral effect of the antagonist was consistent with a role for endogenous hormone in maintaining avoidance behavior. On the other hand, the results obtained with Brattleboro rats have been more confusing. Although de Wied originally reported that these animals learned normally but had memory deficiencies (de Wied *et al.*, 1975), Celestian *et al.* (1975) found that the homozygous Brattleboro rats had a slower rate of *acquisition* (learning) of a conditioned avoidance response but actually showed enhanced retention of the response as compared to heterozygous controls or normal animals. In both laboratories, it was shown that the Brattleboro rats and controls did not differ in their sensitivity to footshock (de Wied *et al.*, 1975; Celestian *et al.*, 1975). Miller *et al.* (1976) subsequently reported that Brattleboro rats differed from controls in acquisition *and* extinction of an active avoidance response, but that the faster rate of extinction was actually an artifact due to a lower performance level at the end of the training period. In a later study, however, Carey and Miller (1982) reported that Brattleboro rats had a slight but statistically significant increase in sensitivity to shock, compared to controls, and again found enhanced retention of a passive avoidance response, as well as better performance in an active avoidance response by the Brattleboro rats. In contrast, Bailey and Weiss (1979) found that Brattleboro rats displayed poorer passive avoidance behavior than normal rats, but that there was no indication of a complete lack of retention as had previously been reported. Thus, it is difficult to reach conclusions regarding the role of AVP on learning or memory based on experiments with Brattleboro rats. It has been pointed out that these animals differ from normal rats in a large number of behaviors, and they appear to be more timid and cautious than normal rats (Brito *et al.*, 1981; Brito, 1983). In the aversive conditions that prevail during avoidance training and testing, these behavioral characteristics would be expected to complicate the results. It has also been suggested that the Brattleboro rat has subtly altered motivation, emotion, and/or attention, as compared to normal rats (Williams *et al.*, 1983a), and that vasopressin may influence performance of various tasks in these rats by affecting emotional or motivational factors, rather than memory (Williams *et al.*, 1983b). Furthermore, the Brattleboro rats differ from normal rats with respect to several physiological parameters other than lack of vasopressin (see Sokol and Valtin, 1982) and are therefore not an ideal model for evaluating the role of the endogenous hormone in memory or behavior.

B. Neurohypophyseal Peptides and Appetitively Motivated Responses

The investigations of peptide effects on active and passive avoidance behavior described above have often been criticized because they deal only with aversively motivated (i.e., by footshock) behaviors. Since these tasks are quite stressful, and vasopressin is known to be released by stress, it is possible that the interpretation of results using vasopressin may be confounded. Animals (and humans) can also be motivated to perform tasks by positive reinforcement, and such tasks are referred to as appetitively motivated. In fact, the performance of positively reinforced behaviors is, presumably, a more natural situation for the animal and provides an opportunity to explore the generality of the effect of vasopressin on motivation and/or memory.

Bohus (1977) reported that DGLVP enhanced the retention by male rats of a sexually motivated choice behavior in a T maze. It was of interest that, in this study, copulation reward after a correct choice was essential in order for the animals to acquire the response, as well as for retention to be affected by DGLVP. Bohus (1977) interpreted his results as indicating that DGLVP facilitates memory consolidation. When rats were trained to discriminate black and white arms of a T maze for a food reward, subcutaneous administration of LVP during extinction (but not during acquisition) retarded the extinction of the appetitive behavior (Hostetter *et al.*, 1977). However, this inhibition of extinction was only observed in animals trained to find food in the black arm. These results again indicated that vasopressin could influence the memory of an appetitively motivated task, and also that the nature of the reward, or cues associated with the reward, were important for observation of the effect. Sara *et al.* (1982) showed that LVP (10 μg) enhanced acquisition of an appetitive task (finding food in a Y maze) when animals were trained to find the food at the end of the lighted arm of the maze. This finding is consistent with an effect of LVP on learning. When the animals were then retrained to find the food on the dark side of the maze (''reversal training''), LVP treatment *impaired* learning of the response. It was suggested that LVP both facilitates learning and increases the tendency to ''perseverate'' (Sara *et al.*, 1982); this latter activity would be consistent with the findings that vasopressin inhibits extinction of avoidance responses. Similarly, Couk and Beckwith (1982) found that, in rats trained in a brightness discrimination task, an analog of AVP ([1-deamino,8-D-arginine]vasopressin, DDAVP) given prior to learning, enhanced acquisition of the task but impaired reversal learning. In this study, apparently only learning or short-term memory was affected, since DDAVP did not affect retention of the task at 5 days after training (Couk and Beckwith, 1982). Messing and Sparber (1983) also found that DGAVP (5 or 10 μg/kg given before training and extinction sessions) facilitated acquisition and delayed extinction of a food-reinforced autoshaped lever-touching response. This latter paradigm represents another appetitively motivated task in which vasopressin was concluded to have specific effects on learning and/or memory. Similarly, Ettenberg *et al.* (1983a,b), as described earlier, found that AVP (1 μg per rat), given immediately after training, facilitated retention in a water-seeking task. It should be pointed out that in these appetitive tasks, animals are food- or water-deprived, as necessary, in order to facilitate learning of the task. There is no information as to whether the response to vasopressin is altered by such treatment, although one might postulate that depriving the animals would affect their motivational state, as has also been suggested for vasopressin.

If it is assumed that the rewards obtained by the animals in the appetitive paradigms are positively reinforcing, then it might be postulated that peptides that enhance learning or memory of these tasks also enhance the reward perceived by the animals. Dorsa and van Ree (1979) studied the effects of DGAVP on intracranial self-stimulation behavior, in order to directly assess the rewarding properties of the peptides. They found that when the stimulating current was "around threshold," DGAVP (1 μg) decreased the number of responses. Unfortunately, studies of this sort are difficult to interpret, since a compound with rewarding properties could conceivably either increase self-stimulation (by potentiating its reinforcing properties) or decrease self-stimulation (by providing reinforcement itself). In any case, the results do appear to indicate that DGAVP treatment can affect the systems in the CNS that mediate reinforcement, and this finding may provide an explanation for the effects of vasopressin on learning or memory of appetitive, as well as aversive, tasks.

In this context, it is of interest that rats given subcutaneous injections (1–4 μg) of DGLVP increased their daily ethanol intake, as compared to controls, and eventually accepted higher concentrations of ethanol than controls (Finkelberg *et al.*, 1978; Mucha and Kalant, 1978). Whether ethanol normally serves as a positive reinforcer in the rats or as an aversive stimulus is open to question; however, the results suggest that DGLVP treatment may modulate either the reinforcing or aversive properties of ethanol. DGLVP treatment also maintained a more lasting pattern of alcohol drinking behavior than that seen in saline-treated controls, suggesting that the peptide affected both acquisition and extinction of alcohol drinking behavior (Finkelberg *et al.*, 1978). On the other hand, Mello and Mendelson (1979) found no effect of DGAVP (25 or 125 μg/kg) on self-administration of food by monkeys, a measure that would also be expected to reflect the reinforcing properties of the peptide.

Not all investigators have found positive effects of vasopressin on appetitively motivated tasks. Sahgal (1983), for example, found that 1 μg of AVP, administered to homozygous Brattleboro rats, *impaired* acquisition of a food-rewarded autoshaping task. Although his interpretation of these results was that AVP caused overarousal of the animals, it must be remembered that the Brattleboro rats differ in a number of behavioral and biochemical aspects from normal rats, and these differences complicate the interpretation of such experiments. Alliot and Alexinsky (1982), using normal rats, also found no effect of LVP (220 μg administered subcutaneously immediately after training) on acquisition or extinction of a light–dark discrimination task where animals were trained to bar-press for food reward. These authors did find that LVP treatment *decreased* bar-press responding for food. In this case, the dose of LVP used was quite high, and the decreased response could reflect sedative properties of high doses of peptide (Krejčí and Kupkova, 1978).

In addition to utilizing the appetitive tasks discussed above, several investigators have used specialized learning paradigms with the intention of localizing the effect of vasopressin to some particular aspect of learning or memory. For example, the influence of vasopressin on short-term memory has been examined and, in pigeons, Teal and Evans (1982) found no effect of DDAVP on this aspect of memory. In this study, animals were trained in a task in which the reward was food. Burešová and Škopková (1980, 1982) reported variable and weak effects of vasopressin and analogs on spatial short-term memory in rats, in a paradigm using a radial arm maze with food reward. The most significant effect seen in these studies was a *worsening* of memory after DDAVP administration (Burešová and Škopková, 1980). The authors stressed, however, that short-term, "working" memory requires that a particular stimulus be forgotten as soon as the trial is over, and inhibition of extinction by vasopressin might be expected to worsen performance in this particular test. Vasopressin has also been tested in other paradigms that measure conditioned responses other than avoidance. For example, rabbits were subjected to classical conditioning and extinction of heart-rate and eyeblink responses (i.e., autonomic responses), and vasopressin (DDAVP, 5 μg/kg) inhibited the extinction of the heart-rate response, but not the eyeblink response (Hernández and Powell, 1983).

C. Summary: Effects of Vasopressin on Aversive and Appetitively Motivated Tasks and Effects on Long-Term Memory

The weight of evidence, from both aversively and appetitively motivated behavioral measures, favors the hypothesis that vasopressin does enhance both learning and long-term memory in animals. Most of the obvious pitfalls—e.g., peptide effects on performance of various tasks—have been controlled in the various studies. Whether vasopressin directly influences a component of memory or whether arousal or motivation are primarily affected remains to be clarified, as does the extent of the aversive or reinforcing properties of the peptides themselves, which may modify behavioral responses. In general, the role of vasopressin in memory and learning seems to be modulatory. A construct previously used to describe the neurobiology of memory suggests that there exist "extrinsic" and "intrinsic" neuronal systems that affect memory (see Squire and Davis, 1981). The intrinsic systems encode within themselves, presumably by changes in synaptic efficacy, information that represents learning or memory. The extrinsic systems are those that modify acquisition, retention, or retrieval of memory. Since vasopressin seems to affect learning and memory of many different kinds of tasks, one may speculate that vasopressin modifies the activity of the *extrinsic* neuronal systems, or that vasopressin-containing neurons represent, in themselves, an extrinsic system. However, the nature of the memory being affected—i.e., the intrinsic systems—may influence the efficacy of the hormone or the duration of the response to vasopressin.

D. Neurohypophyseal Peptides and Retrieval—Studies on Amnesia

The postulate that vasopressin acts on extrinsic neuronal systems associated with memory, or is a component of such a system, is supported by studies that indicate that vasopressin and related peptides, in addition to modifying acquisition and retention of memory, can also affect the retrieval of memory. In studies of active avoidance behavior, de Wied and co-workers found that vasopressin or antiserum to vasopressin would affect retention of "memory" when animals were treated 1 hr after training, but not when they were treated 6 hr after training (see van Wimersma Greidanus *et al.*, 1983). Retention testing took place at 24 hr after training, and it was therefore concluded that vasopressin had to be present during the "consolidation" phase of memory in order for retention to be affected. However, these workers also noted that hormone or antiserum given 23 hr after training—i.e., 1 hr prior to retention testing—would affect retention similarly to treatments given 1 hr after training (see van Wimersma Greidanus *et al.*, 1983). These results were interpreted to indicate that vasopressin, in addition to modulating the retention of memory, also can modify its retrieval.

More direct analysis of peptide actions on this aspect of memory involves studies of amnesia. Flexner and co-workers (Flexner *et al.*, 1963, 1967) used a paradigm in which mice were trained in a Y maze to choose the right or left arm. Footshock was given for failure to move from the stem of the Y or for errors of right–left discrimination. Intracerebral injections of puromycin were given 24 hr after training, and mice were tested for retention of the Y-maze task at 1 and at 2–3 weeks after training. Puromycin causes amnesia in this paradigm, and Lande *et al.* (1972) found that DGLVP, administered subcutaneously immediately after training (i.e., before puromycin treatment), protected the mice against the amnesia. This study was later extended to show that AVP and several analogs of AVP and LVP also protected against puromycin-induced amnesia in mice, although oxytocin was inactive (Walter *et al.*, 1975). Arginine vasopressin was the most potent hormone tested, and dose–response studies showed that all active peptides attenuated amnesia in a dose-related manner; in this particular paradigm, there was no evidence of a biphasic effect (Walter *et al.*, 1975; Flexner *et al.*, 1977). The peptide hormones attenuated puromycin-induced amnesia when administered up to 24 hr prior to training, as well as immediately after training. However, protection against amnesia diminished as the interval between training and peptide treatment increased (Flexner *et al.*, 1978; Walter *et al.*, 1980b). It is of interest that peptides consisting of the C-terminal portion of the hormones, and derivatives of these peptides, also attenuated puromycin-induced amnesia (Walter *et al.*, 1975, 1980a,b; Flexner *et al.*, 1978). The action of some of these smaller peptides was of a longer duration than that of the hormones (Flexner *et al.*, 1978; Walter *et al.*, 1980b), supporting the hypothesis that vasopressin may serve as a precursor to produce longer-acting, more metabolically stable di- and tripeptides [and possibley cyclic derivatives; see Flexner *et al.* (1978) and Walter *et al.* (1980a)] that can affect the retrieval or expression of memory. The implications of the activity of such peptides in terms of the mechanism of neuropeptide action is discussed below.

Another antibiotic, anisomycin, also induces amnesia in mice, and Judge and Quartermain (1982) showed that lysine vasopressin, given sc 1 hr after training, reversed amnesia of a passive avoidance response produced by anisomycin. Similarly, in rats, amnesia for a passive avoidance response was induced by exposure of the animals to CO_2, and this amnesia was reversed by sc administration of DGLVP either 1 hr prior to the acquisition trial or 1–6 hr prior to the test trial (Rigter *et al.*, 1974, 1975). Bookin and Pfeifer produced amnesia for a passive avoidance response in rats with either pentylenetetrazole-induced seizures or electroconvulsive shock. In each case, treatment of animals with LVP (1 μg, sc) 1 hr prior to training or 1 hr prior to retention testing attenuated the amnesia (Bookin and Pfeifer, 1976; Pfeifer and Bookin, 1978). Asin (1980), however, reported that LVP prevented diethyldithiocarbamate-induced amnesia of a passive avoidance response in rats when injected prior to retention testing, but not if the LVP was given prior to training.

It is difficult to distinguish in these experiments whether the peptides modify the processes of memory consolidation in such a way that the memory becomes insensitive to the effects of amnesia-producing agents, or whether the peptides directly affect the retrieval or expression of memory. This issue was addressed in a study by van Wimersma Greidanus (1982). It was shown first that $ACTH_{4-10}$ enhanced retention of passive avoidance behavior when it was injected prior to retention testing, but not when it was administered immediately after training. Intracerebroventricular administration of *antivasopressin* serum, either prior to retention testing, or immediately after training, led to *deficits* in passive avoidance behavior. Injection of $ACTH_{4-10}$ prior to retention testing reversed the effects of the antivasopressin serum, but only when the antiserum was also administered prior to retention testing. $ACTH_{4-10}$ had no effect in animals that received antivasopressin serum immediately after learning. These results were interpreted to indicate that vasopressin may affect both consolidation of memory *and* retrieval of memory: when consolidation is blocked by vasopressin antiserum administered after training, $ACTH_{4-10}$ is ineffective because there is no memory to retrieve. As a caveat, it should be pointed out that, in at least one paradigm [appetitive discrimination learning in rats (Sara *et al.*, 1982)], vasopressin administered immediately prior to retention testing did not affect performance; this finding was interpreted to indicate that vasopressin did not affect retrieval of memory. However, in most instances, vasopressin does appear to enhance retrieval, and it is possible that attenuation of amnesia by vasopressin may reflect actions of the hormone on both aspects of memory. The structure–

activity studies of peptide effects on avoidance behavior and on amnesia shed further light on this issue and are discussed in Section VII.

III. HUMAN STUDIES—NEUROHYPOPHYSEAL PEPTIDES AND MEMORY

The ability of vasopressin to enhance learning or memory in animals has led a number of investigators to evaluate the effects of the peptide in humans, in the hope that this naturally occurring hormone will prove to be a benign means of improving memory or reversing amnesia. Ehrensing *et al.* (1982) performed studies in humans that paralleled animal studies when they evaluated the ability of vasopressin (LVP) to modulate the acquisition and extinction of a conditioned avoidance response to smoking. They found that LVP facilitated the acquisition of the conditioned avoidance response, but apparently accelerated its extinction. In this study, vasopressin was administered intranasally 30 min prior to the aversive conditioning.

The effects of vasopressin on learning and memory in normal, as well as in cognitively impaired, patients have also been studied using more direct measures. Treatment of patients with DDAVP (intranasally, three times a day, for 2–3 weeks) led to more efficient learning of information, as measured by completeness, organization, and reliability of recall (Weingartner *et al.*, 1981a). This enhanced learning was seen in normal volunteers, as well as in patients with depression, and lasted for 4 weeks after peptide treatment was terminated. In this study, it was also found that DDAVP partially reversed the amnesia induced by electroconvulsive therapy (ECT) in two depressed patients. In another investigation, it was found that DDAVP enhanced cognitive performance in patients with early-stage progressive idiopathic dementia (Weingartner *et al.*, 1981b). After an evaluation period and practice with various tests of memory, patients were given placebo or DDAVP 1 hr prior to testing. The tests used were chosen to measure access to semantic memory structures and to learning and memory. The results indicated that DDAVP facilitated access to semantic memory structures that are part of long-term memory and, thus, facilitated learning or memory.

Several other groups of investigators have concluded that vasopressin may enhance attention in human subjects. Legros *et al.* (1978) reported that LVP, given intranasally three times daily for 3 days, induced improvements in tests involving attention, concentration, and motor rapidity, as well as in tests of visual and verbal memory. Beckwith *et al.* (1983) designed a study that could distinguish between attention and memory and found that DDAVP administered intranasally, combined with experience in performing the assigned task, improved attentional processes but not memory in normal volunteers. In another study, Beckwith *et al.* (1982) showed that DDAVP improved the performance of healthy volunteers in a visual discrimination test that required that subjects distinguish relevant from irrelevant information, but peptide treatment did not affect visual memory. In this study, it was shown that DDAVP did not affect blood pressure, heart rate, or anxiety, and it was concluded that vasopressin enhanced selective attention.

The results of studies of the effects of vasopressin on amnesia in humans have been more variable. Oliveros *et al.* (1978) reported that vasopressin treatment (intranasal administration) improved memory in three patients with posttraumatic amnesia and was less active in one patient with amnesia that was "probably of alcoholic origin." Koch-Henriksen and Nielsen (1981), on the other hand, found no effect of intranasal LVP on posttraumatic amnesia in five patients. Similarly, Jenkins *et al.* (1979, 1981) found no effect of DDAVP or DGAVP on memory, using a number of tests, in patients who had impairments of memory arising from severe head injury. Fewtrell *et al.* (1982) also found that LVP, given over a 14-day period, did not improve memory in seven head-injured patients. On the other hand, in a rather detailed study by Legros and his colleagues, lysine vasopressin was found to improve attention and motivation, as well as to induce a more normal electroencephalogram (EEG), in seven subjects suffering from posttraumatic amnesia (Timsit-Berthier *et al.*, 1980). It was felt, in this study, that the improvement in the patients was mainly in their general level of activity.

In other types of amnesia, vasopressin treatment has also produced mixed results. LeBoeuf *et al.* (1978) reported that vasopressin improved memory in a patient with Korsakoff's syndrome. This patient showed a dramatic increase in test scores after 2 weeks of vasopressin treatment, and the improvement was maintained at 2 months after the end of treatment. However, Laczi *et al.* (1983c) found no effect of DGAVP in tests designed to measure attention or short-term or long-term memory in 14 patients with Korsakoff's syndrome. Similarly, Franceschi *et al.* (1982) found no effect of LVP in six patients with Korsakoff's syndrome. These investigators also saw no effect of LVP in patients with Alzheimer's disease (10 patients) or multiinfarct dementia (eight patients). The tests used in this study measured visual, verbal, and spatial long-term memory, verbal long-term memory and retrieval, and verbal and spatial learning. Attention and concentration were also measured. Blake *et al.* (1978) found no effect of vasopressin on memory in two patients with Korsakoff's syndrome, and Jenkins *et al.* (1982) reported no effect of DDAVP, administered intranasally, in patients with Alzheimer's disease (three patients) or Korsakoff's syndrome (one patient) or in normal volunteers. In contrast, Partap *et al.* (1983) found that LVP mitigated memory impairment induced by ECT in one patient, and Koizumi *et al.* (1981) found that vasopressin improved memory in one patient with postencephalitic amnesia.

If endogenous vasopressin is involved in memory or attentional processes in humans, then patients with central diabetes insipidus might be expected to demonstrate memory or learning impairments that would be reversed by vasopressin administration. Gilot *et al.* (1980) reported generally low scores in several tests of attention and memory in five patients with diabetes insipidus but no specific profile of memory disturbances. These patients showed an improvement in their test scores after 3 days of treatment with DDAVP. The improvement was most marked in the patients who initially had the lowest scores. The authors noted that the metabolic side-effects of DDAVP in these patients could influence cognitive function, and that conclusions about a central effect were premature. Burish *et al.* (1981) reported similar results in five patients with diabetes insipidus and found that improvements were greatest in tests requiring visual coordination, short-term memory, learning of new material, and problem-solving abilities. Laczi *et al.* (1982, 1983d) also found impairments in long- and short-term memory in patients with diabetes insipidus. DDAVP or LVP normalized these memory disturbances, when given as a single injection, or when given intranasally several times daily for seven days. DGAVP also improved memory, and was more effective when administered intranasally. The effects of DGAVP persisted after discontinuation of treatment, and DGAVP did not produce endocrinological effects. Thus, it was suggested that DGAVP produced its effects on memory via a central site of action.

In addition to being evaluated for its effects on memory in humans, possible therapeutic effects of vasopressin treatment have been evaluated in various disease states in which there is cognitive impairment. Gold and his colleagues reported that DDAVP, given daily intranasally for 3–7 weeks, induced significant improvement in three out of four patients with major affective illness (Gold *et al.*, 1979). This improvement was noted in tests designed to measure the formation, encoding, and organization of long-term trace events in memory. The effects of DDAVP on long-term memory were more pronounced than those on attentional processes or short-term memory in these patients. The same group of researchers also noted that CSF vasopressin levels were significantly lower in patients with unipolar or bipolar depression, as compared to controls, and the response of plasma vasopressin to an osmotic stimulus was lower in patients in the depressed phase of bipolar depression than in the manic phase (Gold *et al.*, 1981). Raskind *et al.* (1979), on the other hand, found no differences in plasma AVP levels between depressed patients and controls, and no change in plasma AVP levels in patients recovering from depression. Thus, the possible role of endogenous AVP in depressive illness (Gold *et al.*, 1978) remains questionable.

Vasopressin (LVP) has also been tested in patients with chronic anergic schizophrenia (Korsgaard *et al.*, 1981). While there appeared to be some therapeutic effect, overall vasopressin had little influence on the schizophrenic syndrome. LVP also was reported to be ineffective in ameliorating the symptoms of Parkinson's disease (Jensen, 1980), and there were no changes in levels of extrahypothalamic (locus coeruleus or substantia nigra) vasopressin, measured at autopsy, in patients with Parkinson's disease or Huntington's chorea, as compared to controls (Rossor *et al.*, 1982).

In summary, there is evidence that vasopressin may enhance attention or motivation in humans. It is interesting that in all of the studies reviewed, mild cognitive impairment was observed in humans with diabetes insipidus, and improvement was obtained with vasopressin treatment. These results suggest that in humans, as in animals, endogenous vasopressin can enhance memory processes, perhaps by influencing arousal. On the other hand, in patients with serious brain injury or severe lesions, such as Korsakoff's syndrome or Parkinson's disease, vasopressin was ineffective in altering memory or behavior. Such results seem to be consistent with a modulatory role of vasopressin on the neuronal systems that are involved in the formation and/or expression of memory. It may be hypothesized that if these systems are destroyed or severely disrupted, the substrate for vasopressin action is eliminated and the peptide becomes ineffective. In situations where neuronal activity is altered, however—e.g., affective illness?—vasopressin treatment can modulate neuronal activity and, in some instances, return it to a more nearly normal level.

IV. NEUROHYPOPHYSEAL PEPTIDES AND DRUG TOLERANCE AND DEPENDENCE

Tolerance is defined as a reduced response to a drug in an organism that has been previously exposed to the drug (see Tabakoff and Hoffman, 1983; Tabakoff *et al.*, 1982). Functional (CNS) tolerance to the effects of drugs (i.e., tolerance that represents an adaptation at the cellular level, rather than a change in drug metabolism) has been postulated to share underlying mechanisms with learning and/or memory processes, since both tolerance and learning represent adaptive responses of the CNS to external stimuli. Furthermore, under some circumstances, tolerance includes a learned component (Tabakoff *et al.*, 1982; Hoffman and Tabakoff, 1981; LeBlanc and Cappell, 1976). For this reason, several investigators have examined the effects of vasopressin and related peptides on tolerance, and such peptides have been found to influence not only tolerance to different types of drugs, but also other drug-related behaviors.

In initial studies of ethanol tolerance, the ability of vasopressin to maintain tolerance, in analogy to its ability to maintain avoidance behaviors, was examined. Mice were made tolerant to ethanol by feeding them a liquid diet containing ethanol for 7 days (Ritzmann and Tabakoff, 1976). This treatment also produced physical dependence in the animals. After withdrawal from the ethanol-feeding regimen (and after overt withdrawal signs had dissipated), the animals were injected once daily with AVP (10 μg, subcutaneously) or saline. Tolerance to the hypnotic and hypothermic effects of a challenge dose of ethanol was ascertained at 3-day intervals after withdrawal. Tolerance in animals treated with saline dissipated over 3–7 days after withdrawal, while in animals treated with AVP, tolerance was maintained for as long as the peptide was administered (up to 18 days after withdrawal) (Hoffman *et al.*, 1978). AVP treatment had no effect on the response of the control mice to the challenge dose of ethanol, and the peptide did not affect ethanol metabolism (Hoffman and Tabakoff, 1979), but appeared to maintain the central adaptation that results in ethanol tolerance. Tolerance could be maintained by daily injections of as little as 0.4 nmol/kg (approximately 0.1 μg/mouse) of AVP (Hoffman and Tabakoff, 1979). In contrast to the various types of dose–response relationships observed for other behavioral effects of neurohypophyseal peptides, there appeared to be an all-or-none effect of AVP on ethanol tolerance; however, it is difficult to measure a graded response in the tests used for determining tolerance (Ritzmann and Tabakoff, 1976). Oxytocin was found to have no effect on ethanol tolerance, while LVP, DGLVP, and DGAVP all maintained tolerance (Hoffman *et al.*, 1978; Hoffman, 1982). Since, as discussed above, the latter two peptides are essentially devoid of peripheral endocrinological activity, these results supported the hypothesis that neurohypophyseal peptides maintain ethanol tolerance by acting at a central site. This postulate is also supported by a more recent demonstration that low doses of AVP, given intracerebroventricularly, can maintain tolerance to ethanol (Hung *et al.*, 1984).

Since testing for ethanol tolerance involves measuring the response of an animal to a challenge dose of ethanol, it is possible that the peptides might directly influence the response to ethanol, rather than modulate tolerance *per se*. However, the peptides, given prior to acute ethanol injection, did not alter the hypnotic or hypothermic response to ethanol in mice (Hoffman, 1982; Hoffman *et al.*, 1980). Furthermore, AVP, LVP, and DGLVP were found to maintain tolerance to the low-dose incoordinating effects of ethanol, both in rats and mice (Hoffman and Tabakoff, 1984; Lê *et al.*, 1982), as well as to high-dose effects (i.e., hypothermia, hypnosis), and such findings support the idea that vasopressin and related peptides modulate ethanol tolerance, rather than affecting particular responses to ethanol. The fact that the peptides can maintain tolerance both in rats and mice is also indicative of the generality of the effect.

The effects of neurohypophyseal peptides on *acquisition* of ethanol tolerance have also been studied. Crabbe *et al.* (1980) reported that, in mice, DGAVP blocked the development of tolerance to the hypothermic effect of ethanol. In this study, mice were injected daily with ethanol and received DGAVP injections immediately following the ethanol injections. In another study, mice injected twice daily with ethanol developed tolerance to its hypnotic and hypothermic effects, and injection of AVP or DGLVP daily at 2 hr after the second ethanol injection delayed the development of tolerance (Melchior *et al.*, 1983). If it is postulated that tolerance and learning share common mechanisms, then the delay in acquisition of tolerance caused by the neurohypophyseal peptides is surprising, since the peptides in general appear to facilitate acquisition of learned responses. Furthermore, the type of ethanol tolerance measured in these studies is thought to include a prominent learned component. That is, when ethanol is administered in a distinct environment, animals have been shown to develop a conditioned (learned) compensatory response that reduces the effect of ethanol (produces tolerance). The animals display this compensatory response (e.g., hyperthermia, to counteract ethanol-induced hypothermia) *only* in the environment in which they have previously received ethanol. If they are tested in an environment in which they do not "expect" ethanol, the conditioned compensatory response is not displayed, and the animals are less tolerant. This type of tolerance has been termed "environment-dependent" (Tabakoff *et al.*, 1982; Melchior and Tabakoff, 1981). In both studies of the effects of vasopressin on the acquisition of tolerance, ethanol was administered by injection, a procedure that has been shown to produce environment-dependent tolerance (Melchior and Tabakoff, 1981). Thus, the peptides blocked or delayed the acquisition of tolerance in a paradigm where, at least in part, tolerance is thought to reflect a learned response. These results may be interpreted to suggest that the effects of neurohypophyseal peptides on the maintenance of adaptive responses and on the acquisition of these responses may be mediated by different mechanisms and may, therefore, differ depending on the response (e.g., tolerance versus "memory") that is measured.

Rigter *et al.* (1980) also examined the effects of DGAVP on ethanol withdrawal symptomatology. Physical dependence on ethanol is normally ascertained by the appearance of withdrawal symptoms (e.g., see Goldstein, 1972; Mello and Mendelson, 1977), and this study was meant to determine whether DGAVP affected physical dependence. When DGAVP was infused throughout the period of ethanol exposure and withdrawal or was injected during withdrawal, withdrawal seizure activity was exacerbated (Rigter *et al.*, 1980). Although this result was interpreted as modulation of physical dependence by DGAVP, it should be pointed out that AVP has been reported to induce seizure activity in rats (Kasting *et al.*, 1980; and see below), and the effect reported by Rigter *et al.* (1980) may represent an enhancement in the severity of a particular withdrawal symptom, rather than an effect on physical dependence.

It has been accepted for some time that the diuresis observed after ethanol ingestion results from inhibition of the release of vasopressin during the time that plasma ethanol levels are rising (Beard and Sargent, 1979) [although there is some disagreement in this area (e.g., Linkola *et al.*, 1978)]. There is also some evidence for a rebound increase in vasopressin release during the time that blood ethanol levels are falling (Beard and Sargent, 1979). If ethanol changes the secretion of vasopressin, it seems possible that the *endogenous* hormone may contribute to the development or maintenance of ethanol tolerance. However, there are apparently no data available regarding the effect of ethanol treatment on brain or CSF levels of vasopressin, and there are indications that vasopressin in CSF may be regulated differentially from vasopressin that is released into the systemic circulation (e.g., Perlow *et al.*, 1982). Thus, it is not clear whether ethanol ingestion would alter the levels or turnover of vasopressin in the CNS, where the hormone would be most likely to influence tolerance. Nevertheless, some studies do suggest that endogenous vasopressin may modulate ethanol tolerance. For example, Pittman *et al.* (1982) reported that Brattleboro rats did not develop tolerance to the hypothermic effect of ethanol. Although this finding seems to demonstrate a requirement for vasopressin in the development of ethanol tolerance, it should be noted that Sprague-Dawley rats were used as controls in these experiments. It has been shown that different strains of rat develop ethanol tolerance at different rates (Tabakoff and Culp, 1984), and the appropriate control for the Brattleboro rats would have been Long-Evans rats. The fact that heterozygous Brattleboro rats also did not develop ethanol tolerance is in line with the idea that factors other than vasopressin contribute to the lack of tolerance in these animals [although it was stated that the heterozygotes had reduced levels of hypothalamic vasopressin (Pittman *et al.*, 1982)]. Finally, the results in the study by Pittman *et al.* (1982) must be compared to studies described above, in which vasopressin was found to *delay* the acquisition of tolerance.

Another study relevant to the role of endogenous vasopressin in ethanol tolerance development was that in which the effect of ethanol on urinary excretion of vasopressin was examined in selectively bred alcohol-preferring (AA) and nonpreferring (ANA) rats (Linkola *et al.*, 1977). The AA rats, which also display a greater degree of tolerance than ANA rats, were found to excrete twice as much AVP as the ANA rats following a challenge dose of 2.4 g/kg of alcohol. This study provides circumstantial evidence for a role of endogenous vasopressin in tolerance, but does not directly address the question. In preliminary studies in rats, it has been found in our laboratories that a specific peptide antagonist of AVP, $d(CH_2)_5Tyr(OEt)VAVP$ (Manning *et al.*, 1981), facilitated the extinction of ethanol tolerance (C.-R. Hung, B. Tabakoff, and P. L. Hoffman, unpublished). If these results are substantiated, they will indicate an important role for endogenous vasopressin, under normal physiological conditions, in maintaining ethanol tolerance.

As discussed earlier, vasopressin has also been implicated in other alcohol-related behaviors [i.e., intake of ethanol (Finkelberg *et al.*, 1978; Mucha and Kalant, 1978)], which are not easily compared to processes involved in learning or memory but which may reflect an influence of vasopressin on the neuronal systems involved in mediating reinforcing properties of various drugs (Fibiger, 1978; German and Bowden, 1974; Wise, 1980).

The effects of vasopressin on tolerance to other drugs, notably morphine, have also been studied, and the influence of vasopressin and related peptides on morphine withdrawal (physical dependence) has been investigated, since tolerance to and physical dependence on opiates are generally thought to result from the same underlying change in CNS function (Way *et al.*, 1969). In contrast, ethanol tolerance and physical dependence appear to be dissociable phenomena (Tabakoff and Ritzmann, 1977). Krivoy *et al.* (1974) initially demonstrated that DGLVP enhanced the development of tolerance to the analgesic effect of morphine in mice. De Wied and Gispen (1976) reported that Brattleboro rats showed a delayed development of morphine tolerance, and intracerebroventricular treatment of rats with antiserum to vasopressin was also reported to inhibit the development of morphine tolerance (van Wimersma Greidanus *et al.*, 1978). This work supported a role for endogenous vasopressin in morphine tolerance development and also indicated that effects of neurohypophyseal peptides on morphine tolerance are centrally mediated. Van Ree and de Wied (1976) found that the development of physical dependence on morphine was facilitated in rats treated with DGAVP, oxytocin, or the C-terminal tripeptide of oxytocin, prolylleucylglycinamide (PLG). Thus, initial work was consistent with most of the reported effects of vasopressin and related peptides on learning and memory, in that the peptides appeared to facilitate the development of a central adaptive response to morphine. However, it should be remembered that while DGAVP maintained ethanol tolerance that had already developed, it appeared to block or delay the *acquisition* of this tolerance (Melchior *et al.*, 1983; Crabbe *et al.*, 1980). Similar results have been found in some studies of peptide effects on morphine tolerance. Thus, Walter *et al.* (1978b, 1979, 1980a) reported that PLG, as well as derivatives of this peptide, *blocked* the development of tolerance to and physical dependence on morphine in mice, although the peptides did not affect withdrawal symptoms *per se*. This work was supported by that of Kovács *et al.* (1983). There are no obvious differences between the studies in which peptides related to vasopressin facilitated tolerance development, compared to those in which they blocked tolerance development. However, one possibly important difference is that, in the studies of Walter and colleagues, morphine pellets were implanted, while in those of de Wied, morphine was injected. As described above for ethanol tolerance, morphine tolerance can also, under some circumstances, include a learned (conditioned) component (Siegel, 1975), and the method of drug administration (i.e., discrete injection versus continuous exposure) may influence the contribution of learned or conditioned tolerance. Peptide effects on various forms of tolerance may differ, leading to the observed discrepancies. Another possible reason for such discrepancies is a species difference. In any case, it appears that peptides related to the C-terminal portion of oxytocin can modify opiate tolerance and dependence. The effectiveness of vasopressin itself is not as clear, since at least one group of workers reported that neither vasopressin nor oxytocin had any effect on morphine tolerance or dependence in mice (Schmidt *et al.*, 1978).

Other aspects of opiate-related behavior have also been reported to be affected by vasopressin. Van Ree and de Wied (1977) found that heroin self-administration in rats was attenuated by DGAVP, while intracerebroventricular administration of vasopressin antiserum facilitated heroin self-administration. On the other hand, Mello and Mendelson (1979) found no effect of DGAVP on intravenous morphine administration by morphine-dependent monkeys. Later, Post *et al.* (1982) reported that Brattleboro rats showed deficient onset and persistence of cocaine-induced behavioral sensitization (i.e., progressive and long-lasting increases in hyperactivity and stereotypy in response to repeated treatment with cocaine), which has been suggested to represent a model for learning or memory. The study of van Ree and de Wied (1977) could be interpreted to reflect an effect of vasopressin on brain reward systems, while the study of Post *et al.* (1982) was believed to indicate the importance of endogenous vasopressin in processes related to memory or tolerance.

In summary, the results of studies of the effects of vasopressin on the maintenance of ethanol tolerance seem consistent with those of studies of peptide effects on memory, and they suggest that vasopressin and related peptides may influence the activity of neuronal pathways that subserve both of these central adaptive processes. The effects of peptides on acquisition of tolerance need to be further investigated, but most evidence indicates that vasopressin and related peptides can delay the onset of ethanol and opiate tolerance, and perhaps physical dependence on opiates, although the peptides do not appear to directly block the actions of the drugs or to alter drug metabolism. The apparent differences in the effects of peptides on learning compared to those on acquisition of tolerance may eventually facilitate an understanding of the neuronal pathways involved in these processes. One of the most interesting aspects of studies of peptide modulation of drug-related behaviors is the finding that vasopressin and related peptides may modulate the activity of brain reward systems, thus altering the reinforcing properties of drugs. As discussed earlier, peptide effects on brain reward systems may also contribute to effects on learning and memory, and may, in fact, represent a unifying mechanism for a number of seemingly unrelated aspects of neurohypophyseal peptide action in the CNS.

V. EFFECTS OF OXYTOCIN IN THE CNS

Most of the studies of the effects of neurohypophyseal peptides on memory processes have focused on vasopressin and its analogs, although in many instances oxytocin has also been tested, often as a control substance. In initial studies of inhibition of extinction of avoidance behavior, oxytocin was found to be ineffective or, at high doses, to have the same effect as vasopressin (de Wied, 1971; Walter *et al.*, 1978a). However, in later studies, lower doses of oxytocin administered peripherally or oxytocin given intracerebroventricularly were reported to *facilitate* extinction of active and passive avoidance behaviors (van Ree *et al.*, 1978; Kovács *et al.*, 1978; Bohus *et al.*, 1978a,b; Kovács and Telegdy, 1982). In addition, intracerebroventricular administration of antiserum to oxytocin was reported to *enhance* retention of avoidance behaviors (van Ree *et al.*, 1978), and an oxytocin antagonist also enhanced avoidance behavior (Krejčí *et al.*, 1981). Thus, it has been suggested that oxytocin has effects opposite to those of vasopressin in modulating avoidance behavior or memory consolidation (van Ree *et al.*, 1978; Kovács and Telegdy, 1982) and that the endogenous hormone may play a role in amnesia (Kovács and Telegdy, 1982). Part of the reason for failure to observe an effect of oxytocin in earlier experiments may have been the experimental conditions: that is, in order to observe inhibition of extinction, it is desirable to obtain a rapid rate of extinction in controls. This experimental protocol could lead to a "ceiling" effect, which would make an enhanced rate of extinction difficult to detect (Kovács and Telegdy, 1982). Oxytocin has also been reported to have effects opposite to those of vasopressin on self-stimulating behavior (see Kovács and Telegdy, 1982) and on memory retrieval. For example, Bohus *et al.* (1978a) reported that oxytocin attenuated passive avoidance behavior if injected immediately prior to retention testing, consistent with a disruption of retrieval by oxytocin (i.e., oxytocin had an effect opposite to that of vasopressin on retrieval).

However, in some studies, oxytocin and vasopressin have been reported to have similar effects. Urban (1981) found that oxytocin and vasopressin, injected intraseptally at the same dose, altered EEG parameters in a similar manner. Flexner *et al.* (1977) found that high doses of oxytocin attenuated puromycin-induced amnesia, as did vasopressin. As discussed above, attenuation of puromycin-induced amnesia is also thought to represent an effect on memory retrieval or expression. Furthermore, when memory of a conditioned taste aversion was measured in chicks, oxytocin administered intracerebrally caused a dose- and time-dependent *enhancement* of retention (Davis *et al.*, 1983), like that caused by vasopressin. Oxytocin was also reported to have the same effect as DGAVP on development of physical dependence on morphine: both peptides facilitated the development of dependence and, in fact, oxytocin was *more* potent than vasopressin (van Ree and de Wied, 1976).

One possible explanation for different results regarding the relative effects of oxytocin and vasopressin may be that oxytocin can have opposite effects, depending on the brain area in which it acts. For example, oxytocin injected into the rat dorsal septal nucleus facilitated retention of an avoidance response, while the same dose of oxytocin attenuated the response if the peptide was injected into the hippocampus or dorsal raphe nucleus (Kovács and Telegdy, 1982). It is possible that high doses of oxytocin interact with vasopressin receptors to produce responses similar to vasopressin, while low doses may interact with more specific oxytocin receptors, which produce other behavioral responses. Alternatively, oxytocin may, in some cases, simply act like a partial agonist at vasopressin receptors and block responses to endogenous vasopressin. However, neither postulate can fully account for all of the behavioral responses to oxytocin, since in some paradigms in which vasopressin is active, oxytocin has no effect. For example, oxytocin has been generally ineffective in modulating acquisition ("learning") of avoidance responses (Kovács and Telegdy, 1982).

It seems premature to label oxytocin an "amnestic peptide," since its reported effects in various paradigms are conflicting. However, in the few studies of the effects of oxytocin on *human* memory, it was reported that oxytocin did not alter learning but did impair subsequent recall. In one study, subjects were tested for recall of lists of words. Oxytocin, LVP, or placebo was given at the end of a training session, and then 24 hr and 1 hr prior to a retention session. Although LVP and placebo were equally effective in enhancing memory, the subjects receiving oxytocin showed a decreased recall (Fehm-Wolfsdorf *et al.*, 1984). In a second study, oxytocin was taken intranasally three times per day for 3 days, while in a third study, tests were carried out in women receiving intravenous infusions of oxytocin to induce therapeutic abortion. In the latter two cases, word-pair recall and picture-matching tests were used, and oxytocin decreased recall (Ferrier *et al.*, 1980; Kennett *et al.*, 1982). It was speculated by Ferrier and colleagues (Ferrier *et al.*, 1980) that the amnesic action of oxytocin could have the effect of focusing maternal attention on the maternal–infant unit by isolating the mother from external influences.

Another important aspect of the response to oxytocin, which will be discussed in more detail below, is the finding that the C-terminal tripeptide of this hormone, prolylleucylglycinamide (PLG), is active in a number of behavioral paradigms, in addition to influencing morphine tolerance and physical dependence as described in Section IV. In studies related to learning or memory, this tripeptide usually has actions similar to those of oxytocin (Walter *et al.*, 1978a), but it has also been found to mimic the effects of vasopressin in some instances (Walter *et al.*, 1980a; Davis *et al.*, 1982b). It has been suggested that oxytocin may act as a precursor in the CNS to produce this more stable and longer-acting tripeptide, as well as derivatives of the tripeptide (Walter *et al.*, 1980a), which may influence memory processes.

Oxytocin has also been reported to influence behaviors other than those previously found to be affected by vasopressin. Pedersen and Prange (1979) reported that oxytocin, administered intracerebroventricularly to virgin female rats, induced maternal behavior in these animals. In this study, it was found that elevated levels of estrogen were required in order for oxytocin to exert its effect. In a later study, it was found that the ring structure of oxytocin, tocinoic acid, also induced maternal behavior, and that vasopressin produced the same behavior, but with a later onset of action (Pedersen *et al.*, 1982). The effect of oxytocin on maternal behavior is intriguing, but it has not been reproducible in all studies

(Rubin *et al.*, 1983). It is probable that the experimental conditions and handling of the animals are important for observing this behavior, and the influence of these conditions, as well as peptide dose, on maternal behavior needs further investigation.

VI. CNS EFFECTS OF NEUROHYPOPHYSEAL PEPTIDES NOT RELATED TO LEARNING OR MEMORY

There are a number of behavioral and physiological effects of vasopressin and related peptides that seem to be centrally mediated but that appear to be unrelated to effects of the hormone on learning or memory. Some of these actions are only observed with high doses of peptides, while others may represent physiological functions subserved by the hormone.

A. Effects on Spontaneous Behaviors in Mice

Rees *et al.* (1976) and Delanoy *et al.* (1978, 1979) originally reported that relatively high doses of vasopressin (LVP, AVP), as well as oxytocin and arginine vasotocin (AVT), administered intracerebroventricularly, caused hyperactivity, grooming, scratching, foraging, and squeaking in mice. These actions were thought to be centrally mediated, although DGLVP was less active than the other hormones (Delanoy *et al.*, 1978). Meisenberg and Simmons (1982, 1983) reported similar findings using lower doses of the hormones and reported that, in stressful situations, the hormones produced escape-directed activity (Meisenberg, 1982). A study of structure–activity relationships for these activities revealed that the structural requirements for induction of spontaneous behaviors in mice were closely related to those necessary for the effects of vasopressin on blood pressure. This result suggested to the authors that the behavioral effects of vasopressin were mediated by receptors similar to those that mediate the peptide's effects on blood pressure (and, incidentally, different from the receptor involved in effects of vasopressin on memory) (Meisenberg and Simmons, 1982).

B. Convulsive and Hypothermic Effects

Kasting *et al.* (1980) reported that AVP, administered intracerebroventricularly (1.0 μg) to rats, induced an immediate drop in body temperature. This same dose of AVP caused periods of immobility and staring. However, upon a second injection of the same dose of AVP (2 days later), rats developed myoclonic–

myotonic seizures. Thereafter, a dose of 10 ng of AVP was sufficient to induce seizure activity. These results suggested that AVP could induce seizures by a kindling-like process, whereby the animals became more sensitive to the seizure-producing effect of AVP upon repeated treatment.

Several other groups have also reported that AVP, injected either subcutaneously or intracerebroventricularly, causes a drop in body temperature in rodents (Lee and Lomax, 1982; Crine *et al.*, 1981; Hoffman, 1982). It has been suggested that AVP may play a role as an endogenous antipyretic, since plasma AVP levels are increased in febrile animals. Furthermore, since repeated exposure to AVP induced seizure activity, it was postulated that the increase in AVP levels during fevers might be involved in the onset of febrile convulsions (Lederis *et al.*, 1981; Ruwe *et al.*, 1983). Thus, Kasting *et al.* (1981) found that, in the absence of AVP (i.e., in Brattleboro rats or in animals given AVP antiserum), the threshold body temperature at which convulsions occurred was significantly increased. However, no elevations in AVP in the CSF were observed in children who had had a recent febrile convulsion (Lederis *et al.*, 1981), and the role of AVP in producing seizures, especially in humans, needs further investigation. In this context, it is of interest that AVP, as well as some of its analogs, given intracerebroventricularly to rats, induces a characteristic type of convulsion known as "barrel rotation" (rotation along the sagittal body axis) (Yamada and Furukawa, 1981; Abood *et al.*, 1980; Kruse *et al.*, 1977). This effect is observed at relatively high doses (>40 ng/rat, icv), and may represent a toxic effect. In one instance, oxytocin and vasopressin were found to produce similar convulsive activity (Kruse *et al.*, 1977), while in another experiment, oxytocin blocked the effect of vasopressin (Abood *et al.*, 1980). While "barrel rotation" is an unusual behavior and may be somewhat nonspecific, it may also be indicative of the potential of vasopressin for inducing or enhancing seizure activity via a central site of action.

C. Corticotropin-Releasing Activity

Vasopressin has been shown to enhance release of ACTH from the anterior pituitary (Lutz-Bucher *et al.*, 1981), and at one time the hormone was suggested to be the hypothalamic corticotropin-releasing factor (CRF). Although a different peptide is now believed to be CRF (Vale *et al.*, 1981), there is evidence that vasopressin can potentiate the actions of CRF. A number of investigators have reported that AVP enhances the activity of synthetic CRF in *in vitro* assays using anterior pituitary cells (Beny and Baertschi, 1982; Culler *et al.*, 1983; Lutz-Bucher and Koch, 1983), although a similar effect was not observed in the intermediate lobe of the pituitary (Lutz-Bucher and Koch, 1983). In general, the experimental data indicate that endogenous vasopressin may play a role in mediating the release of both ACTH, as well as beta-endorphin, from the anterior pituitary.

D. Other Effects

Vasopressin has been reported to produce a non–opiate-mediated analgesia, both in rats and in mice. In rats, analgesia has been observed after either subcutaneous or intracerebroventricular administration (Berntson and Berson, 1980; Kordower *et al.*, 1982), and in mice, only after subcutaneous administration (Berkowitz and Sherman, 1982). Thus, it is not clear if the analgesia produced by vasopressin is a result of a central or peripheral action; however, it should be noted that high doses (1–10 μg/kg) of the peptide were used for the intracerebroventricular injections in mice. Oxytocin, in contrast to vasopressin, was not reported to produce analgesia (Berkowitz and Sherman, 1982).

By measuring open-field activity or exploratory behavior following systemic administration of hormones, it has been concluded that oxytocin affects arousal (Crine *et al.*, 1983) and that LVP enhances habituation (Höglund and Meyerson, 1982), respectively. Krejčí and Kupková (1978) reported that an analog of vasopressin, [deamino,6-carba,8-ornithine]vasopressin, reduced exploratory behavior and induced sleep in rats; in mice, only a small depression of locomotor activity occurred. It is not clear whether the behavior observed was natural sleep or perhaps a preconvulsive state, although it was reported that sleep was maintained for 2–4 hr. In this study, peptides were administered subcutaneously. AVP has also been reported to increase water intake in dogs after intracerebroventricular administration (Szczepańska-Sadowska *et al.*, 1982), to inhibit sexual behavior in female rats by acting at a central site (Södersten *et al.*, 1983), and to regulate brain water permeability (Raichle and Grubb, 1978; Raichle *et al.*, 1978).

E. Conclusion

While the studies described in this section do not deal directly with effects of neurohypophyseal peptides on learning or memory, there are certain aspects of these results that might influence the results obtained in tests of memory or in tests of drug tolerance. For instance, peptide effects on arousal, locomotor activity, or seizure activity (e.g., immobility) could seriously interfere with the performance of a learned task used to measure "memory." The analgesia induced by vasopressin should be considered and controlled when aversive tests are used for assessing memory. In many cases, tolerance to a drug is assessed by measuring the hypothermic response to the drug; it must be kept in mind that the peptides also influence thermoregulation. Finally, the interpretation of effects of vasopressin on various behaviors should include an awareness that, especially under stressful conditions, vasopressin may increase the release of ACTH and beta-endorphin from the pituitary. Thus, behavioral responses to neurohypophyseal peptides may be secondary effects due to release of other hormones. Although in many investigations such confounding factors have been considered, they may contribute to discrepancies among various studies and should be kept in mind, both in performing experiments and while interpreting results.

VII. POSSIBLE SITES AND MECHANISMS OF ACTION OF NEUROHYPOPHYSEAL PEPTIDES IN THE CNS

In the periphery, neurohypophyseal hormones interact with specific receptors in the cell membranes of their target organs (Roy *et al.*, 1975a,b; Jard, 1983) as a first step in producing endocrine effects. If these peptides produce behavioral responses by their actions in particular areas of brain, it seems likely that, like other neuropeptides (Robberecht *et al.*, 1978; Uhl *et al.*, 1977; Olson *et al.*, 1980), neurohypophyseal peptides would also interact with specific binding sites in neuronal cell membranes. The identification and localization of the CNS site(s) of action of vasopressin and related peptides has been approached in several ways. De Wied and his colleagues examined the ability of LVP to inhibit extinction of an active avoidance response when the hormone was injected directly into particular brain areas. They found that LVP was active when injected into the posterior thalamic region, including the parafascicular nucleus, but was inactive when injected into the ventromedial or anteromedial thalamus, posterior hypothalamus, substantia nigra, reticular formation, substantia grisea, putamen, or dorsal hippocampus (van Wimersma Greidanus *et al.*, 1973, 1975c). In contrast, Stark *et al.* (1978) reported that intrahippocampal injection of LVP *did* result in retention of an active avoidance response. Similarly, Kovács *et al.* (1982) reported that injection of antiserum to AVP into the dorsal hippocampus *attenuated* passive avoidance behavior, an effect opposite to that of AVP. Bilateral injections of AVP (25 pg) into the dentate gyrus, dorsal septal nuclei, or dorsal raphe nuclei were reported by Kovács *et al.* (1979c) to enhance passive avoidance behavior. However, Hartmann *et al.* (1979) did not observe this behavioral response to vasopressin that was injected into the dorsal raphe nucleus, locus coeruleus, or pars compacta of the zona incerta. Some possible reasons why the results from different laboratories are not entirely compatible include diffusion of peptides from the site of the microinjection, as well as difficulties in estimating the peptide dose at the site of injection. However, the results in general indicate that vasopressin acts within the limbic system to modify avoidance behaviors (i.e., memory consolidation or arousal). Microinjection studies have also been used to identify a site of action for vasopressin with respect to effects on *retrieval* of memory. Bohus *et al.* (1982) reported that bilateral injections of AVP into the central amygdala or the dentate gyrus, but not the dorsal septal region or the dorsal raphe nucleus, improved passive avoidance behavior in rats that had been made amnesic by pentylenetetrazole treatment. Keeping in mind the caveats mentioned above with respect to microinjection sites, one may postulate that vasopressin's effects on retrieval of memory and "memory consolidation" could be mediated by receptors in different brain areas.

Another means of localizing the site of action of vasopressin involves brain lesioning. Lesions of the parafascicular area reduced, but did not block, the ability of vasopressin to inhibit extinction of active avoidance behavior (van Wimersma Greidanus *et al.*, 1974). These results suggested that the parafascicular area was important, but not essential, for the action of vasopressin in this paradigm. On the other hand, extensive lesions of the antero–dorsal hippocampus, lesions of the amygdaloid complex, and lesions of the preoptic rostral septal area blocked the effect of vasopressin on extinction of a conditioned avoidance response (van Wimersma Greidanus and de Wied, 1976a; van Wimersma Greidanus *et al.*, 1979a, 1983). Because rather extensive lesions were necessary to block the action of vasopressin, the results of these studies were interpreted to mean that an intact limbic system is necessary in order for vasopressin to influence memory consolidation. However, in other studies it was found that transection of the fornix and the stria terminalis—destroying connections between the limbic system and the hypothalamus—did *not* block the ability of vasopressin (DGLVP) to maintain an active avoidance response (van Wimersma Greidanus *et al.*, 1979b). This result suggested to the authors that the septal region and the dorsal hippocampus may represent "restricted and independent anatomical substrates" for vasopressin action and that an intact limbic system was *not* necessarily required.

All of these studies have attempted to determine a CNS site of action for vasopressin that is administered exogenously. The site of action of *endogenous* hormone has also been the subject of a number of studies. Kovács *et al.* (1982) injected antiserum to AVP directly into the dorsal hippocampus and found an attenuation of passive avoidance behavior. A similar effect was observed after microinjection of antiserum to AVP into the dorsal raphe nucleus (Kovács *et al.*, 1980). Following the hippocampal injection, it was shown that the antiserum spread toward the ventral hippocampus and the dorsal septum (Kovács *et al.*, 1982). Therefore, while these studies may not provide an exact localization of the site of action of endogenous vasopressin, they do suggest a role in memory processes of the hormone found within the limbic–midbrain areas.

As mentioned in the introduction, it was originally suggested that vasopressin might be transported to its site(s) of action in the brain via the CSF, particularly since exogenous hormones, injected intracerebroventricularly, were potent modulators of behavior (de Wied, 1976, 1977). Theoretically, the endogenous hormone could be released from the posterior pituitary into the circulation and taken up into the CSF, or the hormone could be secreted directly into the cerebral ventricles (Zaidi and Heller, 1974; Zimmerman *et al.*, 1975). In order to obtain information about the importance of such a pathway, concentrations of vasopressin in blood and CSF have been measured during behavioral training. No correlation was observed between retention of passive avoidance behavior and levels of vasopressin in the peripheral circulation (van Wimersma Greidanus *et al.*, 1979c). Thus, changes in systemic levels of vasopressin are probably not important for the actions of the hormone on behavior. Since the half-life of vasopressin is short (Janáky *et al.*, 1982; Gaziz, 1978; Lauson, 1974), and apparently little of the peptide crosses the blood–brain barrier (Ang and Jenkins, 1982), it is not surprising that endogenous hormone released into the circulation would not play a major role in the behavioral actions of vasopressin. More positive results were reported when vasopressin was measured in eye plexus blood, which has a higher level of AVP than the peripheral circulation in general (Laczi *et al.*, 1983a). Thompson and de Wied (1973) found a positive correlation between the level of antidiuretic activity in eye plexus blood and the level of retention of passive avoidance behavior. However, although Laczi *et al.* (1983a) replicated these results in Wistar rats, they found that heterozygous Brattleboro rats showed maximal retention of passive avoidance behavior, even though the levels of AVP in eye plexus blood were not different from the levels of AVP in eye plexus blood of controls (i.e., nonshocked rats). Therefore, the conclusion was that changes in circulating levels of endogenous AVP are not important for the behavioral effects of vasopressin.

If vasopressin is secreted *directly* into the CSF and transported to its site of action by this route, changes in CSF levels of AVP might be expected during training and testing of avoidance behaviors. Van Wimersma Greidanus *et al.* (1979c), however, found no correlation of CSF levels of AVP with retention of passive avoidance behavior; similarly, Mens *et al.* (1982) reported that levels of vasopressin in CSF did not change during acquisition or testing for retention of passive avoidance behavior. Thus, even though, under some circumstances, it appears that release of vasopressin and oxytocin into CSF is regulated differentially from release into the systemic circulation (Jenkins *et al.*, 1980; Amico *et al.*, 1983; Perlow *et al.*, 1982) [although osmotic regulation of release seems similar (Doris and Bell, 1984; Barnard and Morris, 1982)], suggesting that vasopressin can be secreted directly from the brain into the CSF, there is no good evidence that endogenous hormone is transported to its sites of action in brain via the CSF during behavioral training or testing.

It was originally believed that the only neurons containing neurohypophyseal hormones were the neurosecretory neurons that projected from the supraoptic and paraventricular nuclei of the hypothalamus to the posterior pituitary. More recently, vasopressin and oxytocin have been localized in a number of brain areas (Rossor *et al.*, 1981; Dogterom *et al.*, 1978), and numerous hypothalamic and extrahypothalamic vasopressin-containing neuronal tracts have been identified. For example, vasopressin-containing processes extend from the hypothalamus to various limbic areas: diencephalon, mesencephalon, brainstem, and spinal cord (Sofroniew, 1980; Buijs, 1978; Sofroniew and Schrell, 1981; Sofroniew and Weindl, 1978a). Cell bodies containing vasopressin have also been identified in the suprachiasmatic nucleus (Sofroniew and Weindl, 1978b; Vandesande *et al.*, 1975; van Leeuwen *et al.*, 1978). Projections from this nucleus to the lateral septum and the medial dorsal thalamus and lateral habenula have been described (Sofroniew and Weindl, 1978b). Dendrites of vasopressin-containing neurons have been observed to terminate on cell bodies and dendrites of noradrenergic neurons, and noradrenergic and other synaptic inputs to vasopressin-containing neurons have also been identified (McNeill and Sladek, 1980; Silverman and Oldfield, 1984). Buijs and van Heerikhuize (1982) found that, in brain areas having terminations of vasopressin- or oxytocin-containing fibers, calcium-dependent release of the hormones evoked by potassium or veratridine was demonstrable. Thus, at least in this aspect, the peptides act like neurotransmitters that are released at synapses in the CNS. All of these data suggest that vasopressin (and oxytocin) may be released, as neurotransmitters or neuromodulators, directly at their site(s) of action in the brain, leading to modification of behavior. When AVP content of specific brain regions was measured by radioimmunoassay during passive avoidance testing, there were significant increases in some brain areas and significant decreases in others (Laczi *et al.*, 1983b). These changes are provocative, although their interpretation is difficult. That is, since synthesis of hypothalamic vasopressin appears to occur via a high-molecular-weight precursor (see Verbales and Robinson, 1983; Land *et al.*, 1982), and metabolism of the hormone is very rapid, the meaning of changes in vasopressin levels is somewhat obscure. Changes in peptide levels could reflect, for example, changes in rate of release from neurons, rate of synthesis, rate of processing of precursor, or other factors. Nonetheless, this study provides some evidence for altered activity of vasopressin-containing neurons during passive avoidance testing. The major problem of dissociating changes in vasopressin levels as a result of stress from changes associated with learning or memory (especially in paradigms that are aversively motivated) has yet to be addressed. Although the interpretation of changes in levels of vasopressin is not altogether clear, it is interesting to note that vasopressin levels in brain regions receiving innervation from the supraoptic and paraventricular nuclei were altered in response to osmotic stimuli, while hormone levels in the suprachiasmatic nucleus, the lateral septal nucleus, and the nucleus of the solitary tract were unaffected (Zerbe and Palkovits, 1984). Thus, it may well be possible at some point to isolate the vasopressin-containing neurons that are primarily concerned with modulating behavioral responses from those that are involved in classical homeostatic physiological processes.

Neurohypophyseal hormones produce their effects on peripheral target organs—e.g., kidney, vascular system, uterus, mammary glands—by interacting with specific receptors (Alexandrova and Soloff, 1980; Jard, 1983; Soloff,

1979). As mentioned above, it is reasonable to suppose that hormone effects in the CNS are also initiated by hormone–receptor interactions, especially since specific binding sites in brain have been described for other neuropeptides. It has been technically difficult to characterize the binding of radioactively labeled neurohypophyseal peptides in brain tissue, presumably because of the chemical, as well as metabolic, instability of these peptides. Very recently, however, specific binding sites for AVP in hippocampal synaptosomal membranes have been described (Pearlmutter and Constantini, 1984). One way to determine whether such binding sites represent receptors associated with the behavioral effects of vasopressin is to compare the binding of various analogs to these sites with the behavioral potencies of the analogs. Quite a large number of neurohypophyseal hormone analogs is available (Chapter 2, this volume), and some of these peptides have been used in structure–activity studies of behavioral responses. Walter *et al.* (1978a) examined the effects of a large number of analogs, administered subcutaneously, on inhibition of extinction of active avoidance behavior in rats. In brief, the results indicated that the N-terminal "ring" structure of vasopressin was important for this behavioral activity: that is, changes in amino acid residues in positions 2 or 3 of the hormone structure reduced or eliminated activity, so that oxytocin and vasotocin (in which the phenylalanine residue that is present in position 3 of vasopressin is replaced by isoleucine) were relatively inactive. However, an analog comprising the vasopressin "ring" alone (i.e., residues 1–6), pressinoic acid, was also inactive. In addition, C-terminal fragments of oxytocin and derivatives of these fragments did not inhibit the extinction of active avoidance behavior (Walter *et al.*, 1978a). In another study, de Wied found slightly different results, using the same behavioral paradigm, when peptides were administered intracerebroventircularly: pressinamide (an analog of the N-terminal "ring" structure of vasopressin) was nearly as active as AVP, although the C-terminal tripeptide of AVP was still relatively inactive (de Wied, 1976). He suggested that pressinamide contains the essential characteristics necessary for inhibition of extinction of active avoidance behavior by vasopressin, although a second "active site" must be present in the C-terminal part of the hormone, since the N-terminal peptide was not as active as the parent hormone. Another explanation for such results is that the entire vasopressin molecule is required for optimal activity. De Wied (1976) also suggested that vasopressin might serve as a precursor for more metabolically stable compound(s) that exert effects on behavior; this possibility is discussed in more detail below.

Investigations of the structure–activity relationship for attenuation of puromycin-induced amnesia by neurohypophyseal peptides (i.e., retrieval of memory) have also been performed (Walter *et al.*, 1975, 1980a). In these studies, the peptides were administered subcutaneously. The results of these investigations indicated the importance of the N-terminal vasopressin "ring" structure for effects on memory retrieval, similar to results regarding memory "consolidation," in that changes of amino acid residues in positions 2 or 3 of the hormone-produced analogs with reduced activity or no activity. However, structural requirements for this action of the peptides seemed somewhat less stringent than for peptide effects on extinction of active avoidance behavior, since arginine vasotocin and oxytocin could both attenuate amnesia, albeit at higher doses than AVP (Flexner *et al.*, 1977; Walter *et al.*, 1980a). Another difference between the effects of neurohypophyseal peptides on the two aspects of memory was that peptides related to the C-terminal structure of oxytocin and vasopressin, and derivatives of these peptides, were very potent in attenuating puromycin-induced amnesia (Walter *et al.*, 1975, 1980a; Flexner *et al.*, 1977).

The structural requirements of neurohypophyseal peptides for maintenance of ethanol tolerance have also been examined (Hoffman, 1982). It was found, as for the other behavioral activities, that the N-terminal "ring" structure of vasopressin was important for maintenance of ethanol tolerance. Oxytocin, at all doses tested, did not affect the maintenance of tolerance (Hoffman *et al.*, 1978; Hoffman and Tabakoff, 1979). However, an analog of the N-terminal "ring" structure of vasopressin alone, deaminopressinamide, also did not maintain ethanol tolerance, suggesting that, as for inhibition of extinction of active avoidance behavior, the whole vasopressin molecule may be important for maintenance of tolerance. The C-terminal peptides of oxytocin, and derivatives of these peptides, were relatively inactive in maintaining ethanol tolerance (Hoffman, 1982). In general, the structure–activity relationships for peptide maintenance of ethanol tolerance and peptide inhibition of extinction of active avoidance responding were similar. This may suggest that similar receptors mediate peptide influences on these two adaptive phenomena and/or that common biochemical mechanisms underly these two processes. In contrast, peptide effects on memory retrieval may be mediated by a different set of CNS receptors (i.e., vasopressin receptors and/or those which recognize C-terminal peptide fragments; see below).

The vasopressin binding sites that have recently been identified in brain areas other than hippocampus (Biegon, 1984; Pearlmutter *et al.*, 1983; Dorsa *et al.*, 1983) and in the pituitary (Baertschi and Friedli, 1985; Antoni, 1984) have not yet been defined in enough detail to permit any evaluation of their status as possible receptors mediating behavioral responses. When one compares results of structure–activity studies with those of ligand binding experiments in hippocampal tissue (Pearlmutter and Constantini, 1984), one finds a relatively close relationship between peptide affinity at hippocampal binding sites and the potency of peptide effects on extinction of avoidance behavior and maintenance of ethanol tolerance. There are a greater number of discrepancies when affinities observed in the ligand binding study are compared to the potencies of peptides to attenuate puromycin-induced amnesia.

Such results could be taken as a very preliminary indication that ligand binding sites in hippocampus represent vasopressin receptors that mediate peptide effects on particular behaviors or aspects of memory. However, the characterization of a ligand binding site as a receptor requires that interaction of ligand with the binding site result in some measurable (biochemical or behavioral) response. In the CNS, vasopressin interaction with receptors may eventually result in effects on memory, attention, or tolerance, but it will be necessary to characterize the more proximate response to hormone–receptor interaction if we are to understand the basic mechanisms leading to the behavioral responses (i.e., effects on memory "consolidation" or retrieval, or drug tolerance).

There have been a number of demonstrations of the electrophysiological effects of vasopressin and related peptides. AVP and oxytocin were reported to produce depolarization of motoneurons in an isolated spinal-cord preparation of newborn rat (Suzue *et al.*, 1981). AVP applied iontophoretically to lateral septal neurons has been reported by one group to enhance the excitatory responses induced by amino acids on stimulation of fimbria–fornix fibers (Joëls and Urban, 1984). Marchand and Hagino (1982), however, reported that iontophoretically applied AVP *decreased* the amplitude of the fimbria-evoked field response and decreased the frequency of fimbria-evoked action potentials in lateral septal neurons. Both LVP and AVP were reported to *increase* the firing rate of the majority of neurons tested when the peptides were applied microiontophoretically to the locus coeruleus of the rat (Olpe and Baltzer, 1981). Neurons in the locus coeruleus other than those containing norepinephrine were less sensitive to the effects of LVP. Vasopressin (AVP and LVP) and oxytocin (but *not* DGAVP) were also found to increase the rate of firing of neurons in the CA1 area of the hippocampus (Mühlethaler *et al.*, 1982). In this study, extracellular recordings were obtained, using slices perfused with solutions of the peptides.

Huston and Jakobartl (1977), using EEG recordings, found that LVP injected directly into the hippocampus elicited spreading depression in this brain area, reflecting a depolarizing action. However, LVP did not induce this depression when injected into the cortex. They concluded that neurons in the hippocampus may be particularly sensitive to vasopressin. In other EEG studies, vasopressin deficiency in rats (i.e., Brattleboro rats) was reported to decrease the hippocampal theta rhythm during rapid eye movement (REM) sleep, and subcutaneous administration of vasopressin returned this rhythm to normal (Urban and de Wied, 1975). DGAVP also enhanced the high frequencies of the hippocampal theta activity in normal animals when administered icv (Urban and de Wied, 1978). The site of action for vasopressin for modulating the hippocampal theta rhythm in rats appeared to be the septum, since intraseptal injection of AVP increased peak and mean theta frequencies during REM (paradoxical) sleep episodes (Urban, 1981). It is of interest that oxytocin was reported to have opposite effects on the hippocampal theta rhythm when injected intraseptally

(Urban, 1981) as compared to intracerebroventricularly (Bohus *et al.*, 1978b), although vasopressin's effects were the same in these two studies. As discussed above, oxytocin may have different effects, depending on the sites in brain that are accessible to the peptide. In humans, LVP did not change the EEG pattern associated with REM sleep or auditory evoked potentials (Timsit-Berthier *et al.*, 1982). However, LVP increased the amplitude of contingent negative variation (CNV) and prevented its habituation. CNV is a slow potential that reflects the cortical changes occurring under a conditioning paradigm (Timsit-Berthier *et al.*, 1982). The effect of LVP on the amplitude of CNV was slow in onset, distinguishing it from other stimulants such as amphetamine. The interpretation of these human EEG studies is complex, but they do indicate that vasopressin can alter neuronal activity, in line with the more direct electrophysiological studies in animals. The sensitivity of cortical EEG activity in humans, as opposed to animals, may be indicative of a species difference in the site(s) of action of vasopressin in the CNS.

In the CNS, as in the periphery, increased turnover (i.e., synthesis, release, metabolism) of neurotransmitters is usually associated with increases in neuronal activity (Aghajanian *et al.*, 1973; Gallagher and Aghajanian, 1975; Salzman and Roth, 1980a,b). Therefore, based on the electrophysiological studies suggesting that vasopressin affects neuronal activity, it may be expected that vasopressin would also affect neurotransmitter metabolism. It has been demonstrated that vasopressin increases the turnover of catecholamines [norepinephrine (NE) and dopamine (DA)] in a number of brain areas (Tanaka *et al.*, 1977a,b; Kovács *et al.*, 1979a,b; Delanoy *et al.*, 1982; Telegdy and Kovács, 1980). The effects on NE were primarily localized in midbrain–limbic regions (i.e., dorsal septum, dentate gyrus, parafascicular nucleus, dorsal raphe nucleus, and locus coeruleus) (Tanaka *et al.*, 1977a,b; Kovács *et al.*, 1979a). Telegdy and Kovács (1980) also found that AVP increased the rate of DA metabolism in the septum and striatum and increased the NE metabolism in the hypothalamus. In frontal cortical slices, LVP (which had been administered intracerebroventricularly) was reported to increase the accumulation of radioactively labeled DA, but not NE, from labeled tyrosine. In these studies, LVP inhibited the accumulation of DA in striatum and did not affect NE or DA accumulation in septum (Delanoy *et al.*, 1982). Differences in results among these investigations may be attributable to differences in methodology. In the former set of studies (Tanaka *et al.*, 1977a,b; Kovács *et al.*, 1979a; Telegdy and Kovács, 1980), a tyrosine hydroxylase inhibitor (alpha-methyl-*p*-tyrosine, AMPT) was administered, and the rate of decline of catecholamines was studied. Under these conditions, synthesis of catecholamines is inhibited, and their rate of decline is thought to reflect the activity of catecholamine-containing neurons (i.e., release and metabolism of catecholamines, leading to reduced levels). In the latter study (Delanoy *et al.*, 1982), accumulation of catecholamines reflects their rate of synthesis from the precursor, tyrosine. If neurohypophyseal peptides produce a direct effect on synthesis or release of catecholamines, rather than, or in addition to, an effect on neuronal activity *per se* (in which case both synthesis and release are expected to be affected in a parallel manner), one would expect a different response using the different techniques. It is of interest that LVP did not appear to alter catecholamine synthesis, uptake, or release when incubated *in vitro* with slices of striatal, cortical, nigral, or hippocampal tissue (Delanoy *et al.*, 1982). These findings may indicate simply that some of the effects observed *in vivo* represent indirect responses to the hormone.

In any case, there is good evidence for the involvement of noradrenergic systems in the behavioral effects of vasopressin. Kovács *et al.* (1979c) showed that microinjections of AVP into the hippocampal dentate gyrus, dorsal raphe nucleus, or dorsal septal nucleus facilitated passive avoidance behavior and also altered NE metabolism (measured after alpha-methyl-*p*-tyrosine treatment) in the areas injected, as well as in other brain nuclei. These authors concluded that changes in catecholamine metabolism represent an important factor in peptide influences on behavior. Furthermore, while AVP administered subcutaneously was found to enhance the retention of passive avoidance behavior, prior destruction of the ascending dorsal noradrenergic bundle with the neurotoxin 6-hydroxydopamine (6-OHDA) blocked this effect of AVP (Kovács *et al.*, 1979c). Norepinephrine is not the only neurotransmitter that influences the CNS actions of vasopressin, however. Kovács *et al.* (1979c) have also provided evidence that, in rats, serotonergic systems influence the ability of vasopressin to facilitate the retention of passive avoidance behavior. In their studies, injection of vasopressin directly into the locus coeruleus, which contains the cell bodies of the noradrenergic neurons in the dorsal bundle, did not affect passive avoidance behavior. Since 6-OHDA-induced lesions of the dorsal bundle had blocked the effect of systemically administered vasopressin, it was concluded that vasopressin's actions involved an effect on noradrenergic terminals. This postulate was supported by their findings that vasopressin, when injected directly into the dorsal raphe nucleus (which receives projections from the locus coeruleus), facilitated passive avoidance behavior. This action was blocked by 6-OHDA lesions of the dorsal raphe nucleus. However, this action of vasopressin was also blocked when 5,6-dihydroxytryptamine, a serotonergic neurotoxin, was injected into the dorsal raphe nucleus, which also contains cell bodies of serotonin-containing neurons. Kovács *et al.* (1979c) therefore postulated that vasopressin acts at noradrenergic terminals to alter the release of norepinephrine, which then alters the activity of serotonergic neurons. This noradrenergic–serotonergic interaction mediates the effect of vasopressin on avoidance behavior.

Furthermore, Ramaekers *et al.* (1977) and Leonard *et al.* (1977) provided evidence for a role of serotonin in mediating the effects of vasopressin on memory retrieval. In their studies, a rise in hippocampal serotonin levels was observed in animals that received footshock during passive avoidance training. The increased serotonin was seen at 24 hr after training. Treatment of the animals with CO_2 immediately after training induced amnesia for the passive avoidance response, and, in these animals, no change in serotonin levels was observed at 24 hr after training. DGLVP attenuated the amnesia, and the rise in hippocampal serotonin was observed in the DGLVP-treated animals. Thus, the authors suggested that DGLVP attenuated amnesia through an effect on hippocampal serotonin metabolism (Ramaekers *et al.*, 1977; Leonard *et al.*, 1977).

The ability of vasopressin to maintain ethanol tolerance in mice may also involve peptide effets on noradrenergic and serotonergic neuronal systems. In mice, pretreatment with intracerebroventricular 6-OHDA, which lowered whole-brain NE levels by about 50%, blocked the ability of AVP to maintain ethanol tolerance (Hoffman *et al.*, 1983). Thus, it was concluded that the presence of intact brain noradrenergic systems is required in order to observe effects of vasopressin on maintenance of ethanol tolerance in mice. In rats, where vasopressin (DGAVP) also maintained tolerance to the hypothermic and motor-impairing effects of ethanol, lesions of the median raphe nucleus or of median plus dorsal raphe nucleus (i.e., destruction of the cell bodies of serotonin-containing neurons) blocked the effect of DGAVP on ethanol tolerance (Lê *et al.*, 1982), as did chemical lesions that depleted serotonin in the hippocampus and parietal cortex (Speisky and Kalant, 1985). In rats, therefore, an intact mesolimbic serotonergic system, and particularly intact serotonergic innervation of hippocampus and parietal cortex, was concluded to be required for vasopressin to maintain ethanol tolerance. It appears that interactions between noradrenergic and serotonergic systems allow optimal development of ethanol tolerance in both mice and rats (Tabakoff and Ritzmann, 1977; Melchior and Tabakoff, 1981; Khanna *et al.*, 1981; Lê *et al.*, 1980), and such interactions may also be important for the ability of vasopressin to maintain ethanol tolerance.

Thus, many of the behavioral effects of vasopressin appear to be mediated via noradrenergic pathways in brain, perhaps, as suggested by Kovács *et al.* (1979c), by vasopressin action at noradrenergic terminals, to facilitate (or inhibit) norepinephrine (NE) release. The findings that LVP did not appear to alter spontaneous or potassium-stimulated release of NE from brain slices *in vitro* (Delanoy *et al.*, 1982; Hagan and Balfour, 1983), however, may, as mentioned, indicate that vasopressin has indirect effects on NE release *in vivo*. Another possible complication is that high doses of peptide were used in these *in vitro* experiments. Considering the bell-shaped dose–response curves that have been noted for vasopressin's effects, a broader dose–response study is warranted. The suggestion that vasopressin may influence "memory" or tolerance by modulating the release of NE in limbic areas is most interesting when one considers the postulated role for these catecholaminergic neurons in brain reward circuitry (Ritter and Stein, 1973; Stein, 1969; Crow *et al.*, 1972; Wise, 1980). As mentioned earlier, it seems possible that modulation of the activity of brain reinforcing systems by vasopressin could account for at least some of the behavioral effects of the peptides, particularly those related to memory of appetitive tasks and drug administration.

If one site of action for vasopressin is at the terminals of noradrenergic neurons, the peptide may interact with presynaptic receptors that modulate neurotransmitter release. Such a hypothesis is supported not only by the behavioral and neurochemical studies described above, but also by the finding that the electrophysiological effects of iontophoretically applied AVP were significantly decreased in animals that had been treated with alpha-methyl-*p*-tyrosine (Marchand and Hagino, 1982). The nature of this hypothetical CNS vasopressin receptor is not clear, but studies of peripheral vasopressin receptors provide a basis for speculation. There appear to be at least two types of peripheral vasopressin receptor (Jard, 1983). The receptor in the kidney, which is designated V_2, is linked to adenylate cyclase and mediates vasopressin-induced antidiuresis (Jard, 1983; Roy *et al.*, 1975a,b). There is also a receptor, designated V_1, that is found on liver and vascular smooth muscle cells and that is not linked to adenylate cyclase (Jard, 1983; Michell *et al.*, 1979; Kirk *et al.*, 1983). This receptor is responsible for vasopressin-mediated Ca^{2+} mobilization, and the initial intracellular event leading to Ca^{2+} flux is most probably the breakdown of polyphosphoinositides (Michell *et al.*, 1979; Kirk *et al.*, 1983). Although the vasopressin binding sites in brain have not yet been characterized as "V_1" or "V_2," the hippocampal binding site shows low affinity for the analog DDAVP, which binds relatively specifically to V_2 receptors (i.e., has little pressor effect) (Jard, 1983). Similarly, in hippocampus, DDAVP was much less active than AVP or LVP in increasing the firing rate of CA1 neurons, and the action of AVP on these neurons was blocked by an antagonist that has relative selectivity for V_1 receptors (Mühlethaler *et al.*, 1982). Assuming that the behavioral effects of vasopressin depend on neurotransmitter release, a Ca^{2+}-requiring process, it seems reasonable to postulate that at least some of the CNS actions of vasopressin will be found to be mediated by V_1 receptors.

On the other hand, systemic injection of mice with DGAVP was reported to alter the content of cyclic adenosine 3′,5′-monophosphate (cAMP) in several brain areas (Schneider *et al.*, 1982). Courtney and Raskind (1983) found no direct effects of AVP on adenylate cyclase activity in any brain area tested (caudate, hippocampus, septum, hypothalamus, thalamus, nucleus accumbens area, olfactory tubercle, locus coeruleus area), but did find that vasopressin potentiated the ability of dopamine to stimulate adenylate cyclase activity in caudate homogenates. Since the studies were performed in homogenates, vasopressin probably did not act by enhancing dopamine release, and the mechanism of this small potentiation is not clear. Church (1983) reported that LVP potentiated norepinephrine-stimulated increases in cyclic AMP accumulation in mouse hippocampal slices, while Hamburger-Bar and Newman (1985) later reported that [1-deamino,8-D-arginine]vasopressin (DDAVP) *inhibited* the increase in cyclic AMP induced by norepinephrine in rat cortical slices. Similarly, Newman (1985) found that AVP and LVP inhibited norepinephrine-stimulated adenylate cyclase activity in rat cortical slices and forskolin-stimulated activity in cortical membranes. These latter reports of vasopressin effects on cAMP production do not necessitate the interaction of the hormone with a V_2 receptor in hippocampus or cortex, since the effects could be indirect—for example, due to effects on Ca^{2+} flux or neurotransmitter release in synaptosomal preparations or slices, respectively.

In rat pituitary and rat hepatocytes, vasopressin has also been reported to stimulate the activity of phospholipid methyltransferase (Alemany *et al.*, 1981; Prasad and Edwards, 1981). This action may be secondary to vasopressin mobilization of Ca^{2+} (Alemany *et al.*, 1981). In brain, phospholipid methylation has been reported to influence neuronal membrane lipid structure and beta-adrenergic receptor sensitivity (Hirata *et al.*, 1979). The possibility that some of the CNS effects of vasopressin may depend on this biochemical pathway deserves attention.

Less work has been done to ascertain a biochemical mechanism of action for oxytocin in the CNS, as compared to vasopressin (Kovács and Telegdy, 1982). Bohus *et al.* (1978b) reported that oxytocin had an effect opposite to that of vasopressin, in that it decreased the peak frequency of the hippocampal theta rhythm generated during REM sleep. This finding was consistent with their observation that oxytocin attenuated passive avoidance behavior (Bohus *et al.*, 1978a,b). Rat pituitary phospholipid methyltransferase was reported to be unaffected by oxytocin but stimulated by vasopressin (Prasad and Edwards, 1981). Since some of oxytocin's actions in the periphery appear to involve Ca^{2+} mobilization, leading to smooth-muscle contraction (Mironneau, 1976), it seems likely that if oxytocin receptors are present in the CNS, they might also affect Ca^{2+} flux. On the other hand, oxytocin may also interact with vasopressin receptors, acting as a partial agonist or antagonist to endogenous vasopressin. Thus, the biochemical effects of oxytocin alone may be difficult to discern.

Two of the recurring questions in studies of the effects of neurohypophyseal peptides on memory-related processes have been (1) how do these peptides get to the brain following systemic injection? and (2) how can peptides with such short half-lives exert rather long-term effects? As discussed in Section II,A,1, the blood–brain barrier appears to effectively prevent intact neurohypophyseal hormones from entering the brain from the systemic circulation. It is possible that the hormones may exert their effects through actions at areas outside the blood–brain barrier (e.g., circumventricular organs) and, in fact, endogenous vasopressin has been located in some of these areas (Laczi *et al.*, 1983b). Another possibility is that the hormones are metabolized to fragments that can more easily enter the brain and that are more metabolically stable than vasopressin, thus allowing long-term actions. In de Wied's study of the effects of various peptides, administered icv, on inhibition of extinction of active avoidance behavior, certain fragments of AVP (i.e., the covalent ring portion and the C-terminal tripeptide) were behaviorally active (de Wied, 1976). Such results suggested that vasopressin could serve as a "prohormone" for production of neuropeptides that influence learning or memory (de Wied, 1976). These peptide fragments, derived from endogenous hormone, could normally reach their site of action either through the CSF or following metabolism of the hormones in brain tissue (de Wied, 1976). For example, both cyclo(Leu-Gly) and cyclo(Arg-Gly), derivatives of the C-terminal dipeptides of oxytocin and arginine vasopressin, respectively, can attenuate puromycin-induced amnesia (Walter *et al.*, 1982; Flexner *et al.*, 1977). These peptides are more resistant to degradation than the parent hormones (Walter *et al.*, 1982; Hoffman *et al.*, 1977), and cyclo(Leu-Gly) was found to easily penetrate the blood–CSF barrier (Hoffman *et al.*, 1977; Rainbow *et al.*, 1979) and to remain intact in brain for up to 96 hr after systemic administration in the mouse (Rainbow *et al.*, 1979). Cyclo(Arg-Gly) was found to form spontaneously from Pro-Arg-GlyNH_2 *in vitro* (Shlank and Walter, 1972), and the formation of the cyclic dipeptide derivatives from vasopressin and/or oxytocin could also occur *in vivo*. Enzymes that release the C-terminal dipeptide from vasopressin have been characterized (Walter and Simmons, 1977), as have hypothalamic enzymes that release the C-terminal tripeptide, prolylleucylglycinamide (PLG), from oxytocin (Walter *et al.*, 1973). PLG is also behaviorally active, as are prolylarginylglycinamide (PAG), the C-terminal tripeptide of AVP, and prolyllysylglycinamide, the C-terminal tripeptide of LVP (Walter *et al.*, 1980a; de Wied, 1976; Davis *et al.*, 1982b). Although the C-terminal peptides and derivatives of these peptides share some of the behavioral activities of the parent hormones (see Walter *et al.*, 1980a), it does not seem likely that the formation of these peptides can account totally for the CNS effects of the hormones. For example, as discussed above, structure–activity studies indicate the importance of the N-terminal ring structure of vasopressin for several behavioral actions (Walter *et al.*, 1975, 1978a; Hoffman, 1982). It might be argued that these structure–activity studies, which were performed for the most part by injecting the peptides systemically, actually demonstrate the structural requirements for uptake of peptides into the brain, rather than for interactions at a CNS receptor. However, it is clear, for example, that cyclo(Leu-Gly) can enter the brain from the circulation more easily than AVP (Rainbow *et al.*, 1979; Zaidi and Heller, 1974), yet cyclo(Leu-Gly) is less active than AVP in many behavioral assays (Walter *et al.*, 1978a; Hoffman, 1982). As another example, cyclo(Arg-Gly) was found to be as potent as AVP in attenuating puromycin-induced amnesia (Walter *et al.*, 1980a, 1982). Cyclo(Arg-Gly) is expected to be formed as readily from arginine vasotocin as from AVP, yet arginine vasotocin was significantly less active than AVP in this assay (Walter *et al.*, 1980b). If the actions of these two hormones were mediated by generation of the same intermediate, their activities should be equal. Thus, the structural requirements for peptide effects that were determined in structure–activity studies most likely reflect the peptide characteristics necessary for efficient interaction with the relevant receptors in the CNS, and they indicate that the N-terminal portion of the hormones plays an important role in this interaction.

Burbach *et al.* (1983) have identified a metabolite of AVP that was formed by enzymes present in rat synaptic membranes. This peptide, pGlu-Asn-Cys-Pro-Arg-GlyNH$_2$, was more potent than AVP in enhancing retention of passive avoidance behavior when both peptides were administered intracerebroventricularly. The peptide was more stable metabolically than AVP and was proposed to represent a possible compound that could mediate the long-term effects of AVP. This is an interesting hypothesis, although, again, it cannot account for many of the results of structure–activity studies, and the peptide's affinity at the hippocampal binding site did not correlate well with its very high behavioral activity (Pearlmutter and Constantini, 1984).

While the fragments of oxytocin and vasopressin appear to share some of the behavioral effects of the parent hormones, certain of these smaller peptides appear also to have other behavioral and biochemical actions. As discussed in Section IV, while the effects of vasopressin and oxytocin on tolerance to and physical dependence on opiates are controversial (van Ree and de Wied, 1976; Schmidt *et al.*, 1978), PLG and a large number of peptides that are structurally related to the C-terminal tripeptide of oxytocin have been reported to block the development of opiate tolerance in mice (Walter *et al.*, 1978b, 1979; Kovács *et al.*, 1983). Treatment with these peptides also blocked some of the signs of morphine withdrawal in these animals (Walter *et al.*, 1978b, 1979; Bhargava *et al.*, 1980). Only certain behavioral signs of morphine tolerance and dependence were blocked by these peptides (Bhargava *et al.*, 1980), and it was shown that treatment with the peptide cyclo(Leu-Gly), which blocked signs of morphine tolerance and dependence, also prevented morphine-induced striatal dopamine receptor supersensitivity (Ritzmann *et al.*, 1979). Similarly, it has been reported that when rats were treated concurrently with neuroleptic drugs (haloperidol or chlorpromazine) and PLG, the peptide antagonized the enhancement of striatal haloperidol binding (i.e., dopamine receptor supersensitivity) that occurred after neuroleptic treatment alone (Chiu *et al.*, 1981a). Cyclo(Leu-Gly) had a similar effect (Bhargava and Ritzmann, 1980). Neither of these peptides interferes directly with dopamine antagonist binding to striatal receptor sites (Chiu *et al.*, 1981b; Ritzmann *et al.*, 1982), although PLG was reported to enhance agonist binding (Chiu *et al.*, 1981b). Thus, the actual mechanism by which peptides related to the C-terminal portion of oxytocin induce their effects on dopamine receptors is not known. For example, there has been no consistent evidence to indicate that PLG affects striatal dopamine metabolism or postsynaptic responses to dopamine (Singhal and Rastogi, 1982; Kostrzewa *et al.*, 1979). It is possible, however, that PLG and related peptides may interact with specific receptors in the CNS, initiating reactions that influence the interaction of opiates with dopaminergic systems. Thus, it is possible that the peptide fragments of neurohypophyseal hormones may influence behavior, not only by interacting weakly at vasopressin binding sites, but also by acting at receptors more specific for these fragments. Binding sites in brain for PLG have been reported (Chiu *et al.*, 1980), and it would be of interest to compare their localization and affinity for various analogs with those of the AVP binding sites (Pearlmutter and Constantini, 1984; Pearlmutter *et al.*, 1983; Dorsa *et al.*, 1983). In summary, it appears that neurohypophyseal hormones, in addition to influencing behavioral responses, may serve as precursors for peptides that can also affect the same, as well as different, physiological or behavioral processes.

VIII. CONCLUSIONS

Our knowledge and understanding of peptidergic pathways in the CNS and of their functional significance is still at an early stage. The evidence at this point that neurohypophyseal hormones can influence animal behavior seems quite convincing, in spite of early controversy. Whether these hormones will be as effective in humans, in whom learning and memory may be more localized to the cerebral cortex, in contrast to limbic areas in animals (Squire and Davis, 1981; Squire, 1982), remains to be seen. However, understanding the mechanism of action of these peptides in the CNS can provide significant insight into the neurochemistry of learning and memory, as well as other CNS adaptive functions, including the development of drug tolerance and/or physical dependence. As the characteristics of vasopressin receptors in the CNS are defined, it may be possible to design analogs that will have potent effects on memory and/or tolerance with few endocrinological side-effects. Such analogs would be invaluable for their ability to alleviate certain incapacitating human disorders.

REFERENCES

Abood, L. G., Knapp, R., Mitchell, T., Booth, H., and Schwab, L. (1980). *J. Neurosci. Res.* **5,** 191–199.

Ader, R., and de Wied, D. (1972). *Psychon. Sci.* **29,** 46–48.

Aghajanian, G. K., Bunney, B. S., and Kuhar, M. J. (1973). *In* "New Concepts in Neurotransmitter Regulation" (A. J. Mandel, ed.), pp. 115–134. Plenum, New York.

Alemany, S., Varela, I., and Mato, J. M. (1981). *FEBS Lett.* **135,** 111–114.

Alexandrova, M., and Soloff, M. (1980). *Endocrinology* **106,** 739–743.
Alliot, J., and Alexinsky, T. (1982). *Physiol. Behav.* **28,** 525–530.
Amico, J. A., Tenicela, R., Johnston, J., and Robinson, A. G. (1983). *J. Clin. Endocrinol. Metab.* **57,** 947–951.
Ang, V. T. Y., and Jenkins, J. S. (1982). *J. Endocrinol.* **93,** 319–325.
Antoni, F. A. (1984). *Neuroendocrinology* **39,** 186–188.
Asin, K. E. (1980). *Pharmacol. Biochem. Behav.* **12,** 343–346.
Baertschi, A. J., and Friedli, M. (1985). *Endocrinology* **116,** 499–502.
Bailey, W. H., and Weiss, J. M. (1979). *Brain Res.* **162,** 174–178.
Barnard, R. R., Jr., and Morris, M. (1982). *Neurosci. Lett.* **29,** 275–279.
Beard, J. D., and Sargent, W. Q. (1979). *In* "Biochemistry and Pharmacology of Ethanol" (E. Majchrowicz and E. P. Noble, eds.), vol. 2, pp. 3–16. Plenum, New York.
Beckwith, B. E., Petros, T., Kanaan-Beckwith, S., Couk, D. I., Hang, R. J., and Ryan, C. (1982). *Peptides* **3,** 627–630.
Beckwith, B. E., Couk, D. I., and Till, T. S. (1983). *Peptides* **4,** 707–709.
Beny, J.-L., and Baertschi, A. J. (1982). *Experientia* **38,** 1078–1079.
Berecek, K. H., Olpe, H. R., Jones, R. S. G., and Hofbauer, K. G. (1984). *Am. J. Physiol.* **247,** H675–H681.
Berkowitz, B. A., and Sherman, S. (1982). *J. Pharmacol. Exp. Ther.* **220,** 329–334.
Berntson, G. G., and Berson, B. S. (1980). *Life Sci.* **26,** 455–459.
Bhargava, H. N., and Ritzmann, R. F. (1980). *Pharmacol. Biochem. Behav.* **13,** 633–636.
Bhargava, H. N., Walter, R., and Ritzmann, R. F. (1980). *Pharmacol. Biochem. Behav.* **12,** 73–77.
Biegon, A., Terlou, M., Voohuis, T. D., and de Kloet, E. R. (1984). *Neurosci. Lett.* **44,** 229–234.
Blake, D. R., Dodd, M. J., and Evans, J. G. (1978). *Lancet* **1,** 608.
Bohus, B. (1977). *Horm. Behav.* **8,** 52–61.
Bohus, B., Ader, R., and de Wied, D. (1972). *Horm. Behav.* **3,** 191–197.
Bohus, B., Gispen, W. H., and de Wied, D. (1973). *Neuroendocrinology* **11,** 137–143.
Bohus, B., Kovacs, G. L., and de Wied, D. (1978a). *Brain Res.* **157,** 414–417.
Bohus, B., Urban, I., van Wimersma Greidanus, T. B., and de Wied, D. (1978b). *Neuropharmacology* **17,** 239–247.
Bohus, B., Conti, L., Kovács, G. L., and Versteeg, D. H. G. (1982). *In* "Neuronal Plasticity and Memory Formation" C. A. Marsan and H. Matthes, eds.), pp. 75–87. Raven, New York.
Bookin, H. B., and Pfeifer, W. D. (1977). *Pharmacol. Biochem. Behav.* **7,** 51–54.
Brito, G. N. O. (1983). *Physiol. Behav.* **30,** 29–34.
Brito, G. N. O., Thomas, G. J., Gingold, S. I., and Gash, D. M. (1981). *Brain Res. Bull.* **6,** 71–75.
Buijs, R. M. (1978). *Cell Tissue Res.* **192,** 423–435.
Buijs, R. M., and van Heerikhuize, J. J. (1982). *Brain Res.* **252,** 71–76.
Burbach, J. P. H., Kovács, G. L., de Wied, D., van Nispen, J. W., and Greven, H. M. (1983). *Science* **221,** 1310–1312.
Burešová, O., and Škopková, J. (1980). *Peptides* **1,** 261–263.
Burešová, O., and Škopková, J. (1982). *Peptides* **3,** 725–727.
Burish, T. G., Slonim, A. E., Davidson, E. S., Moore, G., and Tower, A. (1981). *Clin. Neuropsychol. III,* 3–15.
Carey, R. J., and Miller, M. (1982). *Behav. Brain Res.* **6,** 1–13.
Celestian, J. F., Carey, R. J., and Miller, M. (1975). *Physiol. Behav.* **15,** 707–711.
Chiu, S., Wong, Y. W., and Mishra, R. K. (1980). *Pharmacologist* **22,** 205.
Chiu, S., Paulose, C. S., and Mishra, R. K. (1981a). *Science* **214,** 1261–1262.
Chiu, S., Paulose, C. S., and Mishra, R. R. (1981b). *Peptides* **2,** 105–111.
Church, A. C. (1983). *Peptides* **4,** 261–263.
Cooper, R. L., McNamara, M. C., and Thompson, W. G. (1980). *Neurobiol. Aging* **1,** 53–57.

Couk, D. I., and Beckwith, B. E. (1982). *Peptides* **3,** 521–526.
Courtney, N., and Raskind, M. (1983). *Life Sci.* **32,** 591–596.
Crabbe, J. C., Rigter, H., and Kerbusch, S. (1980). *Behav. Genet.* **10,** 139–152.
Crine, A. F. (1982). *Prog. Neuro-Psychopharmacol. Biol. Psychiatry* **6,** 265–268.
Crine, A. F. (1984). *Prog. Neuro-Psychopharmacol. Biol. Psychiatry* **8,** 379–384.
Crine, A. F., Bredart, S., and Legros, J. J. (1981). *Horm. Behav.* **15,** 226–231.
Crine, A. F., Boulanger, B., and Nizet, G. (1983). *Reg. Peptides* **5,** 145–152.
Crow, T. J., Spear, P. J., and Arbuthnott, G. W. (1972). *Brain Res.* **36,** 275–287.
Culler, M. D., Turkelson, C. M., Thomas, C. R., and Arimura, A. (1983). *Proc. Soc. Exp. Biol. Med.* **173,** 264–269.
Davis, J. L. (1984). *IRCS Med. Sci.* **12,** 865.
Davis, J. L., Pico, R. M., and Cherkin, A. (1982a). *Behav. Neural Biol.* **35,** 242–250.
Davis, J. L., Pico, R. M., and Cherkin, A. (1982b). *Pharmacol. Biochem. Behav.* **17,** 893–896.
Davis, J. L., Pico, R. M., and Cherkin, A. (1983). *Brain Res.* **266,** 355–358.
Dekanski, J. (1952). *Br. J. Pharmacol. Chemother.* **7,** 567–572.
Delanoy, R. L., Dunn, A. J., and Tintner, R. (1978). *Horm. Behav.* **11,** 348–362.
Delanoy, R. L., Dunn, A. J., and Walter, R. (1979). *Life Sci.* **24,** 651–658.
Delanoy, R. L., Kramarcy, N. R., and Dunn, A. J. (1982). *Brain Res.* **231,** 117–129.
de Wied, D. (1965). *Int. J. Neuropharmacol.* **4,** 157–167.
de Wied, D. (1971). *Nature (London)* **232,** 58–60.
de Wied, D. (1976). *Life Sci.* **19,** 685–690.
de Wied, D. (1977). *Life Sci.* **20,** 195–204.
de Wied, D., and Bohus, B. (1966). *Nature (London)* **212,** 1484–1486.
de Wied, D., and Gispen, W. H. (1976). *Psychopharmacology* **46,** 27–29.
de Wied, D., Greven, H. M., Lande, S., and Witter, A. (1972). *Br. J. Pharmacol.* **45,** 118–122.
de Wied, D., Bohus, B., and van Wimersma Greidanus, T. B. (1975). *Brain Res.* **85,** 152–156.
de Wied, D., van Wimersma Greidanus, T. B., Bohus, B., Urban, I., and Gispen, W. H. (1976). *Prog. Brain Res.* **45,** 181–194.
de Wied, D., Gaffori, O., Van Ree, J. M., and de Jong, W. (1984). *Nature (London)* **308,** 276–278.
Dogterom, J., Snijdewint, F. G. M., and Buijs, R. M. (1978). *Neurosci. Lett.* **9,** 341–346.
Doris, P. A., and Bell, F. R. (1984). *Neuroendrinology* **38,** 290–296.
Dorsa, D. M., and van Ree, J. M. (1979). *Brain Res.* **172,** 367–371.
Dorsa, D. M., Majumdar, L. A., Petracca, F. M., Baskin, D. G., and Cornett, L. E. (1983). *Peptides* **4,** 699–706.
Ehrensing, R. H., Michell, G. F., and Baker, R. P. (1982). *Peptides* **3,** 527–530.
Ettenberg, A. (1984). *Behav. Brain Res.* **14,** 201–211.
Ettenberg, A., van der Kooy, D., LeMoal, M., Koob, G. F., and Bloom, F. E. (1983a). *Behav. Brain Res.* **7,** 331–350.
Ettenberg, A., LeMoal, M., Koob, G. F., and Bloom, F. E. (1983b). *Pharmacol. Biochem. Behav.* **18,** 645–647.
Fehm-Wolfsdorf G., Born, J., Voigt, K.-H., and Fehm, H.-L. (1984). *Psychoneuroendocrinology* **9,** 285–292.
Ferrier, B. M., Kennett, D. J., and Devlin, M. C. (1980). *Life Sci.* **27,** 2311–2317.
Fewtrell, W. D., House, A. O., Jamie, P. F., Oates, M. R., and Cooper, J. E. (1982). *Psychol. Med.* **12,** 423–425.
Fibiger, H. C. (1978). *Annu. Rev. Pharmacol. Toxicol.* **18,** 37–56.
Finkelberg, F., Kalant, H., and LeBlanc, A. E. (1978). *Pharmacol. Biochem. Behav.* **9,** 453–458.
Flexner, J. B., Flexner, L. B., and Stellar, E. (1963). *Science* **141,** 57–59.
Flexner, L. B., Flexner, J. B., and Roberts, R. B. (1967). *Science* **155,** 1377–1383.
Flexner, J. B., Flexner, L. B., Hoffman, P. L., and Walter, R. (1977). *Brain Res.* **134,** 139–144.

Flexner, J. B., Flexner, L. B., Walter, R., and Hoffman, P. L. (1978). *Pharmacol. Biochem. Behav.* **8,** 93–95.

Franceschi, M., Tancredi, O., Savio, G., and Smirne, S. (1982). *Eur. Neurol.* **21,** 388–391.

Gallager, D. W., and Aghajanian, G. K. (1975). *J. Pharmacol. Exp. Ther.* **193,** 785–795.

Gaziz, D. (1978). *Proc. Soc. Exp. Biol. Med.* **158,** 663–665.

German, D. C., and Bowden, D. M. (1974). *Brain Res.* **73,** 381–419.

Gilot, P., Crabbe, J., and Legros, J. J. (1980). *Acta Psych. Belg.* **80,** 755–761.

Gold, P. E., and van Buskirk, R. (1976). *Horm. Behav.* **7,** 509–517.

Gold, P. W., Goodwin, F. K., and Reus, V. I. (1978). *Lancet* **1,** 1233–1235.

Gold, P. W., Ballanger, J. C., Weingartner, H., Goodwin, F. K., and Post, R. M. (1979). *Lancet* **2,** 992–994.

Gold, P. W., Goodwin, F. K., Ballenger, J. C., Post, R. M., Weingartner, H., and Robertson, G. L. (1981). *Int. J. Mental Health* **9,** 91–107.

Goldstein, D. B. (1972). *J. Pharmacol. Exp. Ther.* **180,** 203–215.

Hagan, J. J. (1982). *Behav. Neural. Biol.* **36,** 204–210.

Hagan, J. J., and Balfour, D. J. K. (1983). *Life Sci.* **32,** 2517–2522.

Hagan, J. J., Bohus, B., and de Wied, D. (1982). *Behav. Neural. Biol.* **36,** 211–228.

Hamburger-Bar, R., and Newman, M. E. (1985). *Pharmacol. Biochem. Behav.* **22,** 183–187.

Hartmann, G., Schwarzberg, H., and Hahn, Z. (1979). *Acta Physiol. Acad. Sci. Hung.* **53,** 174.

Hernández, L. L., and Powell, D. A. (1983). *Peptides* **4,** 37–41.

Hirata, F., Strittmayer, W. J., and Axelrod, J. (1979). *Proc. Natl. Acad. Sci. USA* **76,** 368–372.

Hoffman, P. L. (1982). *Pharmacol. Biochem. Behav.* **17,** 685–690.

Hoffman, P. L., and Tabakoff, B. (1979). *In* "Currents in Alcoholism" (M. Galanter, ed.), Vol. 5, pp. 5–16. Grune and Stratton, New York.

Hoffman, P. L., and Tabakoff, B. (1981). *In* "Currents in Alcoholism" (M. Galanter, ed.), Vol. 8, pp. 359–378. Grune and Stratton, New York.

Hoffman, P. L., and Tabakoff, B. (1984). *Pharmacol. Biochem. Behav.* **21,** 539–543.

Hoffman, P. L., Walter, R., and Bulat, M. (1977). *Brain Res.* **122,** 87–94.

Hoffman, P. L., Ritzmann, R. F., Walter, R., and Tabakoff, B. (1978). *Nature (London)* **276,** 614–616.

Hoffman, P. L., Ritzmann, R. F., and Tabakoff, B. (1980). *Pharmacol. Biochem. Behav.* **13** (Suppl. 1), 279–284.

Hoffman, P. L., Melchior, C. L., and Tabakoff, B. (1983). *Life Sci.* **32,** 1065–1071.

Höglund, A. U., and Meyerson, B. J. (1982). *Physiol. Behav.* **29,** 189–193.

Hostetter, G., Jubb, S. L., and Kozlowski, G. P. (1977). *Life Sci.* **21,** 1323–1328.

Hostetter, G., Jubb, S. L., and Kozlowski, G. P. (1980). *Neuroendocrinology* **30,** 174–177.

Hung, C.-R., Tabakoff, B., Melchior, C. L., and Hoffman, P. L. (1984). *Eur. J. Pharmacol.* **106,** 645–648.

Huston, J. P., and Jakobartl, L. (1977). *Neurosci. Lett.* **6,** 69–72.

Janáky, T., Laszlo, F. A., Sirokman, F., and Morgat, J.-L. (1982). *J. Endocrinol.* **93,** 295–303.

Jard, S. (1983). *Curr. Top. Membr. Transp.* **18,** 255–285.

Jenkins, J. S., Mather, H. M., Coughlan, A. K., and Jenkins, D. G. (1979). *Lancet* **2,** 1245–1246.

Jenkins, J. S., Mather, H. M., and Ang, V. (1980). *J. Clin. Endocrinol. Metab.* **50,** 364–367.

Jenkins, J. S., Mather, H. M., Coughlan, A. K., and Jenkins, D. G. (1981). *Lancet* **1,** 39.

Jenkins, J. S., Mather, H. M., and Coughlan, A. K. (1982). *J. Neurol. Neurosurg. Psychiatry* **45,** 830–831.

Jensen, J. P. A. (1980). *Acta Neurol. Scand.* **62,** 197–199.

Joëls, M., and Urban, I. J. A. (1984). *Brain Res.* **34,** 201–209.

Jones, P. M., and Robinson, I. C. A. F. (1982). *Neuroendocrinology* **34,** 297–302.

Judge, M. E., and Quartermain, D. (1982). *Pharmacol. Biochem. Behav.* **16,** 463–466.

Kasting, N. W., Veale, W. L., and Cooper, K. E. (1980). *Can. J. Physiol. Pharmacol.* **58,** 316–319.

Kasting, N. W., Veale, W. L., Cooper, K. E., and Lederis, K. (1981). *Brain Res.* **213,** 327–333.

Kennett, D. J., Devlin, M. C., and Ferrier, B. M. (1982). *Life Sci.* **31,** 273–275.

Khanna, J. M., Kalant, H., Lê, A. D., and LeBlanc, A. E. (1981). *Prog. Neuropsychopharmacol.* **5,** 459–465.

King, A. R., and de Wied, D. (1974). *J. Comp. Physiol. Psychol.* **86,** 1008–1018.

Kirk, C. J., Creba, J. A., Hawkins, P. T., and Michell, R. H. (1983). *In* "The Neurohypophysis: Structure, Function and Control," *Prog. Brain Res.* **60** (B. A. Cross and G. Long, eds.), pp. 405–411. Elsevier, Amsterdam.

Koch-Henriksen, N., and Nielsen, H. (1981). *Lancet* **1,** 38–39.

Koizumi, H. M., Rowe, M., and Clark, R. (1981). *J. Clin. Psychol.* **42,** 217.

Koob, G. F., LeMoal, M., Gaffori, O., Manning, M., Sawyer, W. H., Rivier, J., and Bloom, F. E. (1981). *Reg. Peptides* **2,** 153–163.

Kordower, J. H., Sikorszky, V., and Bodnar, R. J. (1982). *Peptides* **3,** 613–617.

Korsgaard, S., Casey, D. E., Pederson, N. E. D., Jorgensen, A., and Gerlack, J. (1981). *Psychopharmacology* **74,** 379–382.

Kostrzewa, R. M., Hardin, J. C., Snell, R. L., Kastin, A. J., Coy, D. H., and Bymaster, F. (1979). *Brain Res. Bull.* **4,** 657–662.

Kovács, G. L., and Telegdy, G. (1982). *Pharmac. Ther.* **18,** 375–395.

Kovács, G. L., Vecsei, L., and Telegdy, G. (1978). *Physiol. Behav.* **20,** 801–802.

Kovács, G. L., Bohus, B., and de Wied, D. (1979a). *Neuroscience* **4,** 1529–1537.

Kovács, G. L., Bohus, B., and Versteeg, D. H. G. (1979b). *Brain Res.* **172,** 73–85.

Kovács, G. L., Bohus, B., Versteeg, D. H. G., de Kloet, E. R., and de Wied, D. (1979c). *Brain Res.* **175,** 303–314.

Kovács, G. L., Vecsei, L., Medve, L., and Telegdy, G. (1980). *Exp. Brain Res.* **38,** 357–361.

Kovács, G. L., Buijs, R., Bohus, B., and van Wimersma Greidanus, T. B. (1982). *Physiol. Behav.* **28,** 45–48.

Kovács, G. L., Szontagh, L., Balaspiri, L., and Telegdy, G. (1983). *Peptides* **4,** 417–419.

Krejčí, I., and Kupkova, B. (1978). *Act. Nerv. Suppl.* **20,** 60–61.

Krejčí, I., Kupkova, B., Metys, J., Barth, T., and Jost, K. (1979). *Eur. J. Pharmacol.* **56,** 347–353.

Krejčí, I., Kupkova, B., and Dlabac, A. (1981). *Reg. Peptides* **2,** 285–291.

Krivoy, W. A., Zimmerman, E., and Lande, S. (1974). *Proc. Natl. Acad. Sci. USA* **71,** 1852–1856.

Kruse, H., van Wimersma Greidanus, T. B., and de Wied, D. (1977). *Pharmacol. Biochem. Behav.* **7,** 311–313.

Laczi, F., Valkusz, Z. S., László, F. A., Wagner, A., Járdánházy, T., Szász, A., Szilárd, J., and Telegdy, G. (1982). *Psychoneuroendocrinology* **7,** 185–193.

Laczi, F., Fekete, M., and de Wied, D. (1983a). *Life Sci.* **32,** 577–589.

Laczi, F., Gaffori, O., de Kloet, E. R., and de Wied, D. (1983b). *Brain Res.* **260,** 342–346.

Laczi, F., Van Ree, J. M., Balogh, L., Szász, A., Járdánházy, T., Wágner, A., Gáspár, L., Valkusz, Z., Dobranovics, I., Szilárd, J., László, F. A., and de Wied, D. (1983c). *Acta Endocrinol.* **104,** 177–182.

Laczi, F., Van Ree, J. M., Wagner, A., Valkusz, Z. S., Járdánházy, T., Kovács, G. L., Telegdy, G., Szilárd, J., László, F. A., and de Wied, D. (1983d). *Acta Endocrinol.* **102,** 205–212.

Land, H., Schuts, G., Schmale, H. and Richter, D. (1982). *Nature (London)* **295,** 299–303.

Lande, S., Witter, A., and de Wied, D. (1971). *J. Biol. Chem.* **246,** 2058–2062.

Lande, S., Flexner, J. B., and Flexner, L. B. (1972). *Proc. Natl. Acad. Sci. USA* **69,** 558–560.

Lauson, H. D. (1974). *In* "Handbook of Physiology" (E. Knobil and W. H. Sawyer, eds.), Section 7, Vol. IV, Part 1, pp. 287–393. Waverly, Baltimore.

Lê, A. D., Khanna, J. M., and Kalant, H. (1980). *Psychopharmacology* **67,** 143–146.

Lê, A. D., Kalant, H., and Khanna, J. M. (1982). *Eur. J. Pharmacol.* **80,** 337–345.
LeBlanc, A. E., and Cappell, H. (1976). *In* "Alcohol and Opiates: Neurochemical and Behavior Mechanisms" (K. Blum, ed.), pp. 65–77. Academic, New York.
LeBoeuf, A., Lodge, J., and Eames, P. G. (1978). *Lancet* **2,** 1370.
Lederis, K., Veale, W. L., Kasting, N. W., Cooper, K. E., and Haskam, R. H. A. (1981). *Proc. West. Pharmacol. Soc.* **24,** 97–99.
Lee, R. J., and Lomax, P. (1982). *Proc. West. Pharmacol. Soc.* **25,** 29–33.
Legros, J. J., Gilot, P., Seron, X., Claessens, J., Adam, A., Moeglen, J. M., Audibert, A., and Berchier, P. (1978). *Lancet* **1,** 41–42.
LeMoal, M., Koob, G. F., Koda, L. Y., Bloom, F. E., Manning, M., Sawyer, W. H., and Rivier, J. (1981). *Nature* **291,** 491–493.
Leonard, B. E., Ramaekers, F., and Rigter, H. (1977). *Act. Nerv. Suppl.* **19,** 160–162.
Leshner, A. I., and Roche, K. E. (1977). *Physiol. Behav.* **18,** 879–883.
Linkola, J., Fyhrquist, F., and Forsander, O. (1977). *Acta Physiol. Scand.* **101,** 126–128.
Linkola, J., Ylikhari, R., Fyhrquist, F., and Wallenius, M. (1978). *Acta Physiol. Scand.* **104,** 180–187.
Lutz-Bucher, B., and Koch, B. (1983). *Neuroendocrinol. Lett.* **5,** 111–115.
Lutz-Bucher, B., Briaud, B., Koch, B., and Schmitt, G. (1981). *J. Physiol. (Paris)* **77,** 939–950.
Manning, M., Lammek, B., Kolodziejczyk, A. M., Seto, J., and Sawyer, W. H. (1981). *J. Med. Chem.* **24,** 701–706.
Marchand, J. E., and Hagino, N. (1982). *Exp. Neurol.* **78,** 790–795.
Matsuguchi, H., Sharabi, F. M., Gordon, F. J., Johnson, A. K., and Schmid, P. G. (1982). *Neuropharmacology* **21,** 687–693.
McNeill, T. H., and Sladek, J. R. (1980). *J. Comp. Neurol.* **193,** 1023–1033.
Meisenberg, G. (1982). *Neuropharmacology* **21,** 309–316.
Meisenberg, G., and Simmons, W. H. (1982). *Pharmacol. Biochem. Behav.* **16,** 819–825.
Meisenberg, G., and Simmons, W. H. (1983). *Neurosci. Behav.* **7,** 263–280.
Melchior, C. L., and Tabakoff, B. (1981). *J. Pharmacol. Exp. Ther.* **219,** 175–180.
Melchior, C. L., Hoffman, P. L., and Tabakoff, B. (1983). *In* "NIAAA Research Monograph No. 13" (T. J. Cicero, ed.), pp. 242–249. U.S. Government Printing Office, Washington, D.C.
Mello, N. K., and Mendelson, J. H. (1977). *In* "Handbook of Experimental Pharmacology" (W. R. Martin, ed.), Vol. 45(1), pp. 613–666. Springer-Verlag, Berlin.
Mello, N. K., and Mendelson, J. H. (1979). *Pharmacol. Biochem. Behav.* **10,** 415–419.
Mens, W. B. J., van Dam, A. F., van Egmond, M. A. H., Bakker, E. A. D., Legros, J. J., and van Wimersma Greidanus, T. B. (1982). *In* "Frontiers in Hormone Research" (E. M. Rodriguez and T. B. van Wimersma Greidanus, eds.), Vol. 9, pp. 119–130. Karger, Basel.
Messing, R. B., and Sparber, S. B. (1983). *Eur. J. Pharmacol.* **89,** 43–51.
Messing, R. B., and Sparber, S. B. (1984). *Trends Pharmacol. Sci.* **5,** 149–152.
Michell, R. H., Kirk, C. J., and Billah, M. M. (1979). *Biochem. Soc. Trans.* **7,** 861–865.
Miller, M., Barranda, E. G., Dean, M. C., and Brush, F. R. (1976). *Pharmacol. Biochem. Behav.* **5**(Suppl. 1), 35–40.
Mironneau, J. (1976). *Pfluegers Arch. Eur. J. Physiol.* **363,** 113–118.
Moore, J. W., and Solomon, P. R., eds. (1980). *J. Physiol. Psychol.* **8,** 145–293.
Morris, J. F. (1982). *Ann. NY Acad. Sci.* **394,** 54–71.
Mucha, R. F., and Kalant, H. (1978). *Pharmacol. Biochem. Behav.* **10,** 229–234.
Mühlethaler, M., Dreifuss, J. J., and Gähwiler, G. H. (1982). *Nature (London)* **296,** 749–751.
Newman, M. E. (1985). *FEBS Lett.* **181,** 203–206.
Oliveros, J. C., Jandali, M. K., Timsit-Berthier, M., Remy, R., Benghezal, A., Audibert, A., and Moeglen, J. M. (1978). *Lancet* **1,** 42.
Olpe, H.-R., and Baltzer, V. (1981). *Eur. J. Pharmacol.* **73,** 377–378.

Olson, G. A., Olson, R. D., Kastin, A. J., and Coy, D. H. (1980). *Peptides* **1,** 365–379.
Partap, M., Jos, C. J., and Dye, C. J. (1983). *Am. J. Psychiatry* **140,** 946–947.
Pearlmutter, A. F., and Constantini, M. G. (1984). *J. Biol. Chem.* **259,** 11739–11745.
Pearlmutter, A. F., Constantini, M. G., and Loeser, B. (1983). *Peptides* **4,** 335–341.
Pedersen, C. A., and Prange, A. J., Jr. (1979). *Proc. Natl. Acad. Sci. USA* **76,** 6661–6665.
Pedersen, C. A., Ascher, J. A., Monroe, Y. L., and Prange, A. J., Jr. (1982). *Science* **216,** 648–650.
Perlow, M. J., Reppert, S. M., Artman, H. A., Fisher, D. A., Seif, S. M., and Robinson, A. G. (1982). *Science* **216,** 1416–1418.
Pfeifer, W. D., and Bookin, H. B. (1978). *Pharmacol. Biochem. Behav.* **9,** 261–263.
Pittman, Q. J., Rogers, J., and Bloom, F. E. (1982). *Reg. Peptides* **4,** 33–41.
Post, R. M., Contel, N. R., and Gold, P. G. (1982). *Life Sci.* **31,** 2745–2750.
Prasad, C., and Edwards, R. M. (1981). *Biochem. Biophys. Res. Comm.* **103,** 559–564.
Raichle, M. E., and Grubb, R. L., Jr. (1978). *Brain Res.* **143,** 191–194.
Raichle, M. E., Grubb, R. L., and Eichling, J. O. (1978). *In* "Cerebral Vascular Smooth Muscle and Its Control" (Ciba Foundation Symposium 56), pp. 219–235. Elsevier/North Holland, Amsterdam.
Rainbow, T. C., Flexner, J. B., Flexner, L. B., Hoffman, P. L., and Walter, R. (1979). *Pharmacol. Biochem. Behav.* **10,** 787–793.
Ramaekers, F., Rigter, H., and Leonard, B. E. (1977). *Brain Res.* **120,** 485–492.
Raskind, M., Orenstein, H., and Weitzman, R. E. (1979). *Lancet* **1,** 164.
Rees, H. D., Dunn, A. J., and Iuvone, P. M. (1976). *Life Sci.* **18,** 1333–1340.
Rigter, H. (1982). *Behav. Neural. Biol.* **34,** 337–351.
Rigter, H., van Riezen, H., and de Wied, D. (1974). *Physiol. Behav.* **13,** 381–388.
Rigter, H., Elbertse, R., and van Riezen, H. (1975). *Prog. Brain Res.* **42,** 166–171.
Rigter, H., Rijk, H., and Crabbe, J. C. (1980). *Eur. J. Pharmacol.* **64,** 53–63.
Ritter, S., and Stein, L. (1973). *J. Comp. Physiol. Psychol.* **85,** 443–452.
Ritzmann, R. F., and Tabakoff, B. (1976). *J. Pharmacol. Exp. Ther.* **199,** 158–170.
Ritzmann, R. F., Walter, R., Bhargava, H. N., and Flexner, L. B. (1979). *Proc. Natl. Acad. Sci. USA* **76,** 5997–5998.
Ritzmann, R. F., Lee, J. M., and Fields, J. Z. (1982). *Life Sci.* **30,** 1573–1580.
Robberecht, P., de Neef, P., Lammens, M., Deschodt-Lanckman, M., and Christophe, J.-P. (1978). *Eur. J. Biochem.* **90,** 147–154.
Roche, K. E., and Leshner, A. I. (1979). *Science* **204,** 1343–1344.
Rossor, M. N., Iversen, L. L., Hawthorn, J., Ang, V. T. Y., and Jenkins, J. S. (1981). *Brain Res.* **214,** 349–355.
Rossor, M. N., Hunt, S. P., Iversen, L. L., Bannister, R., Hawthorn, J., Ang, V. T. Y., and Jenkins, J. S. (1982). *Brain Res.* **253,** 341–343.
Roy, C., Barth, T., and Jard, S. (1975a). *J. Biol. Chem.* **250,** 3157–3168.
Roy, C., Rajerison, R., Bockaert, J., and Jard, S. (1975b). *J. Biol. Chem.* **250,** 7885–7893.
Rubin, B. S., Menniti, F. S., and Bridges, R. S. (1983). *Horm. Behav.* **17,** 45–53.
Ruwe, W. D., Veale, W. L., and Cooper, K. E. (1983). *Can J. Biochem. Cell Biol.* **61,** 579–593.
Sahgal, A. (1983). *Reg. Peptides* **5,** 317–326.
Sahgal, A. (1984). *Psychopharmacology* **83,** 215–228.
Sahgal, A., and Wright, C. (1983). *Psychopharmacology* **80,** 88–92.
Sahgal, A., Keith, A. B., Wright, C., and Edwardson, J. A. (1982). *Neurosci. Lett.* **28,** 87–92.
Salzman, P. M., and Roth, R. H. (1980a). *J. Pharmacol. Exp. Ther.* **212,** 64–73.
Salzman, P. M., and Roth, R. H. (1980b). *J. Pharmacol. Exp. Ther.* **212,** 74–84.
Sara, S. J., Barnett, J., and Toussaint, P. (1982). *Behav. Proc.* **7,** 157–167.
Schmale, H., Ivell, R., Breindl, M., Darmer, D., and Richter, D. (1984). *EMBO J.* **3,** 3289–3293.

Schmidt, W. K., Holaday, J. W., Loh, H. H., and Way, E. L. (1978). *Life Sci.* **23,** 151–158.
Schneider, D. R., Felt, B. T., and Goldman, H. (1982). *Pharmacol. Biochem. Behav.* **16,** 139–143.
Shlank, H., and Walter, R. (1972). *Proc. Soc. Exp. Biol. Med.* **141,** 452–455.
Siegel, S. (1975). *J. Comp. Physiol. Psychol.* **89,** 498–506.
Silverman, A.-J., and Oldfield, B. J. (1984). *Peptides* **5,** 139–150.
Singhal, R. L., and Rastogi, R. B. (1982). *Pharmacol. Biochem. Behav.* **16,** 229–233.
Södersten, P., Henning, M., Melin, P., and Ludin, S. (1983). *Nature (London)* **301,** 608–610.
Sofroniew, M. (1980). *J. Histochem. Cytochem.* **5,** 475–478.
Sofroniew, M. V., and Schrell, U. (1981). *Neurosci. Lett.* **22,** 211–217.
Sofroniew, M. V., and Weindl, A. (1978a). *Endocrinology* **102,** 334–337.
Sofroniew, M. V., and Weindl, A. (1978b). *Am. J. Anat.* **153,** 391–430.
Sokol, H. W., and Valtin, H., eds. (1982). *Ann. NY Acad. Sci.* **394,** 535–654.
Soloff, M. S. (1979). *Life Sci.* **25,** 1453–1460.
Speisky, M. B. and Kalant, H. (1985). *Brain Res.* **326,** 281–290.
Squire, L. R. (1982). *Annu. Rev. Neurosci.* **5,** 241–273.
Squire, L. R., and Davis, H. P. (1981). *Annu. Rev. Pharmacol. Toxicol.* **4,** 323–356.
Stark, H., Bigl, H., and Sterba, G. (1978). *In* "Neurosecretion and Neuroendocrine Activity" (W. Bargmann, A. Okschi, A. Polenov, and B. Scharrer, eds.), p. 292. Springer-Verlag, Berlin.
Stein, L. (1969). *In* "Reinforcement and Behavior" (J. T. Tapp, ed.), pp. 328–355. Academic, New York.
Strupp, B. J., and Weingartner, H. (1984). *Drug Dev. Res.* **4,** 501–516.
Suzue, T., Yanaihara, N., and Ostuka, M. (1981). *Neurosci. Lett.* **26,** 137–142.
Swaab, D. F. (1982). *In* "Chemical Transmission in the Brain," (Progress in Brain Research Vol. 55) (R. M. Buijs, P. Pévet, and D. F. Swaab, eds.), pp. 97–122. Elsevier Biomedical Press, Amsterdam.
Szczepańska-Sadowska, E., Sobocińska, J., and Sadowski, B. (1982). *Am. J. Physiol.* **242,** R372–R379.
Tabakoff, B., and Culp, S. (1984). *Alcohol. Clin. Exp. Res.* **8,** 495–499.
Tabakoff, B., and Hoffman, P. L. (1983). *In* "The Biology of Alcoholism" (B. Kissin and H. Begleiter, eds.), Vol. 7, pp. 199–252. Plenum, New York.
Tabakoff, B., and Ritzmann, R. F. (1977). *J. Pharmacol. Exp. Ther.* **203,** 319–331.
Tabakoff, B., Melchior, C. L., and Hoffman, P. L. (1982). *Alcohol Clin. Exp. Res.* **6,** 252–259.
Tam, W. K., Chen, C., Alpert, J. E., and Iversen, S. D. (1985). *Brain Res.* **37,** 133–137.
Tanaka, M., Versteeg, D. H. G., and de Wied, D. (1977a). *Neurosci. Lett.* **4,** 321–325.
Tanaka, M., de Kloet, E. R., de Wied, D., and Versteeg, D. H. G. (1977b). *Life Sci.* **20,** 1799–1808.
Teal, J. J., and Evans, H. L. (1982). *Pharmacol. Biochem. Behav.* **17,** 1123–1127.
Telegdy, G., and Kovács, G. L. (1980). *Neurosci. Behav. Physiol.* **11,** 429–432.
Thompson, E. A., and de Wied, D. (1973). *Physiol. Behav.* **11,** 277–280.
Timsit-Berthier, M., Mantanus, H., Jacques, M. C., and Legros, J. J. (1980). *Acta Psychiatr. Belg.* **80,** 728–747.
Timsit-Berthier, M., Mantanus, H., Devos, J. E., and Spiegel, R. (1982). *Neuropsychobiology* **8,** 248–258.
Uhl, G. R., Bennett, J. P., Jr., and Snyder, S. H. (1977). *Brain Res.* **130,** 299–313.
Urban, I. J. A. (1981). *Exp. Neurol.* **74,** 131–147.
Urban, I., and de Wied, D. (1975). *Brain Res.* **97,** 362–366.
Urban, I., and de Wied, D. (1978). *Pharmacol. Biochem. Behav.* **8,** 51–59.
Vale, W., Spiess, J., Rivier, C., and Rivier, J. (1981). *Science* **213,** 1394–1397.
Valtin, H. G., and Schroeder, H. A. (1954). *Amer. J. Physiol.* **206,** 425–430.
Vandesande, F., Dierickx, K., and deMey, J. (1975). *Cell Tissue Res.* **156,** 377–380.

Van Leeuwen, F. W., Swaab, D. F., and de Raay, C. (1978). *Cell Tissue Res.* **193,** 1–10.
Van Ree, J. M., and de Wied, D. (1976). *Life Sci.* **19,** 1331–1340.
Van Ree, J. M., and de Wied, D. (1977). *Life Sci.* **21,** 315–320.
Van Ree, J. M., Bohus, B., Versteeg, D. H. G., and de Wied, D. (1978). *Biochem. Pharmacol.* **27,** 1793–1800.
Van Riezen, H., Rigter, H., and de Wied, D. (1977). *Behav. Biol.* **20,** 311–324.
van Wimersma Greidanus, T. B. (1982). *Peptides* **3,** 7–11.
van Wimersma Greidanus, T. B., and de Wied, D. (1976a). *Biochem. Behav.* **5**(Suppl. 1), 29–33.
van Wimersma Greidanus, T. B., and de Wied, D. (1976b). *Behav. Biol.* **18,** 325–333.
van Wimersma Greidanus, T. B., Bohus, B., and de Wied, D. (1973). *In* "Progress in Endocrinology," International Congress Series 273, pp. 197–201. Excerpta Medica Foundation, Amsterdam.
van Wimersma Greidanus, T. B., Bohus, B., and de Wied, D. (1974). *Neuroendocrinology* **14,** 280–288.
van Wimersma Greidanus, T. B., Bohus, B., and de Wied, D. (1975a). *Prog. Brain Res.* **42,** 135–141.
van Wimersma Greidanus, T. B., Dogterom, J., and de Wied, D. (1975b). *Life Sci.* **16,** 637–644.
van Wimersma Greidanus, T. B., Bohus, B., and de Wied, D. (1975c). *In* "Anatomical Neuroendocrinology" (W. E. Stumpf and L. D. Grant, eds.), pp. 284–289. Karger, Basel.
van Wimersma Greidanus, T. B., Fat-Bronstein, H. T. K., and Van Ree, J. M. (1978). *In* "Characteristics and Function of Opioids" (J. M. Van Ree and L. Terenius, eds.), pp. 73–74. Elsevier/North-Holland, Amsterdam.
van Wimersma Greidanus, T. B., Croiset, G., and Schuiling, G. A. (1979a). *Brain Res. Bull.* **4,** 625–629.
van Wimersma Greidanus, T. B., Croiset, G., Bakker, E., and Bouman, H. (1979b). *Physiol. Behav.* **22,** 291–295.
van Wimersma Greidanus, T. B., Croiset, G., Goedemans, H., and Dogterom, J. (1979c). *Horm. Behav.* **12,** 103–111.
van Wimersma Greidanus, T. B., Van Ree, J. M., and de Wied, D. (1983). *Pharmacol. Ther.* **20,** 437–458.
Vawter, M. P., and Green, K. F. (1980). *Physiol. Behav.* **25,** 851–854.
Verbales, J. G., and Robinson, A. G. (1983). *J. Clin. Endocrinol. Metab.* **57,** 115–123.
Vorherr, H., Bradbury, M. W. B., Hoghoughi, M., and Kleeman, C. R. (1968). *Endocrinology* **83,** 246–250.
Walter, R., and Simmons, W. H. (1977). *In* "Neurohypophysis" (A. M. Moses and L. Share, eds.), pp. 167–188. Karger, Basel.
Walter, R., Griffiths, E. C., and Hooper, K. C. (1973). *Brain Res.* **60,** 449–457.
Walter, R., Hoffman, P. L., Flexner, J. B., and Flexner, L. B. (1975). *Proc. Natl. Acad. Sci. USA* **72,** 4180–4184.
Walter, R., van Ree, J. M., and de Wied, D. (1978a). *Proc. Natl. Acad. Sci. USA* **75,** 2493–2496.
Walter, R., Ritzmann, R. F., Bhargava, H. N., Rainbow, T. C., Flexner, L. B., and Krivoy, W. A. (1978b). *Proc. Natl. Acad. Sci. USA* **75,** 4573–4576.
Walter, R., Ritzmann, R. F., Bhargava, H. N., and Flexner, L. B. (1979). *Proc. Natl. Acad. Sci. USA* **76,** 518–520.
Walter, R., Flexner, L. B., Ritzmann, R. F., Bhargava, H. N., and Hoffman, P. L. (1980a). *In* "Polypeptide Hormones" (R. F. Beers and E. G. Bassett, eds.), pp. 321–336. Raven, New York.
Walter, R., Hoffman, P. L., Flexner, J. B., and Flexner, L. B. (1980b). *Int. J. Pept. Prot. Res.* **16,** 482–486.
Walter, R., Hoffman, P. L., Church, A. C., Flexner, J. B., and Flexner, L. B. (1982). *Horm. Behav.* **16,** 234–244.

Wang, S. (1972). *Biochem. Biophys. Res. Commun.* **48,** 1511–1515.

Way, E. L., Loh, H. H., and Shen, F. (1969). *J. Pharmacol. Exp. Ther.* **167,** 1–8.

Weingartner, H., Gold, P., Ballenger, J. C., Smallberg, S. A., Summers, R., Rubinow, D. R., Post, R. M., and Goodwin, F. K. (1981a). *Science* **211,** 601–603.

Weingartner, H., Kaye, W., Gold, P., Smallberg, S., Peterson, R., Gillin, J. C., and Ebert, M. (1981b). *Life Sci.* **29,** 2721–2726.

Wise, R. A. (1980). *Pharmacol. Biochem. Behav.* **13**(Suppl. 1), 213–223.

Williams, A. R., Carey, R. J., and Miller, M. (1983a). *Peptides* **4,** 711–716.

Williams, A. R., Carey, R. J., and Miller, M. (1983b). *Peptides* **4,** 717–720.

Yamada, K., and Furukawa, T. (1981). *Arch. Int. Pharmacodyn.* **251,** 88–94.

Zaidi, S. M. A., and Heller, H. (1974). *J. Endocrinol.* **60,** 195–196.

Zerbe, R. L., and Palkovits, M. (1984). *Neuroendocrinology* **38,** 285–289.

Zimmerman, E. A., Kozlowski, G. P., and Scott, D. E. (1975). *In* "Brain–Endocrine Interaction II" (K. M. Knigge, D. E. Scott, H. Kobayashi, and S. Ishii, eds.), pp. 123–134. Karger, Basel.

Chapter 7

Renal Effects and Hormonal Interaction with the Neurohypophyseal Hormone, Vasopressin

San-e Ishikawa,* Jin K. Kim, Robert W. Schrier

Department of Medicine
University of Colorado School of Medicine
Denver, Colorado 80262

I. INTRODUCTION

Arginine vasopressin (AVP) is a neurohypophyseal peptide hormone that exerts important effects on the kidney. The primary renal effect of AVP is its hydroosmotic action whereby the hormone decreases water excretion by increasing water permeability of renal collecting tubules.

The concentrating ability of the kidney is regulated by several factors, including (1) glomerular filtration rate and proximal tubular reabsorption, which determine fluid and solute delivery to distal nephron segments; (2) sodium and chloride reabsorption in the thick ascending limb of Henle's loop; (3) papillary plasma flow, which modulates interstitial solute content; (4) the availability of urea, which contributes substantially to interstitial tonicity; (5) secretion of AVP; and (6) tubular responsiveness to AVP. Of these factors, AVP is the final rate-limiting determinant of renal water excretion.

Recent *in vivo* and *in vitro* studies have provided insights about the mechanism of the renal effects of AVP, and the complexity of these effects will now be discussed.

II. NEPHRON SITES OF AVP ACTION

In vitro techniques using microdissection and microperfusion systems have advanced the knowledge of the AVP action sites on renal tubules. Two sites of

*Present address: Jichi Medical School, 3311-1 Yakushiji, Minamikawachi-Machi, Tochigi 329-04, Japan.

THE PEPTIDES, VOLUME 8

the nephron have been demonstrated to respond biochemically to AVP by increasing adenylate cyclase activity (*121, 184*). Specifically, AVP stimulates adenylate cyclase activity in the collecting tubules of humans and other species, and in the medullary thick ascending limb of Henle's loop and the collecting tubule of rat and mouse. However, in humans and rabbits the medullary thick ascending limb of Henle's loop does not seem to be a site of AVP action (*51, 184*). While it is clear that AVP increases water permeability of the collecting tubule (*91*), this effect does not occur in the medullary thick ascending limb of Henle's loop. Results of studies by Burg and Green (*44*) and Rocha and Kokko (204) suggested that AVP might enhance chloride reabsorption in the water-impermeable medullary ascending limb of Henle's loop. This was confirmed by studies of Hall and Varney (*99*), Sasaki and Imai (*208*), and Hebert *et al.* (*111*), showing that transepithelial potential difference and active chloride and sodium transport are stimulated by AVP in this portion of the nephron.

Effects of AVP on chloride transport will be discussed in detail later in this chapter.

III. CELLULAR ACTION OF HYDROOSMOTIC ACTION OF AVP

Since Sutherland introduced the role of cyclic adenosine monophosphate (cAMP) in the hormone action (*231*), it has been well accepted that the action of AVP on the renal tubular cell is mediated through cAMP. The cellular action of AVP is initiated by the binding of AVP to the specific receptors on the basolateral membranes. The AVP receptor complex stimulates adenylate cyclase, and the stimulation of adenylate cyclase leads to formation of cAMP in the cell. The production of intracellular cAMP changes water permeability of the luminal membrane via a complex process(es), which is not yet clearly known.

A. AVP Receptor Binding

Although it is generally accepted that the AVP receptor at the basolateral membranes of renal collecting tubules is functionally coupled to the adenylate cyclase, classified as V2 receptors (*178*), the mechanism of AVP receptor activation has been studied intensively in recent years. A correlation of the specific binding of AVP or lysine vasopressin (LVP) and the stimulation of adenylate cyclase by AVP or LVP has been shown in the renal medullopapillary membranes from porcine (*34, 45*), rat (*45, 46, 140*), bovine (*112*), and human (*98, 228*) studies. In all these studies, the binding of [^{3}H]AVP was reversible, time-dependent, and saturable. It seems that the AVP receptors in the mammalian kidney are homogeneous, since the Scatchard analysis showed linearity in all cases. However, there might be a difference in binding characteristics between renal medullary and papillary membranes. [^{3}H]AVP binding to the rat renal papillary membrane showed more specific binding and less nonspecific binding than the rat renal medullary membranes (unpublished data from this laboratory). This difference might be due to heterogeneity of tubular cells in the renal medullary membranes. In contrast to the studies of renal membranes, [^{3}H]LVP binding to the LLC PK1 cells indicated a heterogeneity in the population of binding sites (*207*). Although differences between intact cells and established cell lines in the kinetics of hormone binding are not known, it is clear that AVP binding to the receptors is involved in the activation of adenylate cyclase and the resultant physiological response to AVP.

B. Adenylate Cyclase

Adenylate cyclase is present in all parts of the mammalian nephron. However, AVP only stimulates adenylate cyclase in the collecting tubules and medullary thick ascending limb of Henle's loop (*121, 140, 184*), which are known as target sites of AVP. In the collecting tubule the papillary portion seems most sensitive to AVP, as maximum stimulation of adenylate cyclase is found with 10^{-7} *M* AVP in the papillary collecting tubule compared to 10^{-6} *M* AVP in the medullary collecting tubule (*139*). When response of adenylate cyclase to AVP was studied in the cortical, medullary, and papillary collecting tubules from the same kidney, the papillary-collecting-tubule adenylate cyclase was most sensitive to AVP, followed by the medullary collecting tubule, and the least sensitive to AVP was the cortical collecting tubule (*142*). The physiological significance of this different sensitivities of adenylate cyclase to AVP in the collecting tubules remains to be elucidated. The response of adenylate cyclase to AVP is not different between species in the medullary collecting tubule. The species differences of adenylate cyclase stimulation to AVP may correlate with the species' capacity for maximal urinary concentration; maximal urinary osmolality is 2500–3000 mOsm/kg H_2O in the mouse and rat, 1500 mOsm/kg H_2O in the rabbit, and 800–1200 mOsm/kg H_2O in humans (*68*). These differences in urinary osmolality might relate to the action of AVP on the medullary ascending limb of Henle's loop to increase sodium and chloride reabsorption and thus to enhance interstitial tonicity. Recent studies of two different groups of rats support this concept (*142*). A group of rats in which the adenylate cyclase of the medullary thick ascending limb did not respond to AVP showed significantly lower maximal urinary osmolality than a group of rats in which adenylate cyclase of medullary thick ascending limb responded well to AVP. Collecting tubules from both groups of rats responded comparably to AVP.

C. cAMP

cAMP is generated from adenosine triphosphate (ATP) by adenylate cyclase stimulation and is degraded to the inactive form 5′-AMP by cAMP phosphodiesterase. Increased intracellular cAMP concentration, as modulated by these two enzymes, initiates post-cAMP systems, which culminate in increased water permeability of the luminal membrane of collecting tubules (*65*). In some pathological states, alterations in the above two cAMP modulating enzymes caused an abnormal response to AVP. For example, in mice with hereditary nephrogenic diabetes insipidus, cAMP accumulation was not found in either the medullary ascending limb of Henle's loop or the medullary collecting tubule (*128*). This inability to increase cAMP in response to AVP was found to result, at least in part, from abnormally low activity of adenylate cyclase in the medullary ascending limb of Henle's loop and abnormally high activity of cAMP phosphodiesterase in the medullary collecting tubule of these mice with hereditary nephrogenic diabetes.

D. cAMP-Phosphodiesterase

As suggested by Jackson *et al.* (*129*), in the normal rat kidney cAMP formation in response to AVP does not appear to be by differential rates of cAMP catabolism in the medullary collecting tubule as compared to the medullary ascending limb of Henle's loop. AVP *per se* also does not have any direct effect on cAMP phosphodiesterase. However, activity of cAMP-dependent phosphodiesterase can be altered in some circumstances independent of AVP, including mice with hereditary nephrogenic diabetes insipidus (*128*) and glucocorticoid-deficient rats (*130*). Thus, cAMP catabolism may play an important role in cAMP accumulation in response to AVP.

E. cAMP-Dependent Protein Kinase

Since in several species AVP acts on both the medullary ascending limb of Henle's loop and the collecting tubule, it was important to determine whether post-cAMP systems are related to the action of AVP in these two tubular segments. In this regard, protein kinase has been proposed as one of the post-cAMP systems (*66*). Two relevant experimental findings about renal cAMP dependent protein kinase are (1) that cAMP-dependent protein kinase is present in tissue homogenates from renal medulla and papilla (*70*) and (2) that either AVP or exogenous cAMP has been demonstrated to activate protein kinase in renal medullary and papillary tissue slices and isolated tubules of the medullary ascending limb of Henle's loop and medullary collecting tubules (*70*). Thus, AVP-induced cAMP-dependent protein kinase in renal papilla and medulla has been assumed to be associated with physiological function of AVP. Several matters, however, still remain to be resolved. For example, it is not known whether cytosolic or membrane-bound protein kinase is the relevant enzyme in this effect of AVP. There is also no evidence on whether the protein kinase that is activated by AVP regulates transport processes at the membrane of the medullary ascending limb of Henle's loop and collecting tubules. Also, the interrelationship between other post-cAMP events with protein kinase, microtubules, microfilaments, and membrane aggregation remains to be defined.

Calcium and prostaglandin (PG) have also been suggested to alter protein kinase activity in toad urinary bladder (*210, 211*). These observations are discussed in detail later in this chapter.

F. Microtubules, Microfilaments, and Intramembrane Aggregation

In addition to cAMP-dependent protein kinase, microtubules and microfilaments have also been suggested to regulate the effect of AVP to increase water permeability (*234, 236*). Most previous studies, however, have been performed in the amphibian toad urinary bladder. Studies by Taylor *et al.* (*234, 235*) in the toad urinary bladder have implicated the role of microtubules and microfilaments in the hydroosmotic effect of AVP. Colchicine, which reduces the volume density of microtubules in the cytoplasm of granular-type epithelial cells by binding to the microtubule subunit protein (tubulin), disrupts microtubules and inhibits the hydroosmotic effect of AVP without altering transepithelial sodium and urea movement. Cytochalasin B, which functionally and structurally disrupts microfilaments, also inhibits the hydroosmotic effect of AVP without affecting sodium excretion (*136*). These findings may also be present in mammalian kidney (*63, 127*). A dense network of microfilaments in toad urinary bladder granular cells has been found in close association with the luminal membrane (*54*). An interrelationship between microtubules, microfilaments, and luminal intramembranous aggregation has been implicated in the final step of AVP action, and recent studies with freeze-fracture electron microscopy have supported this concept (*135, 244*). In these studies the phenomenon of intramembrane particle aggregation increased with AVP-dependent enhanced water permeability and was not related to changes in urea or sodium permeability (*135*). Intramembranous particle aggregation probably consists of fusion of cytoplasmic vesicles with the luminal membrane (*188*); this effect is dependent, at least in part, on the calcium–calmodulin complex (*158*). The intramembrane aggregation and water-flow responses to AVP are both independently sensitive to colchicine and cytochalasin B. However, after maximal stimulation by AVP, the response is found to be sensitive to cytochalasin B but not to colchicine (*136*). These findings indicate that microtubules and microfilaments are involved in the initiation of water flow and intramembrane aggregation by AVP, but only microfilaments are required for the maintenance effect. These studies were performed in the toad urinary bladder, but studies by Harmanci *et al.* have demonstrated intramembrane particle aggregates in the mammalian kidney (*107, 108*). When AVP was administered to rats, there was a significant correlation between increased urinary osmolality and increased intramembranous particle aggregation of collecting-duct luminal membrane (*40*). Thus, the findings in the amphibian toad bladder are consistent with those in the mammalian kidney.

Recent studies (*108*) of intramembrane particle aggregates in the abdominal skin of toad reported that the intramembrane particle aggregates might be a general response in epithelia undergoing cAMP-mediated changes in water permeability, since correlation between water-flow and intramembrane particle-aggregate responses to AVP and isoproterenol were the same. Taking together all this information, changes of water permeability in the final step of AVP action involve complex interrelationships of (1) changes of cytoskeleton structures, (2) fusion of cytoplasmic tubules to the cytoplasmic surface of the luminal membrane, and (3) delivery of aggregates of intramembrane particles to the luminal membrane (*109*). All of these recent new studies have added more information to the mechanism of AVP action; however, many questions remained unanswered. In addition, the cytoskeletal system has been suggested to be involved in steps of many hormone actions, such as receptor binding, adenylate cyclase, cAMP phosphodiesterase, and protein kinase (*252*). This possibility might also be applicable to AVP action.

IV. MODULATING FACTORS OF AVP ACTION

A. Calcium and Calmodulin

The calcium ion has been suggested to modulate the hydroosmotic effect of AVP at multiple sites in the renal tubular epithelium. An *in vivo* study by Berl (*28*) demonstrated that the PG inhibitory action of collecting tubules on the hydroosmotic effect of AVP is attenuated by the treatment with the calcium membrane blockers verapamil or proadifen. The author therefore suggested that PG inhibits the *in vivo* effect of AVP primarily by impairing calcium membrane transport (*28*). The effect of PTH on enhanced renal response to AVP has also been suggested to be dependent on the enhanced calcium membrane transport into the cell (*2, 116*). These *in vivo* studies are consistent with *in vitro* studies in toad urinary bladder (*104, 105*) and therefore suggest that intracellular calcium is an important determinant of the action of AVP.

However, two recent studies in toad urinary bladder suggest that calcium membrane blockers may alter cellular enzyme activities without blocking transcellular calcium movement (*117, 159*). Verapamil and its analog (d-600) and diltiazem inhibited AVP-induced water flow across the toad bladder but actually enhanced cAMP-induced water flow. Studies have suggested that these calcium membrane blockers inhibit cAMP generation in response to AVP but enhance the response to exogenous cAMP by inhibiting cAMP phosphodiesterase activity (*159*). These results suggested that the effect of calcium entry blockers on water flow may occur by a direct effect on the cellular enzymes of the membranes in which the enzymes are present. The calcium ion may modulate the action of AVP at the site between the AVP receptor and adenylate cyclase stimulation (*46*).

Intracellular calcium has also been suggested to play a critical role in the action of AVP. Whether the extracellular calcium is a major source of the intracellular calcium for the action of AVP is not clear at the present time. In frog urinary bladder, calcium ionophore A23187 enhanced the hydroosmotic effect of a submaximal dose of AVP, and the serosal calcium was required to sustain the effect (*104*). However, another study in the toad urinary bladder was not consistent with this interpretation (*105*). In addition, in a study using the isolated perfused rabbit cortical collecting tubule, increased bath calcium enhanced AVP but not cAMP-stimulated water flow. These results suggested a role for extracellular calcium in the cellular action of AVP (*88*). Recent studies with primary cell culture of rat renal papillary collecting tubule (*237*) also suggested that AVP-induced intracellular cAMP accumulation is calcium-sensitive. The calcium-sensitive cAMP system in the medullary collecting tubule was markedly more sensitive than in the medullary thick ascending limb (*190*).

Calmodulin, the calcium-binding protein, has been suggested to modulate the hydroosmotic action of AVP (*158*). Trifluoperazine, which binds to calmodulin and thereby prevents activation of enzymes by calcium–calmodulin complex, inhibited water flow induced by either AVP or cAMP in toad urinary bladder (*158*). This effect was not associated with decreased cAMP content or cAMP-dependent protein kinase activity, thus implying a post-cAMP site of AVP action. A decrease in the number of luminal membrane aggregates was found in trifluoperazine-treated tissues in the absence of a reduction in fusion sites, a finding that is consistent with a site of trifluoperazine action at the microfilament level.

As discussed earlier, AVP stimulates PG production in the perfused kidney, renal medullary slices, and renal interstitial cells (*58, 251, 254*). A study by Craven and deRubertis (*58*) has shown that either a calcium-free medium or a medium containing trifluoperazine inhibits AVP-stimulated PG production from the slices of rat inner medulla. This effect was mediated by altering calcium-responsive arachidonate release. AVP *per se* did not alter acyl hydrolase activity, but trifluoperazine suppressed calcium-induced increases in acyl hydrolase activity. Thus, AVP may acutely stimulate inner medullary PGE production by activating the calcium–calmodulin-dependent acyl hydrolase.

B. Interstitial Tonicity

Previous studies by Dousa (*62*) and deRubertis and Craven (*61*) have shown that changes in medium solute concentration influence enzyme activity of cAMP metabolism in homogenates or membrane fractions from renal papilla. The *in vivo* papillary collecting tubule is located in an environment of hyperosmolality that fluctuates widely in different physiologic and pathophysiologic states (*11, 153*). Moreover, the cellular action of AVP in the isolated papillary collecting tubule was demonstrated to be profoundly altered by environmental osmolality (*71*). Stimulation of adenylate cyclase by AVP was markedly enhanced in medium osmolality from 200 to 800 mOsm/kg H_2O and then was decreased at 2000 mOsm/kg H_2O. On the other hand, cAMP-phosphodiesterase activity is gradually reduced by increasing medium osmolality from 200 to 2000 mOsm/kg H_2O. The basal and AVP-stimulated cAMP content increase as the medium osmolality is increased from 300 to 2000 mOsm/kg H_2O. Thus, environmental osmolality is an important modulator in cellular action of AVP in the papillary-collecting-tubule cells.

C. Temperature

Cold baths or applications of cold to the skin is associated with an increase in urine flow (*23*). This cold diuresis can be seen in association with the diminished release of AVP. Inhibition of AVP release may result from either the cooling of the preoptic region (*110*) or the increase in mean arterial pressure (*182*). The study by Morgan *et al.* (*182*) showed that the diuresis is independent of a diminished renal response to AVP, changes in renal hemodynamics, solute excretion, or renal perfusion pressure. The authors, however, indicated a role of suppressed AVP in the cold diuresis in the rat. Discrepant results on the effect of temperature have been published using *in vitro* microperfusion studies (*93, 100*). Changes in temperature of the bathing medium have been found to alter the response of the isolated perfused rabbit cortical collecting tubules to AVP. The responsiveness of the cortical collecting tubules to AVP was higher during exposure to a temperature of 37°C than at 25°C. However, the response was unstable and gradually deteriorated. In contrast, a stable increase in hydraulic water permeability was found at 25°C in response to either AVP or cAMP. The inhibiting effect of physiological temperature (37°C) on the response to AVP is probably due to a decreased responsiveness to cAMP.

D. Extracellular pH

The effect of extracellular pH on the action of AVP is not known yet. However, it has been shown that metabolic acidosis decreases negative free water clearance in rats after administration of AVP (*18*). Recent studies in the isolated perfused rabbit cortical collecting tubule demonstrated that acidification of the peritubular fluid (induced by increasing the CO_2 tension of the peritubular medium at constant [HCO_3]) inhibited the response to AVP while markedly augmenting of the response to 8-CPT-cAMP (*164*). The mechanism of this effect is not known at this time; however, pH changes may well have a complex effect on the action of AVP.

V. INTERACTIONS OF OTHER HORMONES AND AVP

A. Prostaglandins and Thromboxanes

The role of prostaglandins (PG) in the renal function has been studied extensively for the last decade. There have been a considerable number of studies implicating a role of PG on the action of AVP. PGE_2, PGI_2, PGF_{2a}, PGD_2, and thromboxane A_2 (TxA_2) are primarily produced and inactivated in the medulla. Therefore, measurement of urinary PG was suggested to reflect renal PG synthesis rather than alteration in systemic PG synthesis (*83, 151*). It has been suggested, however, that PG excretion is altered by changing urine flow rate, as suggested by Kirschenbaum and Serros (*145*) and by Wright *et al.* (*249*) in dog studies.

The collecting-tubule epithelial cells and interstitial cells both produce PG. Vascular beds are another site of PG synthesis in the kidney. The recent study of Smith *et al.* (*225*) using an immunofluorescence technique demonstrated cyclooxygenase in the collecting-tubule cells. Also, further studies by the same group have revealed the production of PGE_2 and PGI_2 in papillary-collecting-tubule cells in culture (*95*). This finding is consistent with previous observations by Bohman (*35*) and Jackson *et al.* (*131*). In contrast, the evidence regarding AVP stimulation of PGE synthesis in renal interstitial cells is conflicting because of species differences. Zusman and Keiser reported that the cultured rabbit renomedullary interstitial cells produce PGE (*253, 254*). The cultured interstitial cells could be stimulated to release PGE in response to angiotensin II, AVP, and bradykinin (*254*). But Smith and Wilkin (*225*) and Janszen and Nugteren (*133*) have failed to demonstrate cyclooxygenase in the rabbit medullary interstitial cells. Smith and Wilkin, however, have shown cyclooxygenase in the rat interstitial cells (*225*). Differences in methodology may explain some of these discrepancies.

A great deal of studies has accumulated to indicate that PG can antagonize the hydroosmotic effect of AVP in the kidney. The pioneering studies by Orloff, Handler, and Bergstrom using the toad urinary bladder (*194*) and by Grantham and Orloff using the isolated perfused rabbit renal collecting duct (*92*) demonstrated the inhibition by PGE of the water-permeability response to AVP without inhibition of the water-permeability response to exogenous cAMP. These findings were confirmed by other investigators using both *in vitro* (*47, 163, 240*) and *in vivo* techniques (*96, 134*).

The mechanisms whereby PG inhibits the hydroosmotic effect of AVP have still not been fully understood. The studies by Orloff *et al.* (*194*) and Grantham *et al.* (*92*) have suggested that PG may alter a site proximal to cAMP generation (e.g., adenylate cyclase system). PGE has been suggested to inhibit the AVP-induced adenylate cyclase stimulation and other steps unrelated to cAMP generation. Studies performed on the toad urinary bladder suggested multiple sites for interaction between PG and AVP (*210*). More recent studies in which enzymatic activity was measured further support an interaction between PG and AVP. In these studies PGE diminished the effect of AVP on the adenylate cyclase activation and cAMP generation in the toad urinary bladder (*192*), renal slices (*58, 251*), and isolated collecting tubules (*131*). In addition, studies from our laboratory also revealed that indomethacin, an inhibitor of PG biosynthesis, potentiates the ability of AVP to increase the tissue content of cAMP in rats (*167*). These results support the earlier *in vivo* findings from our laboratory (*3, 25*).

Studies of Dousa *et al.* have further examined the effect of PG to inhibit AVP-induced adenylate cyclase activity (*67, 131*). In these studies, PGE_2 did not inhibit adenylate cyclase stimulation by AVP in the isolated nephron segments of medullary thick ascending limb and papillary collecting tubule, but inhibited adenylate cyclase response to AVP in medullary collecting tubules. Recently studies in rat isolated tubules suggested that PGE_2 may suppress intracellular cAMP by inhibiting AVP-induced cAMP formation in the thick ascending limb and that PGE_2 may suppress cAMP in the cortical collecting tubule by acting at a site affected by the phosphodiesterase inhibitor 1-methyl-3-isobutylxanthine (MIX) (*239*). These results indicate that PGE_2 inhibits AVP-induced cAMP accumulation by a different mechanism in different segments of the nephron. Since AVP is known to stimulate chloride transport in the medullary thick ascending limb, PGE_2 may modulate this effect in the medullary thick ascending limb. Culpepper and Andreoli (*59*) reported that PGE_2 reduced the AVP-stimulated net chloride absorption and spontaneous transepithelial voltage in the mouse medullary thick ascending limb. The authors suggested that PGE_2 might compete with AVP at receptor sites, since the inhibitory effect of PGE_2 was overcome by a high concentration of AVP, and PGE_2 exerted no effect on cAMP-stimulated increments in the transepithelial voltage.

High concentrations of exogenous PGE_2 stimulate adenylate cyclase in the toad urinary bladder (*80, 163, 192*), the renal medulla (*61, 67*), the isolated papillary collecting tubule (*71*), and the cultured papillary collecting cells (*95*), using directly or indirectly measured adenylate cyclase activity. In the isolated perfused cortical collecting tubule, a high concentration (100 μM) of PGE alone increases water permeability in a manner similar to that of a low concentration (1 μM) of PGE and theophylline (*71*). However, another study showed that PGE_2 (10^{-5} M) increased the cAMP content and cAMP-dependent protein kinase ratio but resulted in no change in water flow in the toad urinary bladder (*210*). Also, PGE in a high concentration reduced the negative transepithelial voltage of the cortical and medullary collecting tubules and the medullary thick ascending limb; this effect was accompanied by a reduction of sodium transport (*120, 229, 230*). PGE is therefore proposed to have direct effects on these nephron segments independent of an interaction with AVP action.

PG also may alter local *in vivo* environmental factors that modulate water reabsorption across the collecting tubule epithelium. Specifically, exogenous or endogenous PG may increase inner cortical and juxtamedullary renal blood flow (*52, 126*) and "washout" interstitial hypertonicity in the medulla (*220*). Indomethacin, an inhibitor of PG synthesis, decreases inner cortical renal blood flow (*126, 144*), papillary plasma flow (*55, 226*), and erythrocyte velocity in the vasa recta of the renal medulla (*155*), and it increases renal interstitial solute concentration (*85*). PG-mediated alterations in medullary and papillary blood flow may thereby modulate the solute content of the medulla. Thus, the effect of PG to regulate medullary and papillary blood flow and thereby affect maintenance of renal interstitial tonicity would be expected to modulate the hydroosmotic effect of AVP. Taken together, therefore, PG may influence water reabsorption by altering both sodium chloride transport and medullopapillary blood flow.

Although the finding is controversial, AVP has been demonstrated to stimulate PG synthesis in a dose-dependent manner in the rabbit renal medullary interstitial cells (*254*). Further studies suggest that the vascular property of AVP is associated with PG biosynthesis from these interstitial cells. In the study by Beck *et al.* (*19*), AVP, but not 1-deamino-8-D-AVP (dDAVP), which is an AVP agonist without substantial vascular properties, stimulated PGE production in rat renomedullary interstitial cells. This effect of AVP was totally abolished by the administration of the vascular AVP antagonist [1(β-mercapto-β, β-cyclopentamethylene propionic acid), 4-valine]-8-D-arginine vasopressin. Similar results were obtained by Zipser *et al.* (*251*). On the other hand, Grenier *et al.* (*95*) failed to demonstrate that AVP stimulates PG synthesis in the cultured renal papillary-collecting-tubule cells. In the toad urinary bladder AVP also failed to show any significant increase in PGE_2 or thromboxane B_2 (TxB_2) synthesis (*33*). In contrast, recent studies demonstrated that AVP stimulated both PG and thromboxane synthesis in epithelial cells from the toad urinary bladder (*43*). The same results have been reported in studies of rabbit cortical collecting tubules (*146*).

While the PG produced in the tubular epithelial cells is supposed to antagonize the water permeability response to AVP, the role of PG synthesized from the interstitial cells remains to be studied. Although the interstitial cells are located in the interstitium of the medulla, it is not clear that PG released from these cells reaches AVP-sensitive epithelial cells of the collecting tubules and thereby antagonizes the hydroosmotic effect of AVP.

An additional study by Zusman and Keiser (*253*) in rabbit renomedullary interstitial cells suggested that arachidonic acid is released by AVP-activated phospholipase or acyl hydrolase, or both. The AVP-stimulated release of arachidonic acid thus provides the precursor for PG biosynthesis. Indomethacin, an inhibitor of cyclooxygenase, totally abolishes AVP-stimulated PG biosynthesis without affecting arachidonic acid release. Studies by Craven and deRubertis in rat renal inner medullary slices (*57, 58*) suggests that this mechanism of AVP stimulation of arachidonic acid and PGE_2 synthesis is a calcium-dependent process involving a calcium–calmodulin-dependent activation of acyl hydrolase.

Studies by Burch *et al.* (*41, 42*) have suggested that thromboxane (Tx) is also associated with water metabolism in the toad urinary bladder. Tx and the stable PG endoperoxide analogs, (15*Z*)-hydroxy-9a,11a-(epoxymethano)prosta-5*Z*,13*E*-dienoic acid and (15*Z*)-hydroxy-11a,9a-(epoxymethano)prosta-5*Z*,13*E*-dienoic acid, stimulated basal and enhanced AVP-induced osmotic water flow. 7-(1-Imidazolyl)-heptanoic acid (7IHA), also a Tx synthetase inhibitor, inhibited both AVP-stimulated water flow and synthesis of TxB_2; the stable metabolite of TxA_2. 13-Azaprostanoic acid (13APA), a Tx antagonist, also inhibited AVP-stimulated water flow. However, 7IHA and 13APA had no effect on cAMP-stimulated water flow. In addition, as discussed earlier, AVP stimulated synthesis of PGE_2 and TxB_2 in the toad urinary bladder (*33, 41*). Thus, TxA_2 is suggested to be a positive modulator of AVP-stimulated water flow in the toad urinary bladder. Stimulation of TxA_2 synthesis by AVP therefore would cause an enhancement of AVP-induced water flow. This interaction of Tx with AVP would tend to counteract any negative feedback system between AVP and PGE; additional studies are necessary to explore this possibility.

B. Kallikrein–Kinin

Kallikrein has been shown to be located in the cytoplasm of distal convoluted tubules but not in the collecting tubules of rat cortex (*195*). Recent studies in the microdissected rabbit nephrons (*193*), however, reported the localization of kallikrein in the granular portions of the distal and cortical collecting tubule. The kallikrein–kinin system has been suggested to be another modulating factor of the action of AVP (*181*), since intravenous kallikrein increases free water clearance. In humans, alterations in urinary kallikrein secretion are closely related to changes in free water clearance (*1*). Moreover, changes in urinary kallikrein excretion are, at least in part, regulated by AVP (*74, 137*).

It has been shown that AVP-induced osmotic water flow was reduced by exogenous kinins, including bradykinin, kallidin, eleodosin, and physalamine in the toad urinary bladder (*84*). Aprotinin, a kallikrein inhibitor that decreases kinin production, enhances AVP-stimulated and 8-bromo-cAMP-stimulated water flow. Captopril, the kininase II inhibitor, causes kinin accumulation and decreases AVP-stimulated and 8-bromo-cAMP-stimulated water flow in the toad urinary bladder (*93*). On the other hand, aprotinin decreases and captopril increases AVP-stimulated urea permeability. These studies therefore indicate that the kallikrein–kinin system modulates the action of AVP on the water and urea permeability in opposite directions. Also, it may be speculated that the kallikrein–kinin system is associated with AVP as a negative-feedback system. The results of the later study of Greiner *et al.* (*95*) using the cultured rabbit papillary collecting tubule cells have shown that bradykinin stimulates PGE_2 biosynthesis. Since PGE_2 is known to antagonize the hydroosmotic effect of AVP, the interactions among the action of AVP, PG, and the kallikrein–kinin system are in need of study.

C. Catecholamines

Catecholamines have been suggested to modulate the antidiuretic action of AVP. Intrarenal norepinephrine may inhibit the action of exogenous AVP and thus cause a diuresis, and isoproterenol may enhance the action of exogenous AVP and thus result in an antidiuresis *in vivo* (*78, 147, 156*). These findings were observed without affecting any alterations in systemic and renal hemodynamics and were reversed by their respective alpha- and beta-adrenergic antagonists, i.e., phentolamine or propranolol. These findings suggested that catecholamines directly alter the action of AVP in renal tubules through alpha- or beta-adrenergic receptors on the tubular cells (*147, 156*). *In vitro* studies in the toad urinary bladder also demonstrated that norepinephrine antagonizes AVP action (*102*). On the other hand, studies both in humans (*22*) and dogs (*20, 21, 214, 215*) suggested that alpha- and beta-adrenergic stimulation altered renal water excretion primarily through baroreceptor-mediated AVP release. These studies indicated that alpha-adrenergic stimulation suppresses AVP release (*20, 215*) and beta-adrenergic stimulation increases AVP release (*21, 22, 214*) by altering autonomic baroreceptor tone rather than by a direct central effect of the catecholamines in the brain (*213*). A study performed in the isolated perfused papillary collecting duct (*199*) also suggested that neither norepinephrine nor isoproterenol affected the water permeability of collecting ducts. However, recent studies with the rabbit cortical collecting tubules indicated that alpha-2-adrenergic stimulation directly inhibited AVP-mediated water absorption, and this effect was blocked by a specific alpha-2-adrenergic antagonist (*148*). This inhibitory effect of alpha-2-adrenergic stimulation was suggested to be exerted by inhibition of AVP induced cAMP generation (*148*). In a study with toad bladder (*10*), dopamine inhibited AVP- or cAMP-induced water flow, and this inhibitory effect was not due to activation of alpha- or beta-adrenergic or cholinergic receptors. It seems that the effect of catecholamines on the water transport in the tubule is exerted probably both by a direct effect on the tubular cell and by altering AVP release. *In vivo* studies indicate a predominant effect of the latter mechanism.

D. Adrenocortical Hormones

Glucocorticoid and mineralocorticoid deficiency have been demonstrated to impair water excretion by AVP-dependent and AVP-independent effects (*37, 94, 160, 171*). The study of Green *et al.* (*94*) in the adrenalectomized Brattleboro rats with hereditary diabetes insipidus implicated an AVP-independent pathway in the impaired water excretion in both glucocorticoid and mineralocorticoid deficiency (*94*). An increased concentration of cAMP in the renal medulla has been reported in glucocorticoid deficiency (*149*); this effect could, however, be due to increased endogenous AVP concentration. This effect may be due to a diminution in cAMP phosphodiesterase activity, as shown in the isolated rat renal medullary collecting tubules by Jackson *et al.* (*130*). This abnormality may increase intracellular cAMP, and thereby increase water reabsorption during glucocorticoid deficiency. However, an AVP antagonist of the hydroosmotic effect of AVP has been shown to markedly ameliorate water retention in glucocorticoid and mineralocorticoid deficiency (*125*). This latter finding is consistent with the nonsuppressible plasma concentrations of AVP after an acute water load in glucocorticoid-deficient (*125, 160*) and mineralocorticoid-deficient (*38, 125*) animals.

Schwartz and Kokko (*216*) demonstrated that a urinary concentrating defect was present in adrenal insufficiency, and that this defect was corrected by replacement of either deoxycorticosterone or dexamethasone. Similar results were obtained in *in vitro* perfused cortical collecting tubule from rabbits with adrenal insufficiency in which the hydroosmotic response to AVP was blunted. The impaired response to AVP was corrected by exogenous aldosterone, dexamethasone, cAMP, or a phosphodiesterase inhibitor. The authors thus suggested that the absence of adrenal steroids results in diminished cAMP accumulation in collecting-tubule cells. This finding is in contrast to the proposal that adrenal insufficiency enhances water permeability in the mammalian kidney. However, nonsuppressible plasma concentrations of AVP could override this diminution in the tubular effect of AVP.

E. Parathyroid Hormone and Calcitonin

There have been several studies concerning the interaction between AVP and parathyroid hormone (PTH) on urine concentration (*2, 116, 246*). Stimulation of adenylate cyclase by PTH in the rat papillary tip was enhanced by AVP (*191*), indicating an additive effect of these hormones. However, this result was not demonstrated in the isolated tubules (*184*). The study of Humes *et al.* demonstrated that PTH enhanced the renal response to exogenous AVP in the thyroparathyroidectomized rats undergoing water diuresis (*116*). This potentiating effect of PTH in the AVP response was abolished by pretreatment with verapamil, a blocker of cellular calcium uptake (*116*). Similar results were obtained from another laboratory (*2*). However, another study in dogs (*246*) reported that a pharmacological dose of PTH showed ''AVP-like'' action on the tubular water transport.

However, an effect of PTH on adenylate cyclase of medullary and papillary collecting tubules has not been found using a physiologic dose of PTH. PTH stimulates adenylate cyclase in the proximal tubule as well as in two distal nephron sites, the cortical thick ascending limb and the granular portion of the distal and collecting tubules (*184*). Since it has been reported that two or more hormones stimulate adenylate cyclase in the same nephron site, Morel *et al.* (*185*) suggested an additive action of the calcitonin and AVP. In hormone-deprived Brattleboro rats, calcitonin administration restored the corticomedullary concentration gradient and reduced water excretion (*206*), suggesting that calcitonin enhances electrolyte reabsorption along the thick ascending limb and/or the water permeability of the cortical collecting duct. More studies are required to clarify if calcitonin alone has a direct effect on the tubule or modulates the action of AVP at these target sites.

Glucagon is the most potent in stimulating adenylate cyclase of medullary thick ascending limb; however, the physiological role of the hormone on nephron function remains to be clarified. It has been reported that glucagon induces a further stimulation of adenylate cyclase above the maximal stimulatory effect of AVP in the medullary thick ascending limb (*185*). These interactions may have importance relative to renal function, but this remains to be clarified.

F. Somatostatin

Somatostatin, the inhibiting factor for somatotropin release, has been shown to inhibit the release and action of several hormones, including growth hormone, thyroid-stimulating hormone (TSH), prolactin, adrenocorticotropic hormone (ACTH), insulin, and glucagon. Somatostatin was also demonstrated to inhibit the hydroosmotic effect of AVP in the dog and rat by later studies of Reid and Rose (*200*) and Kurokawa *et al.* (*39, 247*). Infusion of somatostatin resulted in an increase in urine volume and free water clearance and a decrease in urinary osmolality, without any alterations in systemic and renal hemodynamics, osmolar clearance, or electrolyte excretion (*200*). The inhibitory effect of somatostatin on the hydroosmotic effect of AVP depends, at least in part, on the diminution of AVP-induced cAMP accumulation in the cortical and medullary collecting tubules (*247*). However, the presence of somatostatin has not been demonstrated in the mammalian kidney (*36*). Thus, it is not clear whether the effect of somatostatin on the hydroosmotic action of AVP is of physiological importance, since somatostatin has been considered to be a locally produced and locally acting hormone (*9*).

G. Thyroid Hormone

The studies in toad urinary bladder have suggested that thyroid hormone potentiates the hydroosmotic effect of AVP (*176*), but this observation could not be confirmed by another laboratory (*233*). An impairment in water excretion, however, has been found in advanced hypothyroidism. In humans, this abnormality has been associated with an elevated concentration of plasma AVP (*224*). An interaction between thyroid hormone and hydroosmotic effect of AVP at the level of the nephron was further studied by Dousa *et al.* (*106*). In these studies, basal and AVP-stimulated activities of renal medullary adenylate cycalse were markedly reduced in hypothyroid rats, and replacement with thyroxine restored the medullary adenylate cyclase activity to levels observed in control rats. The thyroid replacement decreased cAMP phosphodiesterase activity, thus suggesting a cellular interaction between thyroid hormone and AVP on cAMP generation in the renal medulla.

VI. ROLE OF AVP IN PATHOPHYSIOLOGICAL STATES

Renal diluting disorders are frequently associated with the presence of hyponatremia. Hyponatremia may be associated with decreased, increased, or near-normal amounts of total body sodium and extracellular fluid volume (*24*). In general, disorders of renal diluting capacity are due to (1) the continued secretion of AVP in spite of hypoosmolality, which normally would suppress endogenous

AVP release to an undetectable level; (2) intrarenal factors, including a decrease in glomerular filtration rate and/or an increase in proximal tubular reabsorption of sodium and water, which diminishes the delivery of fluid to the distal diluting segment of the nephron; (3) a defect in sodium chloride transport in the medullary thick ascending limb; and (4) an abnormal urea transport and recycling system.

Recent experimental and clinical evidence showed that the nonosmotic, baroreceptor-mediated release of AVP is an important factor in the water retention observed in several pathophysiological states, including diuretic abuse (*77*); glucocorticoid (*37, 125, 160, 171*), mineralocorticoid (*38, 125*), and thyroid hormone deficiency (*224*); liver dysfunction, including cirrhosis (*161*), portal hypertension (*4*), and chronic bile-duct obstruction (*32*); hypoxic (*5*) and hypercapnic states (*32*); low-output (*6*) and high-output cardiac failure (*101*); and the syndrome of inappropriate secretion of ADH (*203*). The persistent release of AVP enhances the water permeability of the collecting tubules, resulting in a concentrated urine. Thus, the excretion of the solute-free water that is necessary to correct the hyponatremia is impossible. In this situation a specific inhibitor of the hydroosmotic action of AVP would be valuable clinically, as well as experimentally (*124, 125, 172, 173*).

Studies by Kurokawa *et al.* (*149*) and Jackson *et al.* (*130*) have shown that glucocorticoid deficiency is associated with increased cAMP contents in medullary tissues; this effect may be due to the diminution in cAMP phosphodiesterase activity. Last, in addition to nonosmotic release of AVP, it should be emphasized that AVP-independent factors, such as a diminution in distal fluid delivery, may be involved in the impaired water excretion associated with pathophysiological states.

A. Nephrogenic Diabetes Insipidus

Defects in urinary concentrating ability are associated with several renal diseases as well as central diabetes insipidus. Studies of mice with hereditary nephrogenic diabetes insipidus showed that AVP-stimulated adenylate cyclase activity is only slightly lower but cAMP-phosphodiesterase activity is markedly higher in the medullary collecting tubule of mice with hereditary nephrogenic diabetes insipidus (*128*). These abnormalities would result in decreased cAMP accumulation in the medullary collecting tubule in response to AVP. AVP-stimulated adenylate cyclase activity in these mice with nephrogenic diabetes insipidus was also markedly lower in the medullary thick ascending limb. cAMP phosphodiesterase activity in the same nephron segment was not different between control and mice with nephrogenic diabetes insipidus. The defect in adenylate cyclase in the medullary thick ascending limb may also contribute to the urinary concentrating defect by decreasing solute transport and thus the medullary solute gradient in the interstitium.

B. Hypokalemia

Hypokalemia is well recognized to produce polyuria and a urinary concentrating defect in humans and experimental animals (*26, 86, 201*). A study from our laboratory has indicated that polydipsia is an important factor in hypokalemia-induced polyuria (*26*). The concentrating defect with hypokalemia also has been suggested to be due either to diminution in renal interstitial tonicity (*73*) or to a decreased cellular response to AVP *in vitro* (*79*). The recent studies with isolated tubules from hypokalemic rats (*139, 141*) further elucidated the impairment in the hydroosmotic effect of AVP. As compared to control rats, adenylate cyclase response to AVP and AVP-stimulated cAMP accumulation were significantly decreased in the papillary collecting tubule from the hypokalemic rats. The adenylate cyclase response to AVP was not different between control and hypokalemic rats in the medullary thick ascending limb or medullary collecting tubules, and cAMP phosphodiesterase was not altered in hypokalemic rats. This study therefore indicates that the abnormal renal tubular cAMP metabolism in the papillary collecting tubule in hypokalemia is one of the factors contributing to impaired urinary concentration.

Recent studies have also shown that potassium may influence the biosynthesis of PG. Potassium inhibits PGE_2 biosynthesis in rabbit renomedullary interstitial cells (*254*). Progressive increases in medium potassium concentration diminish the effect of exogenous arachidonic acid added to the cultured cells to increase PG biosynthesis by lowering the availability of endogenous cellular phospholipids. Increased PGE_2 synthesis and excretion in hypokalemia may suppress the cellular action of AVP (*76, 250*). However, other investigators have failed to demonstrate this effect of hypokalemia, and PG inhibition has not been shown to improve the maximum urinary concentrating ability in the rat (*27, 227*).

C. Hypercalcemia

Hypercalcemia also causes a concentrating defect. Chronic hypercalcemia induced by the oral administration of vitamin D metabolite impairs the maximum urinary concentration in response to either exogenous AVP or 24-hr fluid deprivation in the rat (*157, 217*). This effect has also been suggested to be related to an increase in PG biosynthesis in the hypercalcemic rat. Urinary PGE_2 excretion has been shown to be fivefold greater in hypercalcemic rats than in normocalcemic rats, and PG inhibition has been shown to improve the maximum urinary concentrating ability (*157*). The role of calcium in the biosynthesis of PG appears to be due to enhanced availability of endogenous phospholipids (*57, 58*). However, the increased PG synthesis with acute hypercalcemia does not appear to affect the impaired response to AVP in the dog (*29*). Studies to evaluate the effect of hypercalcemia on the cellular action of AVP remain to be solved.

D. Acute Renal Failure

The mechanism of the renal concentrating defect associated with nonoliguric acute renal failure has been examined in experimental models associated with ischemia, cisplatin, and gentamicin administration (*7, 89, 90*). In all three of these models, a defect in maximum urinary concentration occurred 7–8 days after induction of acute renal failure; this was when glomerular filtration rate had returned to normal. The role of PG in these models can be excluded, since urinary PG excretion and PG synthesis inhibition failed to improve the renal concentrating ability. The papillary interstitial tonicity after fluid deprivation, however, is significantly reduced in the experimental animals as compared to control rats. Thus, the nonoliguric acute renal failure is associated with a defect in generation of renal papillary interstitial solute concentration as the primary cause of the concentrating defect. An end-organ effect on AVP action has yet to be examined.

E. Drug-Induced Nephrogenic Diabetes Insipidus

Drug-induced nephrogenic diabetes insipidus is often found in patients treated with lithium or demeclocycline. Polyuria and polydipsia may occur in 20–70% of lithium-treated patients (*16, 81, 174*). Lithium administration showed urinary and plasma AVP levels within normal ranges or elevated (*14, 16, 87, 168, 180*). Lithium did not increase AVP release from the isolated rat neurohypophysis (*238*). In one patient reported, lithium treatment was associated with central diabetes insipidus (*16*). It has been suggested that lithium may interfere with AVP-induced cAMP metabolism at the level of the renal tubule (*64, 222*). A study by Jackson *et al.* (*132*) showed that lithium causes a dose-dependent inhibition of AVP-stimulated adenylate cyclase in both the medullary thick ascending limb and the medullary collecting tubule; no effect of lithium was observed on cAMP-phosphodiesterase activity. Thus, lithium-induced polyuria *in vivo* may be resulted from inhibition of AVP-stimulated adenylate cyclase; however, an additional site distal to cAMP generation cannot be excluded.

Demeclocycline, (7-chloro-6-demethyltetracycline), a relatively nontoxic antibiotic, is another drug that produces reversible nephrogenic diabetes insipidus (*50, 53, 223*). A direct effect of demeclocycline was supported by *in vitro* studies in toad urinary bladder (*223*). In this study, demeclocycline impaired both AVP- and cAMP-induced water flow. The AVP-induced water flow, however, was more inhibited than the cAMP-induced water flow (*223*). Demeclocycline also has been shown to impair the adenylate cyclase response to AVP in the human renal medullary tissue (*245*). While results in the toad bladder suggest that demeclocycline inhibits both cAMP production and post-cAMP action, further studies will be needed to determine whether demeclocycline acts in a similar manner on renal mammalian tubular cells.

In contrast, drugs such as chlorpropamide, vincristine, and cyclophosphamide produce hyponatremia. The hyponatremia with vincristine and cyclophosphamide appears to depend on the inappropriate secretion of AVP (*60, 202*); the hyponatremia with chlorpropamide is mostly related to the augmentation of antidiuretic action of AVP by chlorpropamide at the level of the renal tubules. This effect of chlorpropamide has been demonstrated in the toad bladder (*122, 177*), rats (*30, 179*), and humans (*189, 197*). It has also been found that cAMP accumulation in response to AVP or dDAVP is increased after the administration of chlorpropamide (*17, 165, 187*). This increase is associated with the enhancement adenylate cyclase response to AVP by chlorpropamide. Chlorpropamide did not alter cAMP phosphodiesterase. Thus, chlorpropamide leads to increased renal cAMP concentration, which may explain the drug's augmentation of the antidiuretic response to AVP. Chlorpropamide has also been shown in humans to increase urinary AVP excretion (*186*).

VII. EFFECT OF AVP ON CHLORIDE AND OTHER ELECTROLYTE TRANSPORT

Although it is well established that AVP increases water permeability of collecting tubules, a separate site of AVP action was suggested when AVP was observed to stimulate adenylate cyclase in the medullary thick ascending limb from the rat (*121*). Studies of mouse medullary thick ascending limb (*99*) clearly indicated that AVP increased electrical potential difference, lumen positive flux, and net chloride flux. There was no effect of AVP on hydraulic water permeability. In rabbit medullary thick ascending limb, no effect of AVP on potential difference, chloride efflux, or sodium flux could be demonstrated (*208*). This is consistent with the biochemical study in which adenylate cyclase of the rabbit medullary thick ascending limb did not respond to AVP (*184*). Results in the rat are somewhat controversial, since only in the hypertonic medium by adding sodium chloride and urea did AVP increase potential difference, chloride efflux, or sodium flux (*208*). Additional studies of the cortical and medullary thick ascending limb of mice confirmed that AVP or cAMP increases the rate of sodium chloride flux only in the medullary thick ascending limb (*111*). Recent studies with Brattleboro rats have confirmed that the medullary thick ascending limb is indeed the site of AVP action on solute transport (*196, 205, 248*). When hormone-deprived Brattleboro rats were used to exclude other hormonal effects (*196, 205*), AVP not only increased sodium chloride reabsorption (*196*) but also stimulated reabsorption of magnesium, calcium, and chloride and total solute reabsorption in the medullary thick ascending limb (*205*). The authors suggested that this increase in solute reabsorption may be due to decreases in tubular flow rates in the distal parts of the tubule consequent to the AVP action.

AVP administration *in vivo* has been known for some time to cause a natriuresis. The *in vivo* studies in rats and dogs have shown that administration of a pharmacological dose of AVP causes an increase in sodium, chloride, and potassium excretion (*8, 75, 118, 150, 166*). In the antidiuretic state, AVP may increase water excretion as a result of the increases in solute excretion. The mechanism of the solute diuresis has been evaluated, and several sites in the nephron have been suggested whereby AVP may decrease solute transport: the proximal tubule (*175*), the loop of Henle (*8, 75, 118, 150*), and the collecting tubule (*166*).

However, *in vitro* studies using frog skin, toad urinary bladder, and isolated nephron segments suggest that AVP enhances sodium transport. AVP stimulates mucosal-to-serosal transport of sodium in the frog skin and the toad urinary bladder (*154, 241*). In the isolated cortical collecting tubule, AVP produces only minor changes in either transepithelial voltage or net lumen-to-bath sodium transport (*82*). The *in vitro* studies of Holt and Lechene, however, produced different results (*114*). In this study, AVP causes a transient increase, followed by a sustained decrease in the potential difference and in the net fluxes of sodium and chloride across rabbit cortical collecting tubules perfused *in vitro*. The sustained inhibitory effects of AVP on sodium chloride reabsorption may be related to PGE_2, since AVP stimulates the biosynthesis of PGE_2 (*146*).

Thus, taken together, the *in vivo* and *in vitro* effects of AVP on sodium transport appear conflicting. This controversy may be related to the different action of AVP on sodium chloride absorption at different segments of the nephron and alterations in environmental factors surrounding the nephrons.

It also has been suggested that AVP may regulate calcium and phosphorus transport in the cortical collecting tubule (*115*). In this study, calcium and phosphorus reabsorption was totally abolished by AVP. Since exogenous PGE_2 mimicked and meclofenamate diminished the inhibitory action of AVP on calcium and phosphorus transport, endogenous PG may modulate the inhibitory action of AVP on calcium and phosphorus transport in the cortical collecting tubule in a mechanism similar to sodium transport (*114*).

Urea transport is also stimulated by AVP in toad urinary bladder (*47*) and in collecting tubules in the rat (*183*). As already discussed, endogenous kinins appear to modulate the action of AVP on the water-transport and urea-transport pathways in the opposite direction (*47*). These results are consistent with the concept that the pathways mediating water and urea transport are independent and distinct from one another (*48*). It seems unclear whether urea transport in the mammalian kidney is directly regulated by AVP.

VIII. EFFECT OF AVP ON RENAL HEMODYNAMICS

Although low doses of AVP exert little effect on renal blood flow and renal vascular resistance, an infusion of a pressor dose of AVP may cause renal vasoconstriction (*123*). Moreover, an antagonist to the vascular effects of AVP has been shown to increase renal blood flow in the fluid-deprived rat (*2a*). Studies in Brattleboro rats also showed that all doses of AVP increased renal vascular resistance and decreased renal blood flow (*13*). The authors also suggested that a physiological dose of AVP may influence the distribution of renal blood flow by causing vasoconstriction in the outer region of renal cortex and a higher dose of AVP may influence vascular resistance throughout the cortex. AVP also has been suggested to alter the process of glomerular ultrafiltration (*111*). AVP probably produces mesangial cell contraction, which is consistent with the *in vitro* finding that AVP stimulates the cultured mesangial cell contraction (*12*). This effect would decrease glomerular capillary surface area and thus the glomerular capillary ultrafiltration coefficient (*212*). However, since glomerular transcapillary hydraulic pressure difference is increased by AVP, these opposing effects may account for the near constancy of the single-nephron glomerular filtration rate and total kidney glomerular filtration rate (*119*).

IX. EFFECT OF AVP ON RENIN SECRETION

AVP also has been suggested to suppress plasma renin in rats and dogs under normal and pathological conditions (*138, 218, 226, 232, 242*). The stimulation of renin release by isoproterenol is inhibited by AVP in the isolated perfused rat kidney (*243*). The study of Malayan *et al.* (*170*) demonstrated that the plasma AVP concentration of 4.2 pg/ml, significantly suppressed renin secretion by 19% from the basal level. When the plasma AVP was 8 pg/ml renin secretion was suppressed by 34%. Since these plasma concentrations of AVP are not higher than the AVP level with dehydration, this inhibitory effect on renin release by AVP is probably of physiological importance. AVP may exert a negative-feedback control over its own release by its inhibitory effect on renin release, since angiotensin II has been suggested to stimulate AVP release (*221*). The suppression of renin secretion by centrally administered angiotensin II may also be mediated by stimulation of AVP release (*169*). The physiological role of these hormonal interactions has, however, been debated.

X. OTHER RENAL EFFECTS AND METABOLISM

AVP-elicited refractoriness to AVP-stimulated water permeability has been suggested (*69, 72*), but the precise mechanism remains to be determined. Loss of apparent receptor number (i.e., "down regulation") is a possible mechanism (*15*), but postreceptor refractoriness may also be involved (*56, 113*). Early refractoriness of the response to AVP is associated with an increase in urinary PGE_2 excretion, and indomethacin treatment prevents the early onset of this refractoriness (*97*). These results are compatible with the well-known action of PG in antagonizing the hydroosmotic action of AVP. Handler and Preston (*103*) recently found reduced AVP-sensitive adenylate cyclase activity in the toad bladder, which might relate to the decreased AVP receptor numbers. A reduced epithelial cellular cAMP accumulation was also observed in response to AVP. This effect occurred without altering adenylate cyclase activity or phosphodiesterase activity, without loss of cAMP into buffer, and without refractoriness of water-permeability response to exogenous cAMP in the toad bladder. Thus, the refractoriness of the response to AVP may be due to decreased receptor number, abnormal metabolism of cAMP, or an interaction between PG and AVP.

Urinary AVP is measurable by specific radioimmunoassay and has been employed as an index of the release of AVP from the posterior pituitary. It was reported that the kidney clears AVP from the circulation via two pathways: glomerular clearance and peritubular clearance (*198*). Urinary AVP concentration generally correlates with plasma AVP but is dependent on several factors that influence the renal clearance of AVP. The following facts appear to be involved in the renal clearance of AVP: (1) the glomerular filtration rate of AVP, (2) proximal reabsorption or degradation of filtered AVP, and (3) the distal secretion of AVP (*143, 152, 198, 219*). However, other investigators have failed to show the renal tubular reabsorption or secretion of AVP (*162*). It should be noted that the AVP excreted into the final urine is probably intact, since it is not only immunoreactive but also biologically active (*152, 198*).

XI. AVP ANTAGONISTS

A recent development of synthesis of AVP antagonists provides a new therapeutic approach for the various pathological states of water retention. These AVP antagonists are analogues of AVP and have a selective inhibitory action for either vasopressor or hydroosmotic action of AVP or both (*172, 173, 209*). $d(CH_2)_5Tyr(Et)VAVP$, [1-(β-mercapto-β,β-cyclopentamethylenepropionic acid),2-*O*-ethyltyrosine,4-valine]AVP, was found to block completely the antidiuretic effect of exogenous and endogenous AVP in the conscious rats (*124*). This antagonist effect was not accompanied by changes in glomerular filtration rate, urinary solute excretion, or mean arterial pressure (*124*). The effect of this antagonist on the AVP stimulation of adenylate cyclase in the isolated tubules from rats was also examined. It has been shown that this antagonist inhibited the stimulation of adenylate cyclase by AVP, but not by any other hormones, thus suggesting the antagonist is a specific inhibitor of AVP action (*140*). Further studies in the isolated, perfused rabbit cortical collecting tubule demonstrated that the enhanced net water flux induced by AVP was completely blocked by this AVP antagonist (*140*). The AVP antagonist also inhibited [^{3}H]AVP binding to the renal papillary membranes (*140*) and medullopapillary membranes (*228*). The recent observation that this antagonist inhibited the adenylate cyclase response to AVP in the isolated papillary collecting tubules from humans (unpublished data from this laboratory) is encouraging for the future clinical application of these antagonists. In this regard, it has been shown that $d(CH_2)_5Tyr(Et)VAVP$ significantly reverses the water retention associated with glucocorticoid and mineralocorticoid deficiency in rats (*125*). Therefore, the AVP antagonists have promising therapeutic implications for the various pathological states of water retention.

REFERENCES

1. Adetuybi, A., and Mills, I. H. (1972). Relations between urinary kallikrein and renal function, hypertension and excretion of sodium and water in man. *Lancet* **1,** 203–207.
2. Agus, Z. S., Chin, P. J. S., and Goldberg, M. (1977). Regulation of urinary calcium excretion in the rat. *Am. J. Physiol.* **233,** F545-F549.

2a. Aisenbrey, G. A., Handelman, W. A., Arnold, P. E., Manning, M., and Schrier, R. W. (1981). Vascular effects of arginine vasopressin during fluid deprivation in the rat. *J. Clin. Invest.* **67,** 961–968.

3. Anderson, R. J., Berl, T., McDonald, K. M., and Schrier, R. W. (1975). Evidence for an in vivo antagonism between vasopressin and prostaglandins in the mammalian kidney. *J. Clin. Invest.* **56,** 420–426.
4. Anderson, R. J., Cronin, R. E.. McDonald, K. M., and Schrier, R. W. (1976). Mechanisms of portal hypertension-induced alterations in renal hemodynamics, renal water excretion, and renin secretion. *J. Clin. Invest.* **58,** 964–970.
5. Anderson, R. J., Pluss, P. G.. Berns, A. S., Jackson, J. T., Arnold, P. E., Schrier, R. W., and McDonald, K. M. (1978). Mechanism of effect of hypoxia on renal water excretion. *J. Clin. Invest.* **62,** 769–777.
6. Anderson, R. J., Cadnapaphornchai, P., Harbottle, J. A., McDonald, K. M., and Schrier, R. W. (1975). Mechanism of effect of thoracic inferior vena cava restriction on renal water excretion. *J. Clin. Invest.* **54,** 1473–1479.

7. Anderson, R. J., Gordon, J. A., Kim, J. K., Peterson, L. M., and Gross. P. A. (1982). Renal concentrating defect following nonoliguric acute renal failure in the rat. *Kidney Int.* **21,** 583–591.
8. Antoniou, L. D., Burke, T. J., Robinson, R. R., and Clapp, J. R. (1973). Vasopressin-related alterations of sodium reabsorption in the loop of Henle. *Kidney Int.* **3,** 6–13.
9. Arimura, A., Sato, H., duPont, A., Nishi, N., and Shally, A. V. (1975). Somatostatin: Abundance of immunoreactive hormones in rat stomach and pancreas. *Science* **189,** 1007–1009.
10. Arruda, J. A. L., and Sabatini, S. (1982). Dopaminergic inhibition of vasopressin-stimulated water flow in the toad bladder: Evidence for local formation of dopamine. *J. Membr. Biol.* **68,** 89–96.
11. Atherton, J. C., Green, R., and Hai, M. A. (1968). The time course of changes in renal tissue composition during water diuresis in rat. *J. Physiol. (London)* **197,** 429–443.
12. Ausiello, D. A., Kreisberg, J. I., Roy, C., and Karnovsky, M. J. (1980). Contraction of cultured rat glomerular cells of apparent mesangial origin after stimulation with angiotensin II and arginine vasopressin *J. Clin. Invest.* **65,** 754–760.
13. Azzawi, S. A., and Shirley, D. G. (1982). The effect of vasopressin in renal blood flow and its distribution in the rat. *J. Physiol. (London)* **341,** 233–244.
14. Balment, R. J., Jones, I. C., and Henderson, I. W. (1977). Time course of lithium-induced alterations in renal and endocrine function in normal and Brattleboro rats with hypothalamic diabetes insipidus. *Br. J. Pharmacol.* **59,** 627–634.
15. Baxter, J. D., and Funder, J. W. (1979). Hormone receptors. *N. Engl. J. Med.* **30,** 1149–1161.
16. Baylis, P. H., and Heath, D. A. (1978). Water disturbances in patients treated with oral lithium carbonate. *Ann. Intern. Med.* **88,** 607–609.
17. Beck, N., Kim, K. S., and Davis, B. B. (1974). Effect of chlorpropamide on cyclic AMP in rat renal medulla. *Endocrinology* **95,** 771–775.
18. Beck, N., and Kim, H. P. (1975). Effect of acute metabolic acidosis on vasopressin dependent cyclic AMP in rat kidney. *Endocrinology* **96,** 1552–1558.
19. Beck, T. R., Hassid, A., and Dunn, M. J. (1980). The effect of arginine vasopressin and its analogs on the synthesis of prostaglandin E_2 by rat renal medullary interstitial cells in culture. *J. Pharmacol. Exp. Ther.* **215,** 15–19.
20. Berl, T., Cadnapaphornchai, P., Harbottle, J. A., and Schrier, R. W. (1974). Mechanism of suppression of vasopressin during alpha-adrenergic stimulation with norepinephrine. *J. Clin. Invest.* **53,** 219–227.
21. Berl, T., Cadnapaphornchai, P., Harbottle, J. A., and Schrier, R. W. (1974). Mechanism of stimulation of vasopressin release during beta-adrenergic stimulation with isoproterenol. *J. Clin. Invest.* **53,** 857–867.
22. Berl, T., Harbottle, J. A., and Schrier, R. W. (1974). Effect of alpha- and beta-adrenergic stimulation of renal water excretion in man. *Kidney Int.* **6,** 247–253.
23. Berl, T., and Kleeman, C. R. (1976). Temperature and fluid balance. *Israel J. Med. Sci.* **12,** 1004–1009.
24. Berl, T., Anderson, R. J., McDonald, K. M., and Schrier, R. W. (1976). Clinical disorders of water metabolism. *Kidney Int.* **10,** 117–132.
25. Berl, T., Raz, A., Wald, H., Horowitz, J., and Czaczkes, W. (1977). *Prostaglandin synthesis inhibition and the action of vasopressin in man and rat. Am. J. Physiol.* **232,** F529–F537.
26. Berl, T., Linas, S. L., Aisenbrey, G. A., and Anderson, R. J. (1977). On the mechanism of polyuria in potassium depletion: the role of polydipsia. *J. Clin. Invest.* **60,** 557–563.
27. Berl, T., Aisenbrey, G. A., and Linas, S. L. (1980). Renal concentration defect in the hypokalemic rat is prostaglandin independent. *Am. J. Physiol.* **238,** F37–F41.

28. Berl, T. (1981). Cellular calcium uptake in the action of prostaglandins on renal water excretion. *Kidney Int.* **19,** 15–23.
29. Berl, T., and Erickson, A. E. (1982). Calcium-prostaglandin interaction on the action of antidiuretic hormone in the dog. *Am. J. Physiol.* **242,** F313–F320.
30. Berndt, W. O., Miller, J., Kettyle, W. M., and Valtin, H. (1970). Potentiation of antidiuretic effect of vasopressin by chlorpropamide. *Endocrinology* **86,** 1028–1032.
31. Berns, A. S., Anderson, R. J., and McDonald, K. M. (1979). Effect of hypercapnic acidosis on renal water excretion in the dog. *Kidney Int.* **15,** 115–125.
32. Better, O. S., Aisenbrey, G. A., Berl, T., Anderson, R. J., Handelman, W. A., Linas, S. L., Guggenheim, S. J., and Schrier, R. W. (1980). Role of antidiuretic hormone in impaired urinary dilution associated with chronic bile duct ligation. *Clin. Sci.* **58,** 493–500.
33. Bisordi, J. E., Schlondorff, D., and Hays, R. M. (1980). Interaction of vasopressin and prostaglandins in the toad urinary bladder. *J. Clin. Invest.* **66,** 1200–1210.
34. Bockaert, J., Roy, C., Rajerison, R., and Jard, S. (1973). Specific binding of [^{3}H] lysine-vasopressin to pig kidney plasma membranes: relationship of receptor occupancy to adenylate cyclase activation. *J. Biol. Chem.* **248,** 5922–5931.
35. Bohman, S. O. (1977). Demonstration of prostaglandin synthesis in collecting duct cells and other cell types of the rabbit renal medulla. *Prostaglandins* **14,** 729–744.
36. Bolaffi, J. L., Reichlin, S., Goodman, D. B. P., and Forrest, J. N., Jr. (1980). Somatostatin: Occurrence in urinary bladder epithelium and renal tubules of the toad, *Bufo marinus*. *Science* **210,** 644–646.
37. Boykin, J., de Torrente, A., Erickson, A., Robertson, G., and Schrier, R. W. (1978). Role of plasma vasopressin in impaired water excretion of glucocorticoid deficiency. *J. Clin. Invest.* **62,** 738–744.
38. Boykin, J., McCool, A., Robertson, G., McDonald, K., and Schrier, R. W. (1979). Mechanisms of impaired water excretion in mineralocorticoid deficient dogs. *Mineral Electrolyte Metab.* **2,** 310–315.
39. Brautbar, N., Levine, B. S., Coburn, J. W., and Kleeman, C. R. (1979). Interaction of somatostatin with PTH and AVP: renal effects. *Am. J. Physiol.* **237,** E428–E431.
40. Brown, D., Grosso, A., and DeSousa, R. C. (1983). Correlation between water flow and intramembrane particle aggregates in toad epidermis. *Am. J. Physiol.* **245,** C334–C342.
41. Burch, R. M., Knapp, D. R., and Halushka, P. V. (1980). Vasopressin-stimulated water flow is decreased by thromboxane synthesis inhibition or antagonism. *Am. J. Physiol.* **239,** F160–F166.
42. Burch, R. M., and Halushka, P. V. (1980). Thromboxane and stable prostaglandin endoperoxide analogs stimulate water permeability in the toad urinary bladder. *J. Clin. Invest.* **66,** 1251–1257.
43. Burch, R. M., and Halushka, P. V. (1982). Vasopressin stimulates prostaglandin and thromboxane synthesis in toad bladder epithelial cells. *Am. J. Physiol.* **243,** F593–F597.
44. Burg, M. B., and Green, N. (1973). Function of the thick ascending limb of Henle's loop. *Am. J. Physiol.* **224,** 659–668.
45. Butlen, D., Guillon, G., Rajerison, R. M., and Jard, S. (1978). Structural requirements for activation of vasopressin-sensitive adenylate cyclase, hormone binding, and antidiuretic actions: Effects of highly potent analogues and competitive inhibitors. *Mol. Pharmacol.* **14,** 1006–1017.
46. Campbell, B. J., Woodward, G., and Borberg, V. (1972). Calcium-mediated interactions between the antidiuretic hormone and renal plasma membranes. *J. Biol. Chem.* **247,** 6167–6175.
47. Carvounis, C. P., Franki, N., Levine, S. D., and Hays, R. M. (1979). Membrane pathways for water and solutes in the toad bladder: I. Independent activation of water and urea transport. *J. Membr. Biol.* **49,** 253–268.

48. Carvounis, C. P., Levine, S. D., Franki, N., and Hays, R. M. (1979). Membrane pathways for water and solutes in the toad bladder: II. Reflection coefficients of the water and solute channels. *J. Membr. Biol.* **49,** 269–281.
49. Carvounis, C. P., Carvounis, G., and Arbeit, L. A. (1981). Role of the endogenous kallikrein-kinin system in modulating vasopressin-stimulated water flow and urea permeability in the toad urinary bladder. *J. Clin. Invest.* **67,** 1792–1796.
50. Castell, D. O., and Spacks, H. A. (1965). Nephrogenic diabetes insipidus due to demethy-chlortetracycline hydrochloride. *JAMA* **193,** 237–239.
51. Chabardes, D., Gagnan-Brunette, M., Imbert-Teboul, M., Gontcharevskaia, O., Montegut, M., Chique, A., and Morel, F. (1980), Adenylate cyclase responsiveness to hormones in various portions of the human nephron. *J. Clin. Invest.* **65,** 439–448.
52. Chang, L. C., Splawinski, J. A., Oates, J. A., and Nies, A. S. (1975). Enhanced renal prostaglandin production in the dog. II. Effect on intrarenal hemodynamics. *Circ. Res.* **36,** 204–207.
53. Cherrill, D. A., Stote, R. M., Birge, J. R., and Singer, I. (1975). Demeclocycline treatment in the syndrome of inappropriate antidiuretic hormone secretion. *Ann. Intern. Med.* **83,** 654–656.
54. Choi, J. K. (1963). The fine structure of the urinary bladder of the toad, *Bufo marinus. J. Cell. Biol.* **16,** 53–72.
55. Chuang, E. L., Reineck, H. J., Osgood, R. W., Kunau, R. T., Jr., and Stein, J. H. (1978). Studies on the mechanism of reduced urinary osmolality after exposure of the renal papilla. *J. Clin. Invest.* **61,** 633–639.
56. Cigorraga, S. B., Dufau, M. L., and Catt, K. J. (1978). Regulation of luteinizing hormone receptors and steroidogenesis in gonadotropin-desensitized Leidig cells. *J. Biol. Chem.* **253,** 4297–4304.
57. Craven, P. A., Briggs, R., and deRubertis, F. R. (1980). Calcium-dependent action of osmolality on adenosine 3′,5′-monophosphate accumulation in rat renal inner medulla: Evidence for a relationship to calcium-responsive arachidonate release and prostaglandin synthesis. *J. Clin. Invest.* **65,** 529–542.
58. Craven, P. A., and deRubertis, F. R. (1981). Effect of vasopressin and urea on Ca^{2+}-calmodulin-dependent renal prostaglandin E. *Am. J. Physiol.* **241.** F649–F658.
59. Culpepper, R. M., and Andreoli, T. E. (1983). Interactions among prostaglandin E_2, antidiuretic hormone, and cyclic adenosine monophosphate in modulating Cl^- absorption in single mouse medullary thick ascending limbs of Henle. *J. Clin. Invest.* **71,** 1588–1601.
60. deFronzo, R. A., Braine, H., Colvin, O. M., and Davis, P. L. (1973). Water intoxication in man after cyclophosphamide therapy. *Ann. Intern. Med.* **78,** 861–869.
61. deRubertis, F. R., and Craven, P. A. (1978). Effects of osmolality and oxygen availability on soluble cyclic AMP-dependent protein kinase activity of rat renal inner medulla. *J. Clin. Invest.* **62,** 1210–1221.
62. Dousa, T. P. (1971). Effect of renal medullary solutes on vasopressin-sensitive adenyl cyclase. *Am. J. Physiol.* **222,** 657–662.
63. Dousa, T. P., and Barnes, L. D. (1974). Effect of colchicine and vinblastine on the cellular action of vasopressin in mammalian kidney. A possible role of microtubules. *J. Clin. Invest.* **54,** 252–262.
64. Dousa, T. P. (1974). Interaction of lithium with vasopressin-sensitive cyclic AMP system of human renal medulla. *Endocrinology* **95,** 1359–1366.
65. Dousa, T. P., and Valtin, H. (1976). Cellular actions of vasopressin in the mammalian kidney. *Kidney Int.* **10,** 46–63.
66. Dousa, T., and Barnes, L. (1977). Regulation of protein kinase by vasopressin in renal medulla in situ. *Am. J. Physiol.* **232,** F50–F57.
67. Dousa, T. P., and Northrup, T. E. (1978). Cellular interactions between vasopressin and prostaglandins in the mammalian kidney. *Cont. Nephrol.* **12,** 106–115.

68. Dousa, T. P. (1981). Cellular action of antidiuretic hormone. *Mineral Electrolyte Metab.* **5,** 144–158.
69. Edelman, I. S., Petersen, M. J., and Gulyassy, P. F. (1964). Kinetic analysis of the antidiuretic action of vasopressin and adenosine-3′,5′-monophosphate. *J. Clin. Invest.* **43,** 2185–2194.
70. Edwards, R. M., Jackson, B. A., and Dousa, T. P. (1980). Protein kinase activity in isolated tubules of rat renal medulla. *Am. J. Physiol.* **238,** F269–F278.
71. Edwards, R. M., Jackson, B. A., and Dousa, T. P. (1981). ADH-sensitive cAMP system in papillary collecting duct: Effect of osmolality and PGE_2. *Am. J. Physiol.* **240,** F311–F318.
72. Eggena, P., Walter, R., and Schwartz, L. L. (1968). Relationship between hydroosmotic flow and the inhibited response of the toad bladder to vasopressin. *Life Sci.* **7,** 59–63.
73. Eigler, J. O., Salassa, R. M., Bahn, R. C., and Owen, C. A., Jr. (1962). Renal distribution of sodium in potassium-depleted and vitamin D-intoxicated rats. *Am. J. Physiol.* **202,** 1115–1120.
74. Fejes-Toth, G., Zahajszky, T., and Filep, J. (1980). Effect of vasopressin on renal kallikrein excretion. *Am. J. Physiol.* **239,** F388–F393.
75. Fejes-Toth, G., and Szenasi, G. (1981). The effect of vasopressin on renal tubular ^{22}Na efflux in the rat. *J. Physiol. (London)* **318,** 1–7.
76. Ferris, T. F. (1978). Prostaglandins, potassium, and Bartter's syndrome. *J. Lab. Clin. Med.* **92,** 663–668.
77. Fichman, M. P., Vorherr, H.\ Kleeman, C. R., and Tefler, N. (1971). Diuretic-induced hyponatremia. *Ann. Intern. Med.* **75,** 853–863.
78. Fisher, D. A. (1968). Norepinephrine inhibition of vasopressin antidiuresis. *J. Clin. Invest.* **47,** 540–547.
79. Finn, A. L., Handler, J. S., and Orloff, J. (1966). Relation between toad bladder potassium content and permeability response to vasopressin. *Am. J. Physiol.* **210,** 1279–1284.
80. Flores, A. G. A., and Sharp, G. W. G. (1972). Endogenous prostaglandins and osmotic water flow in the toad bladder. *Am. J. Physiol.* **223,** 1392–1397.
81. Forrest, J. N., Jr., Cohen, A. D., Torretti, J., Himmelhoch, J. M., and Epstein, F. H. (1974). On the mechanism of lithium-induced diabetes insipidus in man and the rat. *J. Clin. Invest.* **53,** 1115–1123.
82. Frindt, G., and Burg, M. B. (1972). Effect of vasopressin on sodium transport in renal cortical collecting tubules. *Kidney Int.* **1,** 224–231.
83. Frolich, J. C., Wilson, T. W., Sweetman, B. J., Smigel, M., Nies, A. S., Carr, K., Watson, J. T., and Oates, J. A. (1975). Urinary prostaglandins: identification and origin. *J. Clin. Invest.* **55,** 763–770.
84. Furtado, M. R. F. (1971). Inhibition of the permeability response to vasopressin and oxytocin in the toad bladder: Effect of bradykinin, kallidin, elesdosin, and physalamine. *J. Membr. Biol.* **4,** 167–178.
85. Ganguli, M., Tobian, L., Azar, S., and O'Donnell, M. (1977). Evidence that prostaglandin synthesis inhibitors increase the concentration of sodium and chloride in rat renal medulla. *Circ. Res.* **49**(Suppl. I), I135-I139.
86. Giebisch, G., and Lozano, R. (1959). The effects of adrenal steroids and potassium depletion on the elaboration of an osmotically concentrated urine. *J. Clin. Invest.* **38,** 843–853.
87. Gold, P. W., Robertson, G. L., Post, R. M., Kaye, W., Ballenger, J., Rubinow, D., and Goodwin, F. K. (1983). The effect of lithium on the osmoregulation of arginine vasopressin secretion. *J. Clin. Endocrinol. Metab.* **56,** 295–299.
88. Goldfarb, S. (1982). Effects of calcium on ADH action in the cortical collecting tubule perfused in vitro. *Am. J. Physiol.* **243,** F481–F486.
89. Gordon, J. A., Peterson, L.A., and Anderson, R. J. (1982). Water metabolism after cisplatin in the rat. *Am. J. Physiol.* **243,** F36–F43.

90. Gordon, J. A., Grossfeld, P., and Anderson, J. R. (1982). Renal concentration defect (RCD) after gentamycin (G) (abstract). *Clin. Res.* **30,** 78A.
91. Grantham, J. J., and Burg, M. B. (1966). Effect of vasopressin and cyclic AMP on permeability of isolated collecting tubules. *Am. J. Physiol.* **211,** 255–259.
92. Grantham, J. J., and Orloff, J. (1968). Effect of prostaglandin E_1 on the permeability response of the isolated collecting tubules to vasopressin, adenosine 3′,5′-monophosphate and theophylline. *J. Clin. Invest.* **47,** 1154–1161.
93. Grantham, J. J. (1971). Mode of water transport in mammalian renal collecting tubules. *Fed. Proc.* **30,** 14–21.
94. Green, H. H., Harrington, A. R., and Valtin, H. (1970). On the role of antidiuretic hormone in the inhibition of acute water diuresis in adrenal insufficiency and the effects of gluco- and mineralocorticoids in reversing the inhibition. *J. Clin. Invest.* **49,** 1724–1736.
95. Grenier, F. C., Rollins, T. E., and Smith, W. L. (1981). Kinin-induced prostaglandin synthesis by renal papillary collecting tubule cells in culture. *Am. J. Physiol.* **241,** F94–F104.
96. Gross, J. B., and Bartter, F. C. (1973). Effects of prostaglandins E_1, A_1 and F_{2a} on renal handling of salt and water. *Am. J. Physiol.* **225,** 218–224.
97. Gross, P. A., Kim, J. K., and Anderson, R. J.(1983). Studies on the mechanisms of escape from vasopressin in the rat. *Circ. Res.* **53,** 794–804.
98. Guildon, G., Butlen, D., Cantan, B., Barth, T., and Jard, S. (1982). Kinetic and pharmacological characterization of vasopressin membrane receptors from human kidney medulla: Relation to adenylate cyclase activation. *Eur. J .Pharmacol.* **85,** 291–304.
99. Hall, D., and Varney, D. (1980). Effect of vasopressin on electrical potential difference and chloride transport in mouse medullary thick ascending limb of Henle's loop. *J. Clin. Invest.* **66,** 792–802.
100. Hall, D. A., and Grantham, J. J. (1980). Temperature effect on ADH response of isolated perfused rabbit collecting tubules. *Am. J. Physiol.* **239,** F595–F601.
101. Handelman, W., Lum, G., and Schrier, R. (1978). Impaired water excretion in high output cardiac failure in the rat (abstract). *Clin. Res.* **27,** 62A.
102. Handler, J. S., Bensinger, R., and Orloff, J. (1968). Effect of adrenergic agents on toad bladder response to ADH, 3′,5′-AMP, and theophylline. *Am. J. Physiol.* **215,** 1024–1031.
103. Handler, J. S., and Preston, A. S. (1981). Vasopressin-elicited refractoriness of the response to vasopressin in toad urinary bladder. *Am. J. Physiol.* **240,** F551–F557.
104. Hardy, M. A. (1978). Intracellular calcium as a modulator of transepithelial permeability to water in frog urinary bladder. *J. Cell Biol.* **76,** 787–791.
105. Hardy, M. A., and DiBona, D. R. (1982). Extracellular Ca^{++} and the effect of antidiuretic hormone on the water permeability of the toad urinary bladder: An example of flow-induced alteration of flows. *J. Membr. Biol.* **67,** 27–44.
106. Harkcom, T. M., Kim, J. K., Palumbo, P. J., Hui, Y. S. F., and Dousa, T. P. (1978). Modulatory effect of thyroid function on enzymes of the vasopressin-sensitive adenosine 3′,5′-monophosphate system in renal medulla. *Endocrinology* **102,** 1475–1484.
107. Harmanci, M., Kachadorian, W., Valtin, H. and diScala, V. (1978). Antidiuretic hormone-induced intramembranous alterations in mammalian collecting ducts. *Am. J. Physiol.* **235,** F440–F443.
108. Hamanci, M. C., Stern, W., Kachadorian, W. A., Valtin, H., and DiScala, V. A. (1980). Vasopressin and collecting duct intramembraneous particle clusters: A dose-response relationship. *Am. J.Physiol.* **239,** F560–F564.
109. Hays, R. M. (1983). Alteration of luminal membrane structure by antidiuretic hormone. *Am. J. Physiol.* **245,** C289–C296.
110. Hayward, J. N., and Baker, M. A. (1968). Diuretic and thermoregulatory responses to preoptic cooling in the monkey. *Am. J. Physiol.* **214,** 843–850.

111. Hebert, S. C., Culpepper, R. M., and Andreoli, T. E. (1981). NaCl transport in mouse medullary thick ascending limbs. I. Functional nephron heterogeneity and ADH-stimulated NaCl cotransport. *Am. J. Physiol.* **241,** F412–F431.
112. Hechter, O., Terada, S., Nakahara, T., and Flouret, G. (1978). Neurohypophyseal hormone-responsive renal adenylate cyclase. II. Relationship between hormonal occupancy of neurohypophyseal hormone receptor sites and adenylate cyclase activation. *J. Biol. Chem.* **253,** 3219–3229.
113. Ho, R.-J., and Sutherland, E. W. (1975). cAMP-mediated feedback regulation in target cells. *In* "Advances in Cyclic Nucleotide Research" (G. I. Drummond, P. Gıeengard, and G. A. Robinson, eds.), Vol. 5, p. 533–548. Raven, New York.
114. Holt, W. F., and Lechene, C. (1981) ADH–PGE_2 interactions in cortical collecting tubule. I. Depression of sodium transport. *Am. J. Physiol.* **241,** F452–F460.
115. Holt, W. F., and Lechene, C. (1981). ADH–PGE_2 interactions in cortical collecting tubule. II. Inhibition of Ca and P reabsorption. *Am. J. Physiol.* **241,** F461–F467.
116. Humes, H. D., Simmons, C. F., Jr., and Brenner, B. M. (1980). Interaction between antidiuretic and parathyroid hormones on urine concentration. *Am. J. Physiol.* **239,** F244–F249.
117. Humes, H. D., Simmons, C. F., Jr., and Brenner, B. M. (1980). Effect of verapamil on the hydroosmotic response to antidiuretic hormone in toad urinary bladder. *Am. J. Physiol.* **239,** F250–F257.
118. Humphreys, M. H., Friedler, R. M., and Earley, L. E. (1970). Natriuresis produced by vasopressin or hemorrhage during water diuresis in the dog. *Am. J. Physiol.* **219,** 658–665.
119. Ichikawa, I., and Brenner, B. M. (1977). Evidence for glomerular actions of ADH and dibutyryl cyclic AMP in the rat. *Am. J. Physiol.* **233,** F103–F117.
120. Iino, Y., and Imai, M. (1978). Effects of prostaglandins on Na transport in isolated collecting tubules. *Pfluegers Arch. Gesamte Physiol. Menschen Tiere* **373,** 125–132.
121. Imbert-Teboul, M., Chabardes, D., Montegut, M., Clique, A., and Morel, F. (1978). Vasopressin-dependent adenylate cyclase activities in the rat kidney medulla: Evidence for two separate sites of action. *Endocrinology* **102,** 1254–1261.
122. Ingelfinger, J. R., and Hays, R. M. (1969). Evidence that chlorpropamide and vasopressin share a common site of action. *J. Clin. Endocrinol. Metab.* **29,** 738–740.
123. Ishikawa, S., Goldberg, J. P., Schrier, D., Aisenbrey, G. A., and Schrier, R. W. (1984). Interrelationship between subpressor effects of vasopressin and other vasoactive hormones in the rat. *Mineral Electrolyte Metab.* **10,** 184–189.
124. Ishikawa, S., Kim, J. K., and Schrier, R. W. (1983). Further in vivo evidence for antagonist to antidiuretic action of arginine vasopressin. *Am. J. Physiol.* **245,** R713–R719.
125. Ishikawa, S., and Schrier, R. W. (1982). Effect of arginine vasopressin antagonist on renal water excretion in glucocorticoid and mineralocorticoid deficient rats. *Kidney Int.* **22,** 587–593.
126. Itskovitz, H. D., Terrango, N. A., and McGiff, J. C. (1974). Effect of a renal prostaglandin on distribution of blood flow in the isolated canine kidney. *Circ. Res.* **344,** 770–776.
127. Iyengar, R., Lepper, K. G., and Mailman, D. S. (1976). Involvement of microtubules and microfilaments in the action of vasopressin in canine renal medulla. *J. Supramol. Struct.* **5,** 521–530.
128. Jackson, B. A., Edwards, R. M., Valtin, H., and Dousa, T. P. (1980). Cellular action of vasopressin in medullary tubules of mice with hereditary nephrogenic diabetes insipidus. *J. Clin. Invest.* **66,** 110–122.
129. Jackson, B. A., Edwards, R. M., and Dousa, T. P. (1980). Measurements of cyclic AMP and cyclic GMP phosphodiesterase activity in isolated tubular segments. *Kidney Int.* **18,** 512–518.
130. Jackson, B. A., Edwards, R. M., and Dousa, T. P. (1980). Effects of adrenalectomy (ADX) on vasopressin (VP)-sensitive cAMP metabolism in microdissected medullary nephron segments (abstract). *Am. Soc. Nephrol.* **13,** 96A.

131. Jackson, B. A., Edwards, R. M., and Dousa, T. P. (1980). Vasopressin–prostaglandin interaction in isolated tubules from rat outer medulla. *J. Lab. Clin. Med.* **96,** 119–128.
132. Jackson, B. A., Edwards, R. M., and Dousa, T. P. (1980). Lithium-induced polyuria: Effect of lithium on adenylate cyclase and adenosine 3′,5′-monophosphate phosphodiesterase in medullary ascending limb of Henle's loop and in medullary collecting tubules. *Endocrinology* **107,** 1693–1698.
133. Janszen, F. H. A., and Nugteren, D. H. (1971). Histochemical localization of prostaglandin synthetase. *Histochemie* **27,** 159–164.
134. Johnston, H. H., Herzog, J. P., and Lauler, D. P. (1968). Effect of prostaglandin E_1 on renal hemodynamics, sodium and water excretion. *Am. J. Physiol.* **213,** 939–946.
135. Kachadorian, W. A., Levine, S. D., Wade, J. B., deScala, V. A., and Hays, R. M. (1977). Relationship of aggregated intramembranous particles to water permeability in vasopressin-treated toad urinary bladder. *J. Clin. Invest.* **59,** 576–581.
136. Kachadorian, W. A., Ellis, S. J., and Muller, J. (1979). Possible roles for microtubules and microfilaments in ADH action on toad urinary bladder. *Am. J. Physiol.* **236,** F14–F20.
137. Kauker, M. L., Crofton, J. T., Share, L., and Nasjletti, A. (1984). Role of vasopressin in regulation of renal kinin excretion in Long-Evans and diabetes insipidus rats. *J. Clin. Invest.* **73,** 824–831.
138. Khokhar, A. M., Slater, J. D. H., Forsling, M. L., and Payne, N. N. (1976). Effect of vasopressin on plasma volume and renin release in man. *Clin. Sci. Mol. Med.* **50,** 415–424.
139. Kim, J., Summer, S., and Berl, T. (1984). Studies on the cyclic AMP system in the inner medullary collecting duct of the potassium depleted rat. *Kidney Int.* **26,** 384–391.
140. Kim, J. K., Summer, S. N., and Schrier, R. W. (1983). Cellular effects of antagonist of antidiuretic action of arginine vasopressin (abstract). *Am. Soc. Nephrol.* **16,** 167A.
141. Kim, J. K., Jackson, B. A., Richards, R. M., and Dousa, T. P. (1982). Effect of potassium depletion on the vasopressin-sensitive cyclic AMP system in rat outer medullary tubules. *J. Lab. Clin. Med.* **99,** 29–38.
142. Kim, J. K., Summer, S. N., Erickson, A. E., and Schrier, R. W. (1984). Effect of arginine vasopressin (AVP) on the renal medullary thick ascending limb (MAL) in the maximal concentrating ability of the rat (abstract). *Int. Congr. Nephrol.*
143. Kimura, T., and Share, L. (1981). Characterization of the renal handling of vasopressin in the dog by stop-flow analysis. *Endocrinology* **109,** 2089–2094.
144. Kirschenbaum, M. A., White, N., Stein, J. H., and Ferris, T. F. (1974). Redistribution of renal cortical blood flow during inhibition of prostaglandin synthesis. *Am. J. Physiol.* **227,** 801–805.
145. Kirschenbaum, M. A., and Serros, E. R. (1980). Effect of alterations in urine flow rate on prostaglandin E excretion in conscious dogs. *Am. J. Physiol.* **238,** F107–F111.
146. Kirschenbaum, M. A., Lowe, A. G., Trizona, W., and Fine, L. G. (1982). Regulation of vasopressin action by prostaglandins: Evidence for prostaglandin synthesis in the rabbit cortical collecting tubule. *J. Clin. Invest.* **70,** 1193–1204.
147. Klein, L. A., Liberman, B., Laks, M., and Kleeman, C. R. (1971). Interrelated effects of antidiuretic hormone and adrenergic drugs on water metabolism. *Am. J. Physiol.* **221,** 1657–1665.
148. Krothapalli, R. K., and Suki, W. N. (1984). Functional characterization of the alpha-adrenergic receptor modulating the hydroosmotic effect of vasopressin on the rabbit cortical collecting tubule. *J. Clin. Invest.* **73,** 740–749.
149. Kurokawa, K., Azner, E., Descoeuders, C., and Massry, S. (1978). Effects of glucocorticoid deficiency on renal medullary cyclic AMP. *Clin. Sci. Mol. Med.* **54,** 573–577.
150. Kurtzman, N. A., Rogers, P. W., Boonjarern, S., and Arruda, J. A. L. (1975). Effect of infusion of pharmacologic amounts of vasopressin on renal electrolyte excretion. *Am. J. Physiol.* **228,** 890–894.

151. Larsson, C., and Anggard, E. (1974). Regional differences in the formation and metabolism of prostaglandins in the rabbit kidney. *Eur. J. Pharmacol.* **25,** 327–334.
152. Lauson, H. D., Bocanegra, M., and Beuzeville, C. F. (1965). Hepatic and renal clearance of vasopressin from plasma of dogs. *Am. J. Physiol.* **209,** 199–214.
153. Law, R. O. (1975). The inulin space, solute concentrations and weight changes in rat renal medullary slices incubated in isoosmolal media and their modification during anoxia and hypothermia. *J. Physiol. (London)* **247,** 37–54.
154. Leaf, A., and Dempsey, E. (1960). Some effects of mammalian neurohypophysial hormones on metabolism and active transport of sodium by the isolated toad bladder. *J. Biol. Chem.* **235,** 2160–2163.
155. Lemley, K. V., Robertson, C. R., and Jamison, R. L. (1980). Effect of prostaglandin inhibition on erythrocyte velocities in vasa recta of water loaded rats (abstract). *Clin. Res.* **28,** 63A.
156. Levi, J., Grinblat, J., and Kleeman, C. R. (1971). Effect of isoproterenol on water diuresis in rats with congenital diabetes insipidus. *Am. J. Physiol.* **221,** 1728–1732.
157. Levi, M., Peterson, L., and Berl, T. (1983). Mechanisms of concentrating defect in hypercalcemia. Role of polydipsia and prostaglandins. *Kidney Int.* **23,** 489–497.
158. Levine, S. D., Kachadorian, W. A., Levin, D. N., and Schlondorff, D. (1981). Effects of trifluoperazine on function and structure of toad urinary bladder: Role of calmodulin in vasopressin-stimulation of water permeability. *J. Clin. Invest.* **67,** 662–672.
159. Levine, S. D., Levin, D. N., and Schlondorff, D. (1983). Calcium flow-independent actions of calcium channel blockers in toad urinary bladder. *Am. J. Physiol.* **244,** C243–C249.
160. Linas, S. L., Berl, T., Robertson, G. L., Aisenbrey, G. A., Schrier, R. W., and Anderson, R. J. (1980). Role of vasopressin in the impaired water excretion of glucocorticoid deficiency. *Kidney Int.* **18,** 58–67.
161. Linas, S. L., Anderson, R. J., Guggenheim, S. J., Robertson, G. L., and Berl, T. (1981). Role of vasopressin in impaired water excretion in conscious rats with experimental cirrhosis. *Kidney Int.* **20,** 173–180.
162. Lindheimer, M. D., Reinhanz, A., Grandchamp, A., and Vallotton, M. B. (1980). Fate of vasopressin perfused into nephrons of Wistar and Brattleboro (diabetes insipidus) rats. *Clin. Sci.* **58,** 139–144.
163. Lipson, L. C., and Sharp, G. W. G. (1971). Effect of prostaglandin E_1 on sodium transport and osmotic water flow in the toad bladder. *Am. J. Physiol.* **220,** 1046–1052.
164. Lorenzen, M., Taylor, A., and Windhager, E. E. (1983). pH Effect on osmotic response of collecting tubules to vasopressin and 8-CPT-cAMP. *Am. J. Physiol.* **245,** F188–F197.
165. Lozada, E. S., Gouaux, J., Franki, N., Appel, G. B., and Hays, R. M. (1972). Studies of the mode of action of sulfonylureas and phenylactamides in enhancing the effect of vasopressin. *J. Clin. Endocrinol. Metab.* **34,** 704–712.
166. Luke. R. G. (1973). Natriuresis and chloruresis during hydropenia in the rat. *Am. J. Physiol.* **224,** 13–20.
167. Lum, G. M., Aisenbrey, G. A., Dunn, M. J., Berl, T., Schrier, R. W., and McDonald, K. M. (1977). In vivo effect of indomethacin to potentiate the renal medullary cyclic AMP response to vasopressin. *J. Clin. Invest.* **29,** 8–13.
168. MacNeil, S., Jennings, G., Eastwood, P. R., Paschalis, C., and Jenner, F. A. (1976). Lithium and the antidiuretic hormone. *Br. J. Clin. Pharmacol.* **3,** 305–313.
169. Malayan, S. A., Keil, L. C., Ramsay, D. J., and Reid, I. A. (1979). Mechanism of suppression of plasma renin activity by centrally administered angiotensin II. *Endocrinology* **104,** 672–675.
170. Malayan, S. A., Ramsay, D. J., Keil, L. C., and Reid, I. A. (1980). Effects of increases in plasma vasopressin concentration on plasma renin activity, blood pressure, heart rate and plasma corticosteroid concentration in conscious dogs. *Endocrinology* **107,** 1899–1904.
171. Mandell, I. N., deFronzo, R. A., Robertson, G. L., and Forrest, J. N., Jr. (1980). Role of

plasma arginine vasopressin in the impaired water diuresis of isolated glucocorticoid deficiency in the rat. *Kidney Int.* **17,** 186–195.

172. Manning, M., Lammek, B., Kolodziejczyk, A. M., Seto, J., and Sawyer, W. H. (1981). Synthetic antagonists of in vivo antidiuretic and vasopressor responses to arginine vasopressin. *J. Med. Chem.* **24,** 701–706.
173. Manning, M., Klis, W. A., Olma, A., Seto, J., and Sawyer, W. H. (1982). Design of more potent and selective antagonists of the antidiuretic responses to arginine vasopressin devoid of antidiuretic antagonism. *J. Med. Chem.* **25,** 414–419.
174. Marini, J. L., and Sheard, M. H. (1976). Sustained-release lithium carbonate in a double-blind study: Serum lithium levels, side effects and placebo response. *J. Clin. Pharmacol.* **16,** 276–283.
175. Martinez-Maldonado, M., Eknoyan, G., and Suki, W. N. (1971). Natriuretic effects of vasopressin and cyclic AMP: Possible site of action in the nephron. *Am. J. Physiol.* **220,** 2013–2020.
176. Marusic, E., and Torretti, J. (1964). Synergistic action of vasopressin and thyroxine on water transfer on the isolated toad bladder. *Nature (London)* **202,** 1118–1119.
177. Mendoza, S. A., and Brown, C. F. (1974). Effect of chlorpropamide on osmotic water flow across toad bladder and the response to vasopressin, theophylline and cyclic AMP. *J. Clin. Endocrinol. Metab.* **38,** 883–889.
178. Michell, R. H., Kirk, C. J., and Billah, M. M. (1979). Hormonal stimulation of phosphatidylinositol breakdown, with particular reference to the hepatic effects of vasopressin. *Biochem. Soc. Trans.* **7,** 861–865.
179. Miller, M., and Moses, A. M. (1970). Potentiation of vasopressin action by chlorpropamide in vivo. *Endocrinology* **86,** 1024–1027.
180. Miller, P. D., Dubovsky, S. L., McDonald, K. M., Katz, F. H., Robertson, G. L., and Schrier, R. W. (1979). Central, renal and adrenal effects of lithium in man. *Am. J. Med.* **66,** 797–803.
181. Mills, I. H., Macfarlane, N. A. A., Ward, P. E., and Obika, L. F. O. (1976). The renal kallikrein-kinin system and the regulation of salt and water excretion. *Fed. Proc.* **35,** 181–188.
182. Morgan, M. L., Anderson, R. J., Ellis, M. A., and Berl, T. (1983). Mechanism of cold diuresis in the rat. *Am. J. Physiol.* **244,** F210–F216.
183. Morgan, T., Sakai, F., and Berliner, R. (1968). In vitro permeability of medullary collecting ducts to water and urea. *Am. J. Physiol.* **214,** 574–581.
184. Morel, F. (1981). Sites of hormone action in the mammalian nephron. *Am. J. Physiol.* **240,** F159–F164.
185. Morel, F., Chabardes, D., Imbert-Teboul, M., LeBoffant, F., Hus-Citharel, A., and Montegut, M. (1982). Multiple hormonal control of adenylate cyclase in distal segments of the control of adenylate cyclase in distal segments of the rat kidney. *Kidney Int.* **21**(Suppl. 11), 555–562.
186. Moses, A. M., Numann, P., and Miller, M. (1973). Mechanism of chlorpropamide-induced antidiuresis in man: Evidence for release of ADH and enhancement of peripheral action. *Metabolism* **22,** 59–66.
187. Moses, A. M., and Coulson, R. (1980). Augmentation by chlorpropamide of 1-deamino-8-D-arginine vasopressin-induced antidiuresis and stimulation of renal medullary adenylate cyclase and accumulation of adenosine 3′,5′-monophosphate. *Endocrinology* **106,** 967–972.
188. Muller, J., Kachadorian, W. A., and diScala, V. A. (1980). Evidence that ADH-stimulated intramembrane particle aggregates are transferred from cytoplasmic to luminal membranes into toad bladder epithelial cells. *J Cell Biol.* **85,** 83–95.
189. Murase, T., and Yoshida, S. (1973). Mechanism of chlorpropamide action in patients with diabetes insipidus. *J. Clin. Endocrinol. Metab.* **36,** 174–177.
190. Murayama, N., Kusano, E., Werness, J., Yusufi, A. N. K., Christensen, S., and Dousa, T. P.

(1983). Modulating effect of calcium (Ca) on the vasopressin-sensitive cAMP metabolism in medullary tubules (abstract). *Am. Soc. Nephrol.* **16,** 130A.

191. Nagata, N., Araki, N., Ono, Y., and Kimura, N. (1978). The distribution of hormone sensitive adenylate cyclase in rat kidney and the interaction of hormones in activating the enzyme. *Acta Endocrinol. Scand.* **89,** 404.
192. Omachi, R. S., Robbie, D. E., Handler, J. S., and Orloff, J. (1974). Effects of ADH and other agents on cyclic AMP accumulation in toad bladder epithelium. *Am. J. Physiol.* **226,** 1152-1157.
193. Omata, K., Carretero, O. A., Scicli, A. G., and Jackson, B. A. (1982). Localization of active and inactive kallikrein (kininogenase activity) in the microdissected rabbit nephron. *Kidney Int.* **22,** 602–607.
194. Orloff, J., Handler, J. S., and Bergstrom, S. (1965). Effect of prostaglandin (PGE_1) on the permeability response of toad bladder to vasopressin, theophylline and adenosine 3′,5′-monophosphate. *Nature (London)* **205,** 397–398.
195. Ostravik, T. B., Nustad, K., Brandtzaeg, P., and Pierce, J. V. (1976). Cellular origin of urinary kallikrein. *J. Histochem. Cytochem.* **24,** 1037–1039.
196. Peterson, L. N., DeRouffignac, C., and Levine, D. Z. (1983). An in vivo assessment of the stimulatory effect of antidiuretic hormone (dDAVP) on NaCl reabsorption by the thick ascending limb (TAL) in the rat (abstract). *Am. Soc. Nephrol.* **16,** 176A.
197. Piters, K. (1976). Chlorpropamide-induced hyponatremia. *J. Clin. Endocrinol. Metab.* **43,** 1085–1087.
198. Rabkin, R., Shade, L., Payne, P. A., Young, J., and Crofton, J. (1979). The handling of immunoreactive vasopressin by the isolated perfused rat kidney. *J. Clin. Invest.* **63,** 6–13.
199. Rayson, B. M. R., Ray, C., and Morgan, T. (1978). A study of the interaction of catecholamines and antidiuretic hormone on water permeability and the cyclic AMP system in isolated papillae of the rat. *Pfluegers Arch. Gesamte Physiol. Menschen Tiere* **373,** 99–103.
200. Reid, I. A., and Rose, J. C. (1977). An intrarenal effect of somatostatin on water excretion. *Endocrinology* **100,** 782–785.
201. Relman, A. S., and Schwartz, W. B. (1956). The nephropathy of potassium depletion: A clinical and pathological entity. *N. Engl. J. Med.* **255,** 195–203.
202. Robertson, G. L., Bhoopalam, N., and Zelkowitz, L. J. (1973). Vincristine neurotoxicity and abnormal secretion of antidiuretic hormone. *Arch. Intern. Med.* **132,** 717–720.
203. Robertson, G. L., Athar, S., and Shelton, R. L. (1977). Osmotic control of vasopressin function, *In* "Disturbances in Body Fluid Osmolality" (T. E. Andreoli, J. J. Grantham, and F. C. Rector, Jr., ed.), pp. 125–148. American Physiological Society, Bethesda, Md.
204. Rocha, A. S., and Kokko, J. P. (1973). Sodium chloride and water transport in the medullary ascending limb of Henle. Evidence for active chloride transport. *J. Clin. Invest.* **52,** 612–623.
205. deRouffignac, C., Corman, B., and Roinel, N. (1983). Stimulation by antidiuretic hormone of electrolyte tubular reabsorption in rat kidney. *Am. J. Physiol.* **244,** F156–F164.
206. deRouffignac, C., and Elalouf, J. M. (1983). Effects of calcitonin on the renal concentrating mechanism. *Am. J. Physiol.* **245,** F506–F511.
207. Roy, C., and Ausiello, D. A. (1981). Characterization of (8-lysine) vasopressin binding sites on a pig kidney cell line ($LLC\text{-}PK_1$). *J. Biol. Chem.* **256,** 3415–3416.
208. Sasaki, S., and Imai, M. (1980). Effects of vasopressin on water and NaCl transport across the in vitro perfused medullary thick ascending limb of Henle's loop of mouse, rat, and rabbit kidneys. *Pfluegers Arch. Gesamte Physiol. Menschen Tiere* **383,** 215–221.
209. Sawyer, W. H., Pang, P. K. T., Seto, J., McEnroe, M., Lammek, B., and Manning, M. (1981). Vasopressin analogs that antagonize antidiuretic responses by rat to the antidiuretic hormone. *Science* **212,** 49–51.
210. Schlondorff, D., Carvounis, C. P., Jacoby, M., Satriano, J. A., and Levine, S. D. (1981).

Multiple sites for interaction of prostaglandin and vasopressin in toad urinary bladder. *Am. J. Physiol.* **241,** F625–F631.

211. Schlondorff, D., Levin, D. N., and Levine, S. D. (1982). Calcium flow independent actions of calcium entry blockers on water permeability and cyclic AMP metabolism in toad bladder (abstract). *Clin. Res.* **30,** 542A.
212. Schor, N., Ichikawa, I., and Brenner, B. M. (1981). Mechanisms of action of various hormones and vasoactive substance on glomerular ultrafiltration in the rat. *Kidney Int.* **20,** 442–451.
213. Schrier, R. W. (1974). Effects of adrenergic nervous system and catecholamines on systemic and renal hemodynamics, sodium and water excretion and renin secretion. *Kidney Int.* **6,** 291–306.
214. Schrier, R. W., Lieberman, R., and Ufferman, R. C. (1972). Mechanism of antidiuretic effect of beta adrenergic stimulation. *J. Clin. Invest.* **51,** 97–111.
215. Schrier, R. W., and Berl, T. (1973). Mechanism of effect of alpha adrenergic stimulation with norepinephrine on renal water excretion. *J. Clin. Invest.* **52,** 502–511.
216. Schwartz, M. J., and Kokko, J. P. (1980). Urinary concentrating defect of adrenal insufficiency. Permissive role of adrenal steroids on the hydroosmotic response across the rabbit cortical collecting tubule. *J. Clin. Invest.* **66,** 234–242.
217. Serros, E. R., and Kirschenbaum, M. A. (1981). Prostaglandin-dependent polyuria in hypercalcemia. *Am. J. Physiol.* **241,** F224–F230.
218. Shade, R. E., Davis, J. O., Johnson, J. A., Gotshall, R. W., and Spielman, W. S. (1973). Mechanism of action of angiotensin II and antidiuretic hormone on renin secretion. *Am. J. Physiol.* **224,** 926–929.
219. Shade, R. E., and Share, L. (1977). Renal vasopressin clearance with reductions in renal blood flow in the dog. *Am. J. Physiol.* **232,** F341–F347.
220. Shimizu, K., Kurokawa, T., Maeda, T., and Yoshitoshi, Y. (1969). Free water excretion and washout of renal medullary urea by prostaglandin E_1. *Jpn. Heart J.* **10,** 437–455.
221. Simpson, J. B. (1981). The circumventricular organs and the central actions of angiotensin. *Neuroendocrinology* **32,** 248–256.
222. Singer, I., Rotenberg, D. and Puschett, J. B. (1972). Lithium-induced nephrogenic diabetes insipidus: In vivo and in vitro studies. *J. Clin. Invest.* **51,** 1081–1091.
223. Singer, I., and Rotenberg, D. (1973). Demeclocycline-induced nephrogenic diabetes insipidus. In vivo and in vitro studies. *Ann. Intern. Med.* **79,** 679–683.
224. Skowsky, W. R., and Kikuchi, T. A. (1978). The role of vasopressin in the impaired water excretion of myxedema. *Am. J. Med.* **64,** 613–621.
225. Smith, W. L., and Wilkin, G. P. (1977). Immunochemistry of prostaglandin endoperoxide-forming cycloxygenase: The detection of cycloxygenases in rat, rabbit, and guinea pig kidneys by immunofluorescence. *Prostaglandins* **13,** 873–892.
226. Solez, K., Fox, J. A., Miller, M., and Heptinstall, R. H. (1974). Effects of indomethacin on renal inner medullary plasma flow. *Prostaglandins* **7,** 91–97.
227. Stoff, J. S., Rosa, R. M., and Epstein, F. H. (1979). The concentrating defect of acute potassium depletion in man is independent of renal prostaglandins (abstract). *Kidney Int.* **16,** 874.
228. Stassen, F. L., Erickson, R. W., Huffman, W. F., Stefankiewicz, J., Sulat, L., and Wilbelhaus, V. D. (1982). Molecular mechanisms of novel antidiuretic antagonists: Analysis of the effects on vasopressin binding and adenylate cyclase activation in animal and human kidney. *J. Pharmacol. Exp. Ther.* **223,** 50–54.
229. Stokes, J. B., and Kokko, J. P. (1977). Inhibition of sodium transport by prostaglandin E_2 across the isolated, perfused rabbit collecting tubule. *J. Clin. Invest.* **59,** 1099–1104.
230. Stokes, J. B. (1979). Effect of prostaglandin E_2 on chloride transport across the rabbit thick

ascending limb of Henle: Selective inhibition of the medullary portion. *J. Clin. Invest.* **64,** 495–502.

231. Sutherland, E. W. (1970). On the biological role of cyclic AMP. *JAMA* **214,** 1281–1288.
232. Tagawa, H., Vander, A. J., Bonjour, J.-P., and Malvin, R. L. (1971). Inhibition of renin secretion by vasopressin in unanesthetized sodium-deprived dogs. *Am. J. Physiol.* **220,** 949–951.
233. Taylor, R. E., Jr., and Barker, S. B. (1967). Absence of an in vitro thyroxine effect on oxygen consumption and sodium or water transport by anuran skin and kidney. *Gen. Comp. Endocrinol.* **9,** 129–134.
234. Taylor, A., Mamelak, M., Reaven, E., and Maffly, R. (1973). Vasopressin: Possible role of microtubules and microfilaments in its action. *Science* **181,** 347–350.
235. Taylor, A., Maffly, R., Wilson, L., and Reaven, E. (1975). Evidence for involvement of microtubules in the action of vasopressin. *Ann. NY Acad. Sci.* **253,** 723–737.
236. Taylor, A. (1977). Role of microtubules and microfilaments in action of vasopressin. *In* "Disturbances in Body Fluid Osmolality" (T. E. Andreoli, J. J. Grantham, and F. C. Rector, Jr., eds.), pp. 97–124. American Physiological Society, Bethesda, Md.
237. Teitelbaum, I., and Berl, T. (1983). Effect of calcium (Ca) on AVP mediated cyclic AMP (cAMP) synthesis in cultured rat renal papillary collecting tubule (RPCT) cells (abstract). *Am. Soc. Nephrol.* **16,** 181A.
238. Torp-Pederson, C., and Thorn, N. A. (1973). Acute effects of lithium on the action and release of antidiuretic hormone in rats. *Acta Endocrinol.* **73,** 665–671.
239. Torikai, S., and Kurokawa, K. (1983). Effect of PGE_2 on vasopressin-dependent cell cAMP in isolated single nephron segments. *Am. J. Physiol.* **245,** F58–F66.
240. Urakabe, S., Takamitsu, Y., Shirai, S., Yuasa, S., Kimura, G., Orita, Y., and Abe, H. (1975). Effect of different prostaglandins on the permeability of the toad urinary bladder. *Comp. Biochem. Physiol.* **52C,** 1–4.
241. Ussing, M. M., and Zerarn, K. (1951). Active transport of sodium as the source of electric current in the short-circuited frog skin. *Acta Physiol. Scand.* **23,** 110–127.
242. Vander, A. J. (1968). Inhibition of renin release in the dog by vasopressin and vasotocin. *Circ. Res.* **23,** 605–609.
243. Vandongen, R. (1975). Inhibition of renin secretion in the isolated rat kidney by antidiuretic hormone. *Clin. Sci. Mol. Med.* **49,** 73–76.
244. Wade, J., Kachadorian, W., and deScala, V. (1977). Freeze-fracture electron microscopy: Relationship of membrane structural features to transport physiology. *Am. J. Physiol.* **232,** F77–F83.
245. Wilson, D. M., Perry, H. O., and Sam, W., Jr. (1973). Selective inhibition of human distal tubular function by demeclocycline. *Curr. Ther. Res.* **15,** 734–740.
246. Winaver, J., Chen, T. C., Fragola, J., Robertson, G., Slatopolsky, E., and Puschett, J. B. (1982). Alterations in renal tubular water transport induced by parathyroid hormone. *J. Lab. Clin. Med.* **99,** 457–473.
247. Winkler, S. N., Torikai, S., Levine, B. S., and Kurokawa, K. (1982). Effect of somatostatin on vasopressin-induced antidiuresis and renal cyclic AMP of rats. *Mineral Electrolyte Metab.* **7,** 8–14.
248. Work, J., Booker, B., Schafer, J. A., Galla, J., and Luke, R. (1983). In vivo and in vitro effect of ADH on loop of Henle chloride reabsorption in the Brattleboro (DI) rat (abstract). *Am. Soc. Nephrol.* **16,** 185A.
249. Wright, L. F., Rosenblatt, S. G., and Lifschitz, M. D. (1981). High urine flow rate increases prostaglandin E excretion in the conscious dog. *Prostaglandins* **22,** 21–34.
250. Zipser, R. D., Rude, R. K., Zia, P. K., and Fichman, M. P. (1979). Regulation of urinary prostaglandins in Bartter's syndrome. *Am. J. Med.* **67,** 263–267.

251. Zipser, R. D., Myers, S. I., and Needleman, P. (1981). Stimulation of renal prostaglandin synthesis by the pressor activity of vasopressin. *Endocrinology* **108,** 495–499.
252. Zar, U. (1983). Role of cytoskeletal organization in the regulation of adenylate cyclase–cyclic adenosine monophosphate by hormones. *Endocrinol. Rev.* **4,** 1–21.
253. Zusman, R. M., and Keiser, H. R. (1977). Prostaglandin biosynthesis by rabbit renomedullary interstitial cells in tissue culture: Mechanism of stimulation by angiotensin II, bradykinin, and arginine vasopressin. *J. Biol. Chem.* **252,** 2069–2071.
254. Zusman, R. M., and Keiser, H. R. (1977). Prostaglandin biosynthesis by rabbit renomedullary interstitial cells in tissue culture. Stimulation by angiotensin II, bradykinin and arginine vasopressin. *J. Clin. Invest.* **60,** 215–223.

Chapter **8**

Cardiovascular Effects of Neurohypophyseal Peptides

PHILLIP G. SCHMID, KAUSHIK P. PATEL

Veterans Administration Medical Center
Cardiovascular Center
Department of Internal Medicine
University of Iowa
Iowa City, Iowa 52240

I. EFFECTS ON CELLS

A. Membrane Receptors

There is a large variety of tissues and cells that respond to vasopressin. The character of a response is determined by (1) the type of vasopressin receptor localized on the cell membrane, (2) coupling with intracellular processes through second messengers, and (3) an interplay of cellular processes if cells happen to have receptor and coupling systems that allow responses to more than one action of vasopressin.

Vasopressin activates two distinct types of membrane receptors. The V_1 receptors mediate contraction in vascular smooth muscle and glycogenolysis in liver cells. The V_2 receptors mediate increases in water transport across renal tubular epithelium (antidiuresis).

Arginine vasopressin binds to a single class of receptors on rat aortic smooth muscle cells in culture with a K_d of 12 n*M* (Penit *et al.*, 1983). However, a substantially lower concentration, 0.25 n*M*, was calculated to cause half-maximal vasopressor response, so that the effective concentration in terms of producing a response was 50 times lower than the K_d. Assuming a similar K_d for canine vascular muscle, the data of Montani *et al.* (1980) would imply an effective concentration of vasopressin for resistance vessels on the order of 500 times

THE PEPTIDES, VOLUME 8

lower than the K_d. Thus, it would appear that activation of relatively few vasopressin V_1 receptors is responsible for marked vasoconstriction. Also, it would appear that the efficiency of V_1-induced coupling between receptor and effector mechanisms is perhaps even higher in vascular smooth muscle in resistance vessels than it is in aortic vascular muscle.

Competitive displacement of radiolabeled vasopressin by analogs with a wide spectrum of vasopressor activities clearly established the receptor for arginine vasopressin in rat aortic smooth muscle as a V_1 receptor (Penit *et al.*, 1983). Competitive binding studies did not detect a receptor in aortic muscle with any characteristics of a V_2 receptor. Thus, the vasoconstrictor action of vasopressin in animals is attributed to activation of V_1 receptors.

The V_1-receptor occupancy initiates hydrolysis of membrane polyphosphoinositides and results in the generation of two "second messengers": (1) D-inositol 1,4,5-triphosphate and (2) diacylglycerol. These two messengers initiate independent branches of one intracellular response system linked to contraction of smooth muscle and provision of energy substrate for that contraction. D-Inositol 1,4,5-triphosphate triggers release of intracellular calcium and the early phase of responses, whereas diacylglycerol triggers activation of protein kinase C and phosphorylation of proteins responsible for the sustained phase of responses. In contrast, V_2-receptor occupancy initiates activation of adenylate cyclase and results in the generation of cyclic adenosine monophosphate (cAMP), which regulates a separate intracellular response system. The separate response systems associated with membrane polyphosphoinositides and adenylate cyclase are now recognized to interact in a variety of ways that modify cellular responses to vasopressin.

B. Second Messengers

1. Polyphosphoinositides

It is ironic that polyphosphoinositides were discovered in the brain (Folch, 1949), where much of the attention on vasopressin effects on the circulation is now directed. It was quickly established that turnover of these phospholipids in the cell must be rapid (Dawson, 1949), but a significant role in receptor-mediated signal transduction in cells generally was not recognized until later (Michell, 1975).

Polyphosphoinositides refer to a group of membrane phospholipids that are synthesized in sequence by phosphorylating (a) phosphatidylinositol (PI), (b) phosphatidylinositol 4-phosphate (PIP), and (c) phosphatidylinositol 4,5-phosphate (PIP_2) (Fig. 1). These steps are catalyzed by tyrosine kinases, which appear to have a unique role in this regard. Hydrolysis of PIP and PIP_2 by activated phospholipase C, a membrane phosphodiesterase, yields diacylglycerol

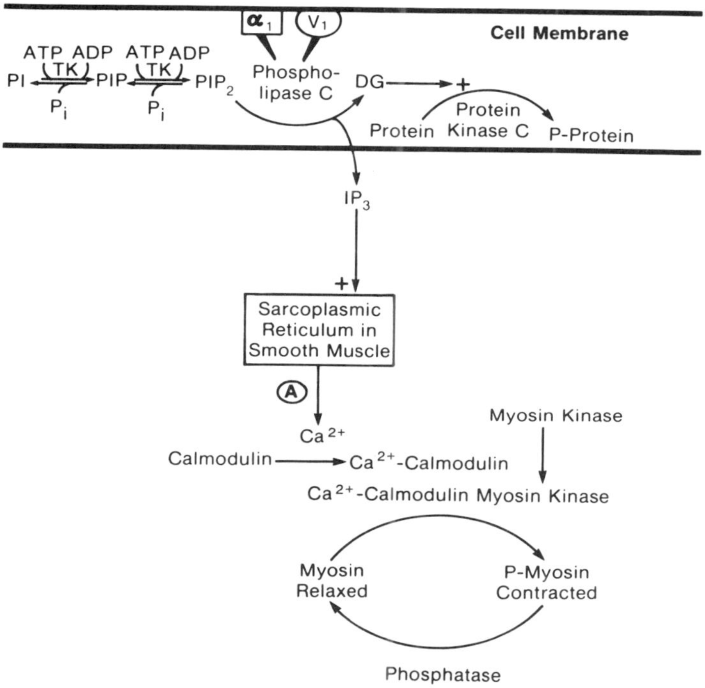

Figure 1. See text for discussion.

(DG) and D-inositol 1,4,5-triphosphate (IP_3) or D-inositol 1,4-biphosphate (IP_2). DG and IP_3 subserve significant and independent functions in message transduction.

Membrane polyphosphoinositides are recycled rapidly by the conversion of myoinositol and phosphatidic acid to PI and the phosphorylation of PI to PIP and PIP to PIP_2. This conversion does not commence until the receptor occupancy by hormone that triggered the response has been terminated. Since the stores of PIP_2 in the membrane may be limited, this raises questions about how responses to neurotransmitter or hormone are sustained. This aspect of the signal response is discussed in the next section.

That IP_3 and DG regulate independent events is indicated by the experiments of Garrison, *et al.* (1984). IP_3 and Ca^{2+}-linked hormones produced an increased phosphorylation state in a profile of 16 proteins. However, when (a) vasopressin, (b) a phorbol ester to mimic DG, and (c) a Ca^{2+} ionophore were tested separately, a different profile of phosphorylated proteins emerged for the phorbol ester and Ca^{2+} ionophore. Thus, DG and IP_3 have distinct roles in cellular regulation.

a. Diacylglycerol (DG). Diacylglycerol, one product of PIP_2 hydrolysis, has independent transducing capability. Diacylglycerol, Ca^{2+}, and phosphotidylserine acting coordinately trigger activation of protein kinase C (PK-C), which catalyzes phosphorylation of a limited number of key enzymes involved in cellular processes. PK-C, a cytosolic enzyme, is postulated to reside near the inner surface of the cell membrane (Rasmussen and Barrett, 1984). In this position, PK-C may have access to a sufficient Ca^{2+} concentration to remain activated even when the acute transient Ca^{2+} release as a result of PIP_2 hydrolysis and IP_3 generation has subsided.

b. D-Inositol 1,4,5-Triphosphate (IP_3). IP_3 stimulates calcium (Ca^{2+}) mobilization intracellularly. Williamson, *et al.* (1985) reviewed evidence that polyphosphoinositide breakdown initiates Ca^{2+} mobilization. Receptor occupancy by vasopressin (V_1 agonist) or a number of other transmitters, catecholamines (alpha-1 agonist), histamine (H-1 agonist), serotonin, and acetylcholine (muscarinic cholinergic agonist), is linked to phospholipase-C–triggered hydrolysis of PIP_2. PIP_2 breakdown occurs in the plasma membrane and precedes Ca^{2+} mobilization. Hormone dose–response curves are parallel for PIP_2 breakdown and IP_3 generation. Finally, IP_3 is capable of mobilizing Ca^{2+} from intracellular stores.

2. *Adenylate Cyclase*

The role of adenylate cyclase in the stimulus–response coupling between vasopressin V_2 receptors and free-water transport in renal tubules has been reviewed by Jard and Bockaert (1975). The occupancy of a V_2 receptor triggers the activation of adenylate cyclase and conversion of ATP to cAMP. cAMP binds to the regulatory unit of a protein kinase, leaving the catalytic subunit as an activated enzyme capable of phosphorylating a number of cellular proteins (see Fig. 2). Intracellular mechanisms provide opportunity for numerous interactions between the two second-messenger systems involving polyphosphoinositides and adenylate cyclase. One of the ways in which interactions occur may involve calmodulin, the enzyme activated by calcium and capable of activating other intracellular processes. The calcium required to activate calmodulin results either from the adenylate cyclase–cAMP–protein kinase system or from the polyphosphoinositide–IP_3 system.

C. Intracellular Mechanisms

Adenylate-cyclase–mediated events seem always to interact with IP_3- and Ca^{2+}-mediated events. Rasmussen and Barrett (1984) refer to this as "synarchic regulation" and identify five patterns of interaction:

1. "coordinate control, in which a hormone (or other extracellular messenger) interacts with two separate receptors with similar affinities for the hormone, and consequently both the [Ca^{2+}] and [cAMP] rise";
2. "hierarchical control, in which one messenger activates the Ca^{2+} messenger system, and either a higher concentration of this first messenger or another first messenger leads to an activation of the cAMP system (the rise in [cAMP] in such a system enhances the responses to the initial first messenger)";
3. "redundant control, in which the two messenger systems are activated by a distinctly different first messenger, and activation of either induces a cellular response";
4. "antagonistic control, in which a rise in [Ca^{2+}] initiates a cellular response, and a rise in [cAMP] inhibits the Ca^{2+}-induced response"; and
5. "sequential control, in which an increase in [Ca^{2+}] leads to a rise in [cAMP] or vice versa."

Receptor-mediated increases in IP_3 and DG trigger separate intracellular events that interact to further modify biological responses. Rasmussen and Barrett (1984) have suggested that Ca^{2+}–calmodulin-mediated events modify the amplitude of responses (i.e., the initial level of stimulus), whereas events mediated by DG–PK-C modify the sensitivity of responses, that is, allow them to continue even as stimulus is diminishing.

1. *Intracellular Mechanisms Affecting Contraction*

Calcium is the factor coupling the stimulus, V_1-receptor occupancy and membrane polyphosphoinositide hydrolysis, with events that comprise the response, contraction of vascular smooth muscle (Takenawa, 1982) and vasoactive mesangial cells of renal glomeruli (Troyer *et al.*, 1985) (Fig. 1). This typifies stimulus–response coupling in vascular smooth muscle (Berridge and Fain, 1979; Bolton, 1979; Loutzenhizer and Van Breeman, 1981). Contraction is thought to result from the Ca^{2+}–calmodulin-activated myosin light-chain kinase and phosphorylation of myosin light-chains (Webb and Bohr, 1981). However, recent observations have indicated that intracellular Ca^{2+} increases transiently (Morgan and Morgan, 1983) and phosphorylation of myosin light chain is not maintained (Aksoy *et al.*, 1983; Chatterjee and Murphy, 1983; Gerthoffer and Murphy, 1983). Therefore, additional events are postulated to account for sustained contraction despite diminishing myosin light-chain phosphorylation. Rasmussen and Barrett (1984) suggest that DG–Ca^{2+}–phosphotidylserine activation of PK-C leads to phosphorylation of a unique protein that promotes myosin light-chain interaction with actin and thus sustains contraction. This exemplifies interaction between the two messengers resulting from polyphosphoinositide hydrolysis: Ca^{2+} triggers contraction, and PK-C activation sustains contraction.

2. *Intracellular Mechanisms Affecting Glucose Production and Glycogen Breakdown to Supply Energy for Contraction*

Vasopressin stimulates glycogenolysis and gluconeogenesis in hepatocytes by increasing cytosolic free Ca^{2+} via mechanisms unrelated to cAMP (Williamson *et al.*, 1981; Exton, 1980, 1981). Studies by Garrison *et al.* (1984) indicate that the Ca^{2+} branch of the polyphosphoinositide second-messenger system induces phosphorylation of phosphorylase kinase, and at least one and possibly more unrecognized enzymes. In addition, Ca^{2+}–calmodulin activates glycogen synthase kinase (Ahmad *et al.*, 1982; Payne *et al.*, 1983). Thus, Ca^{2+} induces activation of two enzymes that promote availability of glucose (Fig. 2). These two enzymes also can be activated by the cAMP second-messenger system (Exton, 1980). This is an example of "redundant control," in which two separate messenger systems are activated by distinct stimuli and either can trigger the response.

Although speculative, vasopressin may influence glycogenolysis and energy substrates in vascular muscle (Webb and Bohr, 1981). It seems very possible that the pathways involving V_1-receptor activation of polyphosphoinositides and Ca^{2+} and/or V_2-receptor activation of cAMP may promote availability of energy substrates in vascular smooth muscle, mesangial cells, or any tissues with contractile behavior.

3. *Intracellular Mechanisms Modifying Contraction*

When vasopressin occupies the V_1 receptor and induces hydrolysis of polyphosphoinositides and generation of IP_3 and DG, it not only activates phosphorylation of myosin light chains and contraction, but it also activates PK-C, which contributes to maintenance of contraction, as pointed out earlier. DG, however, is rapidly metabolized, resulting in the generation of aracadonic acid, a precursor of prostaglandins, prostacyclins, and thromboxanes (Troyer *et al.*, 1985). The prostaglandins are capable of activating adenylate cyclase and generating cAMP, which can induce relaxation as discussed below (Nasjletti and Malik, 1981). This is an example of "sequential control" in which an increase in Ca^{2+} initiates the response and secondarily leads to an increase in cAMP, which inhibits the Ca^{2+}-induced response.

4. *Intracellular Mechanisms Activating Relaxation*

Vasopressin V_2 receptors mediating antidiuresis also may mediate relaxation in vascular smooth muscle, mesangial cells, and other contractile tissues. Forsklin, which activates adenylate cyclase, inhibits the contraction of vascular muscle induced with phorbol ester, mimicking the PK-C branch of the poly-

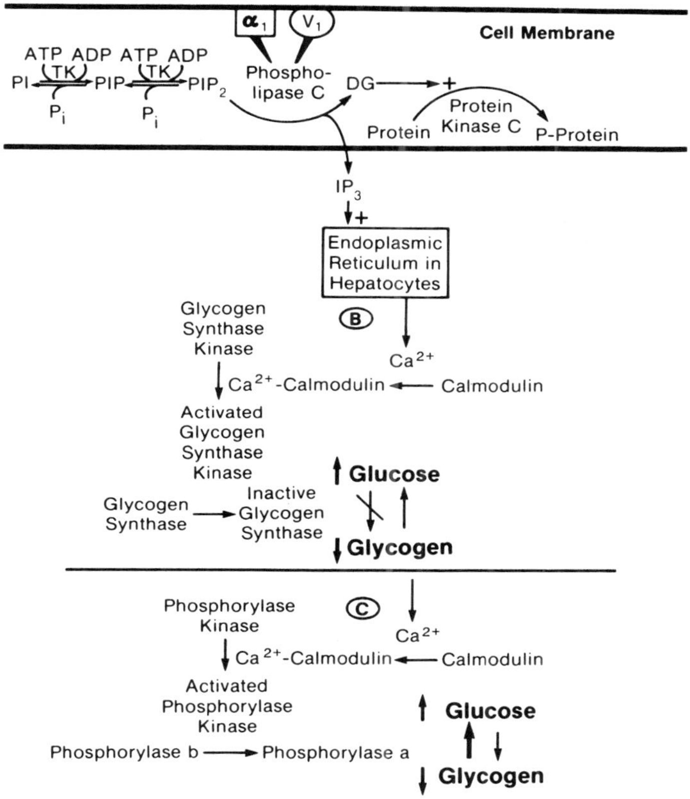

Figure 2. See text for discussion.

phosphoinositide system (Rasmussen and Barrett, 1984). The presumed mechanism of this relaxation is the cAMP-induced phosphorylation of myosin light-chain kinase, which renders the enzyme inactive, thereby inhibiting phosphorylation of myosin light chains and their contractile interaction with actin. Thus, vasopressin activation of V_1 and V_2 receptors on vascular or other smooth muscle may induce "coordinate control," in which the hormone interacts with two receptor systems and both Ca^{2+} and cAMP increase. The responses initiated by the separate systems, contraction versus relaxation, would be antagonistic. The net response expressed by a tissue might depend on the activity of these various messenger systems within the particular tissue. Conceivably, this balance could vary also among regional vascular beds and account for the variability in vascular responses to vasoconstrictor V_1 and antidiuretic V_2 agonists (see Section III for the integrated hemodynamic effects of vasopressin).

II. LOCAL EFFECTS ON REGIONAL VASCULAR TISSUES

Vasopressin is the most powerful vasoconstrictor agent known per weight basis (Altura and Altura, 1977). However, the potent pressor effect of vasopressin was not recognized until the opposing baroreflex mechanisms were removed in dogs (Cowley, 1982) and absent in humans (Mohring *et al.*, 1980). As a result, it should be recognized that plasma levels of vasopressin over the full range from a hydrated state, when they are at a minimum value, to a massive hemorrhage, where they are at the maximum level, would effect the total and regional vascular resistance without any detectable changes in blood pressure (Cowley, 1982; Montani *et al.*, 1980; Schmid *et al.*, 1974). There has been a large number of comprehensive reviews on the circulatory effects of vasopressin (Cowley, 1982; Nakano, 1974; Sawyer, 1961; Somlyo and Somlyo, 1970). They have identified the differences between species (Sawyer, 1961) and differences within a species from one vascular bed to another (Monos *et al.*, 1978; Saameli, 1968). The present section attempts to summarize the effects of vasopressin in various vascular beds with the concept that vasopressin is involved in the redistribution of blood flow during various special conditions (challenges), when such a redistribution may be required for circulatory homeostasis. Differences in vascular sensitivity will determine the pattern or distribution of blood flow in the vascular tree that may be needed for special circumstances such as hemorrhage. It should be noted that vasopressin has different effects from one vascular bed to another (Monos *et al.*, 1978; Saameli, 1968). Furthermore, within a single vascular bed, vasopressin has differing effects on the vascular components in series; for example, topical application of vasopressin on rat mesenteric microvascular produces a greater constriction in arterioles than in veins (Altura, 1978; Mellander and Johansson, 1978; Nakano, 1974).

A. Cerebral Vasculature

The pial vessels are generally very insensitive to vasopressin in rats (Lassoff and Altura, 1980). Angiographic studies in dogs have shown that blood flow in the internal carotid is increased at the expense of decreased blood flow in the external carotid artery (Aronsen and Nylander, 1964, 1966). These studies also have demonstrated vasodilation of cerebral, orbital, and external ophthalmic arteries during administration of vasopressin. Similarly, Sunahara *et al.* (1949) have shown in humans that the blood flow to face and ear is decreased in response to vasopressin. These various studies suggest that vasopressin diverts blood from extracerebral to intracerebral (brain) circulation. In other studies in humans and dogs, responses to vasopressin by basilar and cortical arteries have been shown to be vasoconstrictor in nature. Liard *et al.* (1982) have shown that physiological levels of vasopressin in conscious dogs produce a decrease in blood flow to the brain comparable to the decrease in cardiac output, suggesting a relatively mild vasoconstrictor action of vasopressin on cerebral vasculature.

B. Coronary Vasculature

Vasopressin vasoconstricts the coronary arteries in rabbits (Lu, 1951), cats (Melville, 1936), and dogs (Hausler, 1930). The vasoconstriction of coronary vasculature by vasopressin is responsible for decreased coronary blood flow myocardial contractile force (Nakano, 1967). In addition, the coronary vasoconstriction also induces a myocardial ischemia. Some of the reflex effects of vasopressin via the central nervous system (see Section IV) may be mediated by the cardiac afferents from the myocardial ischemic regions. It should be noted that infusion of vasopressin in patients can be detrimental, as it is conceivable that vasopressin may constrict coronary vasculature in humans and subsequently produce a decrease in myocardial contractility and cause myocardial ischemia, as it does in animals (Nakano, 1967, 1974). Such a contention is supported by a report of myocardial infarction in a patient that received a subcutaneous injection of pitressin (Slotnik and Teigland, 1951).

C. Mesenteric Vasculature

Vasopressin constricts mesenteric vasculature in cats (Barer, 1961), dogs (Delaney *et al.*, 1966), and humans (Schwartz *et al.*, 1959). Mesenteric vasculature resistance is at least three times more sensitive to vasopressin than it is to angiotensin II (Altura and Altura, 1977), another vasoactive peptide. The relative decreases in the gastrointestinal blood flow were greater compared to total cardiac output (Shaldon *et al.*, 1961) during vasopressin infusion. The portal blood flow is decreased (Nakano, 1974) in response to vasopressin infusion, probably as a result of increased vascular resistance upstream in the superior mesenteric and hepatic arteries. Schmid *et al.* (1974) showed in ganglionic-transmission-blocked animals that vasopressin caused a significant reduction in the total mesenteric blood flow. Later, Liard *et al.* (1982) found that vasopressin infusion at levels observed under physiological conditions in conscious dogs causes vasocontriction in the gastrointestinal tract.

Norepinephrine and vasopressin seem to have an interaction at the level of the mesenteric microvessels, because vasopressin potentiates the effects of norepinephrine (Altura and Altura, 1977). A possible explanation for such an observation is that prostaglandin E_2 facilitates the vasocontriction of the mesenteric vasculature by norepinephrine, and vasopressin mediates activation of prostaglandin E_2 synthesis.

D. Renal Vasculature

Vasopressin has been shown to constrict renal vasculature in dogs (Brooks and Pickford, 1958), cats (Barer, 1961, 1963), rabbits (Walker *et al.,* 1937), and humans (Longo *et al.,* 1964). Intravenous infusion of vasopressin, however, did not decrease renal blood flow as measured by clearance methods in dogs (Brooks and Pickford, 1958). It has been suggested that the increase in renal blood flow is a function of redistribution of systemic blood flow as a consequence of vasoconstriction in other regions of the body (Nakano, 1974). Within the kidney, vasopressin decreases medullary blood flow (Fisher *et al.,* 1970) and shifts flow from the outer to the inner cortex (Akatsuka *et al.,* 1977; Johnson *et al.,* 1977), although not consistently (Banks, 1976). Vasopressin infusion in human causes a transient decrease followed by an increase in renal blood flow (Gothlin, 1976), which may reflect renal vasoconstriction with overall redistribution of flow from other beds, such as seen in the dog (Schmid *et al.,* 1974). Liard *et al.,* (1982) have shown that infusion of vasopressin to attain physiological levels in plasma in conscious dogs produced no significant changes in renal blood flow, despite a decrease in cardiac output and increased vasoconstriction in other beds. In conclusion, it appears that the renal vasculature is less responsive to constriction by vasopressin or that the intensity of vasoconstriction is less than other vascular tissues, thus maintaining normal renal blood flow in spite of decreased cardiac output.

E. Muscle Vasculature

The muscle vasculature is the most sensitive of all vascular beds to the vasoconstrictor action by vasopressin. Powerful constriction has been reported in the cat (Barer, 1961) and in conscious (Liard *et al.,* 1982) and anesthetized dogs (Schmid *et al.,* 1974). Liard *et al.* (1982) have reported that infusion of vasopressin (in the physiological range) caused a large drop in the blood flow to the muscle as measured by microspheres. All these studies show that vasopressin has its most profound vasoconstirctor effect on the vasculature of skeletal muscle.

III. INTEGRATED HEMODYNAMIC EFFECTS

A. Heart Rate and Cardiac Output

1. Role of V_1 Receptors

Vasopressin caused myocardial depression secondary to coronary vasoconstriction and bradycardia secondary to efferent parasympathetic activation when administered intravenously in doses varying from bolus injections of 5–125 mU/kg to infusions of 10–20 mU/kg min (Nakano, 1974). Although these amounts of vasopressin clearly produced supraphysiological plasma levels of vasopressin, the results established that vasopressin had significant direct and reflex effects on the hearts of dogs.

In an effort to estimate the threshold levels for direct effects of vasopressin on the heart, Schmid *et al.* (1974) studied dogs with ganglionic blockade. Results seemed to indicate that plasma vasopressin levels of 200 μU/ml were required to reduce cardiac output. However, the studies were carried out in anesthetized dogs with baseline plasma vasopressin levels of 50 μU/ml, so that significant cardiac effects of lower levels of vasopressin were not excluded.

The studies of Montani *et al.* (1980) involving conscious dogs with endogenous plasma levels below 1 μU/ml seem the most definitive to date. Intravenous infusions of vasopressin that increased plasma vasopressin only 1 μU/ml reduced cardiac output; infusions that increased plasma levels 6 μU/ml reduced heart rate. The bradycardia and possibly, in part, the reduced cardiac output appeared to result from a central action of vasopressin, because they were not detected in sinoaortic baroreceptor-denervated dogs (Montani *et al.*, 1980), and, in a separate study, they were produced by intervertebral-artery infusions of vasopressin and were more pronounced than responses produced by intravenous infusions with corresponding plasma levels of vasopressin (Liard *et al.*, 1981).

In conscious rabbits, intravenous administration of vasopressin that increased plasma levels to 56 μu/ml produced decreases in heart rate and cardiac output that were not altered detectably by autonomic blockade with methylscopolamine and propranolol (Elliot *et al.*, 1985). In contrast to the results in dogs, these results support a direct depressant effect of vasopressin on the heart. However, our own studies of rabbits (G. B. Guo, 1985, unpublished data) have demonstrated that sinoaortic baroreceptor denervation and ganglionic blockade progressively attenuated vasopressin-induced bradycardia, so the possibility of centrally mediated bradycardia in response to vasopressin has not been excluded in the rabbit.

The ability of intravenous vasopressin to induce bradycardia and reduce cardiac output in conscious dogs appears to be associated with activation of V_1

(vasoconstrictor) receptors; these actions were reproduced by [phenylalanine^2,ornithine^8]oxytocin, a selective vasoconstrictor agonist (J. Schwartz *et al.*, 1985). These actions also were reversed by the selective V_1 antagonist, [(1-β-mercapto-β, β-cyclopentamethylenepropionic acid), 2-(*O*-methyl)tyrosine]AVP [$d(CH_2)_5$Tyr(Me)AVP].

Thus, V_1 receptor activation appears to initiate reductions in heart rate and cardiac output in dogs. Involvement of V_1 receptors in rabbits and the site of these receptors, central versus peripheral, is unsettled, however.

2. *Role of V_2 Receptors*

The intriguing possibility that activation of vasopressin V_2 receptors might produce opposite effects on heart rate and cardiac output from those of V_1-receptor activation arose when increases in heart rate and cardiac output were noted after administering vasoconstrictor antagonists, dPVDAVP (see Section V) and $d(CH_2)_5$Tyr(Me)AVP, to dogs (Liard and Spadone, 1984). The analog, dPVDAVP, had potent antidiuretic (V_2-receptor) agonist activity, in addition to its vasoconstrictor (V_1-receptor) antagonist activity, strengthening speculation that V_2 receptors were involved. Subsequently, a more thorough evaluation of this possibility produced additional evidence that the antidiuretic component (V_2) of vasopressin's influence induces tachycardia and increases cardiac output in dogs (J. Schwartz *et al.*, 1985).

Appraisal of the data of Elliot *et al.* (1985) concerning responses of heart rate and cardiac output to vasopressin in rabbits after vasopressin antagonist indicate that increases occur in both variables, consistent with the results in dogs.

Thus, V_2-receptor activation appears to initiate increases in heart rate and cardiac output in dogs and rabbits, although the site of these receptors, central versus peripheral, is not yet settled.

B. Peripheral Vascular Resistance and Regional Blood Flow

1. *Role of V_1 Receptors*

Vasopressin administered to dogs in relatively high intravenous doses, up to 125 mU/kg, increased total peripheral vascular resistance even when mean arterial pressure was not detectably altered (Nakano, 1974). Electromagnetic flow meters and simultaneous measurement of multiple regional vascular responses to vasopressin made possible the demonstration that some resistance beds were more sensitive to vasopressin than others (Schmid *et al.*, 1974). In dogs with cardiovascular reflexes inhibited by ganglionic blockade and high baseline vasopressin levels (66 μU/ml, increments of plasma vasopressin of 144 μU/ml and

215 μU/ml caused constriction of hindquarters resistance vessels and redistributed a diminishing cardiac output to kidney and mesentery. This indicated more pronounced vasoconstriction in the iliac bed. Additional studies demonstrated greater sensitivity to vasopressin in skeletal muscle (gracilis muscle), with levels in the range of 20 μU/ml causing significant increases in muscle vascular resistance. However, in view of the high basal level of vasopressin, significant regional vascular responses to lower levels of vasopressin could not be excluded.

Vasopressin administered to conscious, unrestrained rats in intravenous doses of 0.5 and 5 mu/kg min caused more pronounced vasoconstriction in muscle and skin than in other vascular beds, an observation consistent with that in dogs (Hoffman, 1980). Significant vasoconstriction in response to vasopressin was observed in the forebrain, hindbrain, kidney, and intestine, in addition. Curiously, the vasoconstrictor responses to vasopressin in the kidney and intestine were more pronounced with efferent sympathetic neural control intact than with it blocked, either by phenoxybenzamine or by hexamethonium. This has raised speculation that vasopressin and efferent sympathetic neural mechanisms interact selectively in the kidney and intestine of rats. Evidence of such an interaction was not detected in other regional beds of the rat.

The major question that was finally answered by the study of Montani *et al.* (1980) was, how low a level of vasopressin might produce vasoconstriction in a conscious animal? Conscious dogs were given a range of intravenous vasopressin infusions from 0.016 to 2 mU/kg min, which raised plasma vasopressin concentrations above the endogenous level of <1 μU/ml by as little as 1 μU/ml and by as much as 200 μU/ml. These investigators have reported that total peripheral vascular resistance increased significantly with an increment of only 1 μU/ml in vasopressin. They observed also that corresponding increments in plasma vasopressin induced with hypertonic saline infusions produced systemic vasoconstriction. The increases in total peripheral vascular resistance were similar in dogs with intact arterial baroreceptors and those with sinoaortic baroreceptor denervation, a finding interpreted as indicating that low levels of vasopressin exerted a direct peripheral vasoconstrictor influence. The obvious implication of these results is that vasopressin has a physiological role in regulating vascular resistance and distribution of cardiac output as well as arterial pressure.

2. *Role of V_2 Receptors*

In view of the lack of evidence for V_2 receptors in aortic smooth muscle, it is puzzling that antidiuretic agonists seem capable of reducing vascular resistance (Liard and Spadone, 1984; J. Schwartz *et al.*, 1985; Liard, 1986). Future studies will need to determine if V_2 receptors activate release of an endothelial derived relaxation factor or if V_2 receptors are localized to smooth muscle in resistance vessels. Alternatively, antidiuretic agonists might induce vasodilatation by activating within the central nervous system V_2 receptors that regulate neural control of the circulation.

IV. EFFECTS OF VASOPRESSIN ON NEURAL CONTROL OF CIRCULATION

A. Evidence Supporting a Central Action

For a long time vasopressin was thought to have actions apart from directly vasoconstrictor and antidiuretic influences on the circulation and kidney, respectively. However, evidence that vasopressin might also influence neural control of the circulation evolved slowly. Initially, investigators noted that pressor sensitivity to vasopressin was increased substantially by reserpine pretreatment (Nash, 1965) and division of the vagi and sinus nerves (Gardier and Abreu, 1958), procedures that eliminated reflex buffering of responses. At the outset, this increased sensitivity to vasopressin was attributed to increased vascular reactivity by an unstimulated and more reactive vasculature in the reserpine experiments and a facilitative effect of catecholamines on vascular responses to vasopressin in the baroreceptor-denervation experiments.

In the course of demonstrating that vasopressin levels associated with hemorrhage had significant pressor effects, Rocha de Silva and Rosenberg (1969) investigated pressor responses to vasopressin and depressor responses to hypophysectomy in baroreceptor-denervated dogs. Only after elimination of baroreflexes were these investigators able to show that the changes in vasopressin levels associated with hemorrhage or hypophysectomy affected arterial pressure. The significance of these observations seemed to be their support for the concept that vasopressin had an important role in blood pressure regulation in a pathophysiological situation. However, a direct vasoconstrictor influence rather than an influence on neural control of the circulation was emphasized.

Although the work of Cowley *et al.* (1974) is generally credited with establishing an important influence of vasopressin on baroreceptor-mediated reflex neural control of the circulation, their goal was to determine the threshold levels of vasopressin with detectable pressor influence. This threshold seemed to be on the order of 4 μU vasopressin per milliliter plasma in baroreceptor-denervated conscious dogs and substantially higher in conscious dogs with baroreflexes intact. More importantly, the sensitivity to vasopressin in terms of increasing arterial pressure was increased 60- to 100-fold in baroreceptor-denervated animals compared to intact animals. This sensitivity was increased further by eliminating central neural control altogether with decapitation. Again, however, as in the study by Rocha de Silva and Rosenberg (1969), a direct vasoconstrictor influence rather than an influence on neural control of the circulation was emphasized (Cowley *et al.*, 1974).

In their discussion, Cowley *et al.* (1974) speculated that since norepinephrine was pontentiated only threefold by baroreceptor denervation, compared to a 60- to 100-fold potentiation of vasopressin, other factors than elimination of buffering reflexes must have been responsible. Their speculation included the possibilities that (1) pressor responsiveness to vasopressin might be enhanced because plasma levels of vasopressin were depressed in chronically baroreceptor denervated animals and (2) an interaction of high catecholamine levels and vasopressin contributed to augmented pressor responses in the barodenervated state.

In retrospect, this paper by Cowley *et al.* (1974) has proven to be pivotal. As results of other studies became available, the significance of the enhanced pressor effect of vasopressin in baroreceptor-denervated animals seemed totally consistent with an important central action of vasopressin on neural control of the circulation.

Another pivotal report was published by Lohmeier *et al.* (1981). These investigators infused angiotensin II intravenously into dogs to induce hypertension and then added infusions of vasopressin to determine whether this could produce further increases in arterial blood pressure. Vasopressin had no detectable pressor effect, but when the infusion was stopped while infusion of angiotensin II continued, a pronounced decrease in arterial pressure was noted that lasted for more than 24 hr. Withdrawal of vasopressin under these conditions had a sustained depressor influence, and, although the authors focused on the possibility that the results occurred because of water and electrolyte changes, the effects were consistent with yet another action of vasopressin, mediated either via local vasodilator mechanism(s) or via mechanism(s) related to neural control of the circulation.

The work of Cowley *et al.* (1974) and Lohmeier *et al.* (1981) had a major influence on the interpretation of our own observations in desoxycorticosterone- and salt-treated rats (DOC–salt rats) (Matsuguchi and Schmid, 1982a,b). We investigated the role of neurogenic factors and vasopressin in the increased hindquarters vascular resistance of normotensive and hypertensive DOC–salt rats by comparing vasodilator responses to lumbar sympathectomy in the two groups and by comparing vasodilator responses to a vasopressin (V_1 receptor) antagonist in intact and sympathetically denervated hindquarters.

In normal rats, the vasodilator influence of vasopressin antagonist was augmented in sympathetically denervated hindquarters compared to the vasodilator influence in sympathetically innervated hindquarters. Based on the work cited above, we interpreted these results to indicate that vasopressin had a central depressant effect on vasomotor activity, such that vasopressin antagonism produced offsetting effects by blocking the local vasoconstrictor effect of vasopressin and also the central inhibitory effect of vasopressin on sympathetic vasomotor activity. Thus, overall, after vasopressin (V_1 receptor) antagonist, little net change occurred in vascular resistance in the sympathetically intact hindquarters of normal rats. In contrast, in DOC–salt rats with impairment of arterial baroreflexes, the offsetting influence of vasopressin's central inhibition of sympathetic vasomotor activity was not expressed, and the local vasodilator influence of vasopressin antagonist was similar in sympathetically intact and denervated hindquarters. The major conclusions of these studies was that vasopressin had a central action that inhibited sympathetic vasomotor activity and that this action required normal baroreflex function, since it was not expressed in DOC–salt rats with impaired baroreflexes.

Although our results clearly indicated that vasopressin must be acting centrally, there were aspects of our observations and those of others that were puzzling. Chiu and McNeill (1985) noted in spontaneously hypertensive rats (SHR) with impaired baroreflexes that 3-hr infusions of vasopressin (8 mU/kg min), which minimally elevated arterial pressure, when stopped, were followed by pronounced hypotension that was significantly greater than in normotensive WKY rats and greater than the hypotension associated with cessation of equipressor infusions of phenylephrine. This would seem to indicate that vasopressin caused a centrally mediated inhibition of sympathetic vasomotor activity in conjunction with baroreflex impairment in SHR but not in hypertensive DOC–salt rats.

These studies taken together, therefore, implicated a sustained depressor influence of vasopressin when its pressor effects were terminated by stopping administration of the peptide or by a vasopressin (V_1 receptor) antagonist. Moreover, the observations raised important new questions. Did vasopressin have a central action to inhibit peripheral sympathetic vasomotor activity? Were intact baroreflexes necessary for the central action of vasopressin to be expressed? Did vasopressin affect neural control of the circulation by more than one mechanism? Could these depressor responses be mediated by V_2-receptor activation that was unmasked and persisted after V_1-receptor activation had been terminated?

B. Influence on Baroreflex Control of the Circulation

1. *Facilitation of Baroreflex Inhibition of Sympathetic Activity*

Although the work cited above had suggested an interaction between vasopressin and baroreflex mechanisms, a direct effect of vasopressin on baroreflex control of the circulation was first reported in abstracts by Sharabi and Schmid (1982) and Guo *et al.* (1982). These investigators monitored lumbar sympathetic nerve activity in chloralose-anesthetized rabbits during graded, 2-min intravenous infusions of vasopressin (8, 16, and 32 mU/kg min). Compared to the effects of phenylephrine in control and recovery periods, vasopressin caused similar reflex decrements in lumbar sympathetic-nerve activity at significantly lower levels of arterial pressure (Guo *et al.*, 1982, 1986; Sharabi and Schmid, 1982; Schmid *et al.*, 1984, 1985a,b; Sharabi *et al.*, 1985). This established that vasopressin facilitated baroreflex inhibition of peripheral sympathetic nerve activity. Furthermore, sympathetic neural inhibition occurred even as arterial pulse pressure and mean arterial pressure decreased, supporting a central action of vasopressin on baroreflex pathways (Sharabi *et al.*, 1985).

Investigators (Schmid *et al.*, 1985a,b) further implicated a central action on baroreflex pathways with the demonstration that vasopressin given intravenously to rabbits and excluded from isolated carotid sinus baroreceptors (aortic and cardiopulmonary baroreceptors denervated) facilitated reflex inhibition of lumbar sympathetic-nerve activity. However, Undesser *et al.* (1985) established most definitively that vasopressin facilitated baroreflex inhibition of renal sympathetic activity through a central action with the demonstration that lesions of the area postrema blocked this action.

2. *Local Influence on Baroreceptor Afferents*

Guo *et al.* (1986) monitored afferent activity in the aortic depressor nerves and efferent activity in the lumbar sympathetic nerves of chloralose-anesthetized rabbits during aortic occlusion and repeated this while administering vasopressin intravenously. Although vasopressin, in contrast to phenylephrine, caused a decrease in arterial pulse pressure and more pronounced bradycardia, the increase in aortic baroreceptor activity per unit rise in arterial pressure was greater with vasopressin than with phenylephrine. Over a higher range of arterial pressures, 100 and 150 mm Hg, aortic occlusion induced greater afferent discharge, both per unit time and per heart beat, in the presence of vasopressin than in the presence of control saline infusions. Finally, local application of vasopressin to the isolated carotid sinus shifted the pressure stimulus–response curve so that greater reflex inhibition of lumbar sympathetic nerve activity occurred for any carotid sinus pressure. These observations indicate that there is an augmentation of arterial baroreceptor activity with vasopressin that contributes to the facilitation of the baroreflex.

3. *Contrasting Effects of Vasopressin in Plasma and CSF*

Imaizumi and Thames (1986) studied the influence of intravenously and cerebroventricularly administered vasopressin on baroreflex control of renal sympathetic nerve activity. Baroreflexes were evaluated in chloralose-anesthetized rabbits by increasing and decreasing arterial pressure with phenylephrine and nitroglycerin, respectively. Vasopressin given in 1-min intravenous doses of 4 and 40 mU inhibited basal renal sympathetic nerve activity and reduced baroreflex inhibition of this activity. Sinoaortic baroreceptor denervation in rabbits with the vagi sectioned did not eliminate the effects of vasopressin on basal renal sympathetic-nerve activity, whereas it did eliminate inhibitory effects of vasopressin on lumbar sympathetic-nerve activity, confirming reports of Guo *et al.* (1983) and Sharabi *et al.* (1985). Since recorded lumbar and renal sympathetic-nerve activity represented pre- and postganglionic sympathetic traffic, respectively, and since the postganglionic renal nerve traffic was also inhibited by vasopressin in rabbits with spinal-cord transection, the investigators concluded that vasopressin facilitated barereflex inhibition of lumbar preganglionic activity by acting centrally but that it reduced postganglionic renal sympathetic activity by interfering with ganglionic transmission.

In contrast, vasopressin given intracerebroventricularly had no effect on basal renal sympathetic nerve activity or on augmented reflex increases in this activity induced by lowering arterial pressure with nitroglycerin.

Although Imaizumi and Thames (1986) refined the conclusions of Undesser *et al.* (1985) that vasopressin affected baroreflex control of renal sympathetic traffic with the suggestion that, instead, vasopressin might be inhibiting ganglionic transmission, they basically did not challenge the conclusions of Guo *et al.* (1982) and Sharabi *et al.* (1985) that vasopressin facilitated baroreflex inhibition of lumbar preganglionic sympathetic traffic. In view of this, Imaizumi and Thames (1986) concluded that vasopressin augments baroreflex gain whether given intravenously or intracerebroventricularly, but that intravenous vasopressin affects reflex responses to hypertensive stimuli, whereas cerebroventricular vasopressin affects reflex responses to hypotensive stimuli.

4. Vasopressin and Baroreflexes in Rat and Dog

Unlike rabbits, urethane-anesthetized rats did not exhibit facilitated baroreflex inhibition of lumbar sympathetic-nerve activity when given vasopressin intravenously (Sharabi *et al.*, 1985). Unlike rabbits and rats, sodium-pentobarbital-anesthetized, neurohypophysectomized dogs exhibited augmented reflex vasoconstriction during carotid sinus hypotension and not augmented reflex vasodilatation during carotid sinus hypertension when given vasopressin intravenously (Cowley *et al.*, 1984). Thus there appear to be marked species differences with regard to the influence of vasopressin on baroreflex control of the peripheral circulation. However, since baroreflexes were evaluated in both conscious and chloralose-anesthetized rabbits, urethane-anesthetized rats, and pentobarbital-anesthetized dogs, the different anesthetics rather than the different species could account for these variable results.

The effects of vasopressin on reflex control of heart rate also vary among the different species. Vasopressin seemed to augment reflex bradycardia in urethane-anesthetized rats (Schmid *et al.*, 1985a) and in conscious rabbits in one study (Undesser *et al.*, 1985) but not in another (Elliot *et al.*, 1985). In anesthetized, neurohypophysectomized dogs, on the other hand, vasopressin had no detectable effect on baroreflex control of heart rate.

A recent study in our laboratory (K. P. Patel and P. G. Schmid, 1986, unpublished results) clearly has established that vasopressin modifies central baroreflex control of heart rate in anesthetized rabbits. Microinjection of 2% lidocaine into the median preoptic nucleus reversibly blocked the facilitation of reflex bradycardia induced by vasopressin without altering the reflex bradycardia induced by phenylephrine. Undesser *et al.* (1985) also reported that lesions of the area postrema block the facilitation of reflex bradycardia by vasopressin. In view of these results, there is no clear explanation why facilitation of reflex bradycardia during intravenous administration of vasopressin has been detected in most but not all experiments.

5. Role of V_1 and V_2 Receptors

In the studies of Imaizumi and Thames (1986), baroreflex effects of vasopressin were evaluated in the absence and in the presence of a vasopressin (V_1 receptor) antagonist, d(CH_2)$_5$Tyr(Me)AVP. The antagonist reversed the influence of intravenous vasopressin but not the influence of cerebroventricular vasopressin on renal sympathetic nerve activity.

Imai *et al.* (1983) reported that vasopressin and the antidiuretic analog 1-desamino-D-arginine vasopressin (DDAVP) augmented baroreflex sensitivity regarding heart rate in rats with hereditary diabetes insipidus. In addition, they reported that the V_1 antagonist d(CH_2)$_5$Tyr(Me)AVP did not affect baroreflex control of heart rate in normal Long-Evans rats.

These results implicate antidiuretic vasopressin (V_2) receptors in the baroreflex control of heart rate and vasoconstrictor (V_1) receptors in the baroreflex control of peripheral sympathetic-nerve activity in rat and rabbit, respectively.

C. Central Amine Pathways

1. Noradrenergic Pathways

Barereceptor information is carried over afferents in the vagus and glossopharyngeal nerves to the medullary nucleus tractus solitarius. On the basis of neuroanatomical and electrophysiological studies, some of this information is relayed to the forebrain by noradrenergic and adrenergic pathways [for summaries see reviews by Swanson and Sawchenko (1983) and Mancia and Zanchetti (1981)]. Catecholamine neurons of the A_1 group in the ventrolateral medulla terminate mostly on magnocellular neurons of the paraventricular and supraoptic nuclei. In contrast, catecholamine neurons of the A_2 group of the dorso–vagal complex and the A_6 group of the locus ceruleus terminate mostly on parvocellular subnuclei of the paraventricular nucleus. In rabbits, there is evidence that central noradrenergic neurons participate in arterial baroreflex control of the circulation (Chalmers and Reid, 1972). Vasopressin introduced in the brain ventricular system activates noradrenergic neurons in A_2 and locus ceruleus, as indicated by increases in norepinephrine turnover (Tanaka *et al.*, 1979).

In view of these considerations, we investigated the possibility that intravenous administration of vasopressin could affect central and peripheral noradrenergic function. A low dose of vasopressin (4 mU/kg min) given intravenously for 1 hr caused increases in norepinephrine turnover in the median preoptic nucleus and paraventricular nucleus and decreases in norepinephrine turnover in the kidney and gracilis muscle. A higher dose of vasopressin (16 mU/kg min), on the other hand, caused increases in norepinephrine turnover in the locus ceruleus as well as the median preoptic nucleus but no longer caused decreases in peripheral norepinephrine turnover. These results establish that vasopressin in the circulation causes increases in central noradrenergic function and that these central changes are associated (not necessarily by cause and effect) with changes in peripheral noradrenergic function. The results also imply that vasopressin may have biphasic effects on central and peripheral noradrenergic function.

While this study cannot establish that the influence of vasopressin on baroreflex control of the circulation is related to its effects on central noradrenergic function, it is consistent with a hypothesis to that effect and indicates a direction for future studies.

2. Serotonergic Pathways

Vasopressin and the serotonin antagonist methysergide facilitate baroreflex inhibition of peripheral sympathetic nerve activity (Schmid *et al.*, 1985a,b; Sharabi *et al.*, 1985; Axtelle *et al.*, 1985). In contrast, serotonin depresses baroreflex inhibition of peripheral sympathetic activity. There appear to be important interconnections between central noradrenergic neurons originating in the medulla and projecting to serotonergic neurons in the dorsal raphe nucleus (Baraban and Aghajanian, 1981). The dorsal raphe serotonergic neurons project to potential sites of integrative baroreflex function in the median preoptic area and anterior hypothalamus. Manipulation of serotonin at these terminal sites in the forebrain attenuates pressor responses to stimulation of the dorsal raphe nucleus where the pathway originates. These considerations are the basis for a hypothesis that central serotonergic pathways may be activated by vasopressin and may participate in the baroreflex changes induced by the peptide.

D. Autonomic Ganglia

Imiazumi and Thames (1986) have suggested that vasopressin inhibits ganglionic neurotransmission to kidney. Vasopressin has been reported to inhibit ganglionic transmission in isolated superior cervical ganglia of rabbit (Wali, 1984) and to stimulate rapid polyphosphoinositide hydrolysis in superior cervical ganglia of rat (Bone *et al.*, 1984). Vasopressin-like peptide identified by immunohistochemical techniques is widely distributed in the sympathetic nervous system of mammals, both in principal noradrenergic neurons of ganglia and in nerve-fiber terminals in tissues (Hanley *et al.*, 1984). This suggests a regulatory role for vasopressin in peripheral sympathetic neural function, but that role is not known.

V. AGONISTS AND ANTAGONISTS USED TO STUDY CARDIOVASCULAR EFFECTS

A. V_1 Receptors Mediating Vasoconstriction and Glycogenolysis

1. *Vasoconstrictor Agonists*

Vasopressin, of course, is the prototype agonist, but vasopressin is capable of activating antidiuretic (V_2) receptors in addition to V_1 receptors, so responses are not pure. This aspect of vasopressin's influence on the circulation is emphasized throughout this review. The most selective V_1 receptor agonist available for practical use is [2-phenylalanine,8-ornithine]oxytocin. Structures for agonists are presented schematically and activities are indicated as follows (Manning and Sawyer, 1983, 1984).

Vasopressin's vasopressor activity is 369 U/mg; its antidiuretic activity is 323 U/mg; its oxytocic activity is 14 U/mg; and its milk-ejecting activity is 70 U/mg. Thus, vasopressin is not a very selective agonist (Manning and Sawyer, 1984).

```
1   2   3   4   5   6   7   8   9
Cys-Tyr-Phe-Gln-Asn-Cys-Pro-Arg-Gly(NH2)
|                   |
S-------------------S
----------------------------------------
```

The vasopressor activity of [2-phenylalanine,8-ornithine]oxytocin is 124 U/mg; its antidiuretic activity is 0.55 U/mg; its oxytocic activity is 1 U/mg; and its milk-ejecting activity is 7 U/mg. Thus, this compound is a more selective V_1-receptor agonist (Manning and Sawyer, 1984).

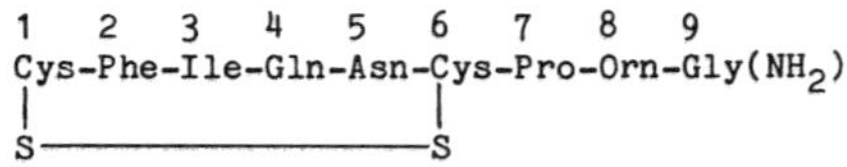

2. Vasoconstrictor Antagonists

Analogs of vasopressin with antivasopressor activity were developed using information derived from investigation of antioxytocic compounds (Manning and Sawyer, 1984). The three antivasopressor antagonists utilized most frequently in physiological and pharmacological studies relating to the circulation are referred to by the abbreviations: dPVDAVP (Manning *et al.*, 1977), d$(CH_2)_5$VDAVP (Lowbridge and Manning, 1978), and d$(CH_2)_5$Tyr(Me)AVP (Kruszynski *et al.*, 1980). The structures for these antivasopressor analogs are presented schematically next and their various activities are indicated (Manning and Sawyer, 1983, 1984).

[1-Deaminopenicillamine,4-valine,8-D-arginine]vasopressin (dPVDAVP) has an antivasopressor pA_2 of 7.83, an antioxytocic pA_2 of 7.23, and antidiuretic (agonist) potency of 123 U/mg. This compound is a relatively weak V_1-receptor antagonist with some antidiuretic (V_2) receptor agonist activity, which must be considered when interpreting its effects *in vivo* (Manning *et al.*, 1977).

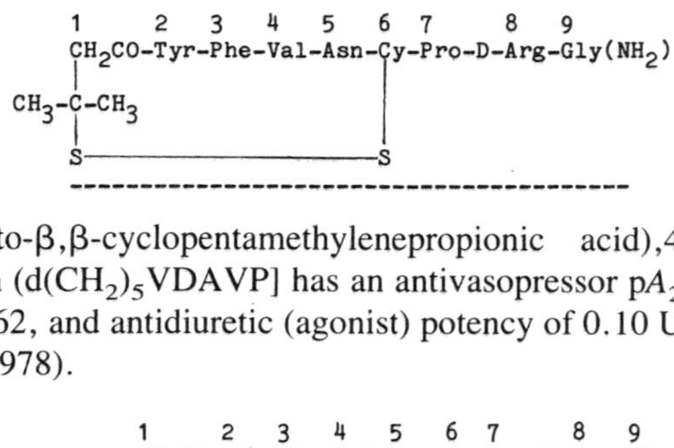

[1-(β-Mercapto-β,β-cyclopentamethylenepropionic acid),4-valine,8-D-arginine]vasopressin (d$(CH_2)_5$VDAVP] has an antivasopressor pA_2 of 7.68, antioxytocic pA_2 of 6.62, and antidiuretic (agonist) potency of 0.10 U/mg (Lowbridge and Manning, 1978).

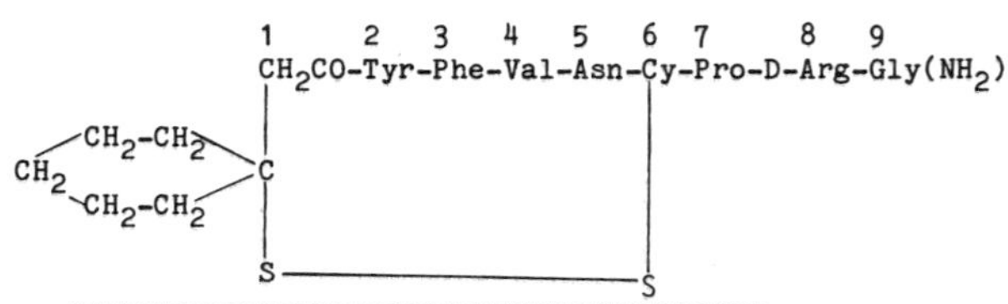

[1-(β-Mercapto-β,β-cyclopentamethyleneproprionic acid), 2-*O*-methyltyrosine]arginine vasopressin [d$(CH_2)_5$[Tyr(Me)^2]AVP] has an antivasopressor pA_2 of 8.62, an antioxytocic pA_2 of 8.13, antidiuretic (agonist) potency of 0.31 U/mg, and pronounced anti-CRF-like (antagonist) potency at 10 μ*M in vitro*. This compound is the most potent available V_1-receptor antagonist, but it also has antioxytocic activity and anti-CRF-releasing activity, although no apprecia-ble antidiuretic (V_2) receptor agonist or antagonist activity (Kruszynski *et al.*, 1980; Aizawa *et al.*, 1982; Knepel *et al.*, 1984).

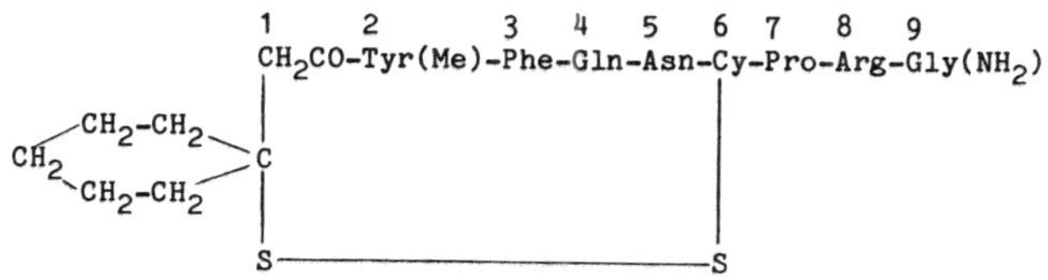

B. V_2 Receptors Mediating Vasodilatation and Antidiuresis

1. *Antidiuretic Agonists*

1-Deamino[8-D-arginine]vasopressin (dDAVP) and 1-deamino[4-valine,8-D-arginine]vasopressin (dVDAVP) have relatively selective antidiuretic (V_2) receptor agonist activity (Manning and Sawyer, 1984).

1-Deamino[8-D-arginine]vasopressin (dDAVP) has antidiuretic activity of 1200 U/mg, vasopressor activity of 0.4 U/mg, oxytocic activity of 1.5 U/mg, and milk-ejecting activity of 0.4 U/mg.

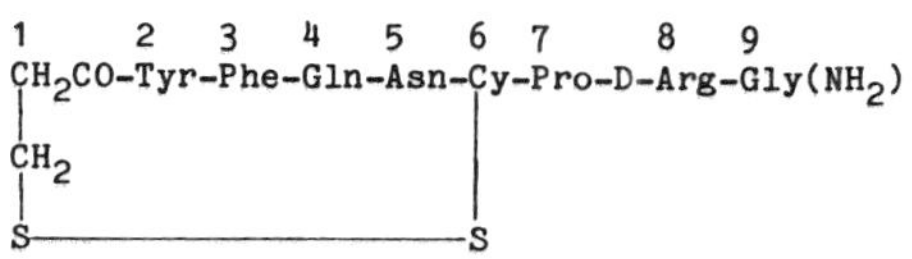

1-Deamino[4-valine,8-D-arginine]vasopressin (dVDAVP) has antidiuretic activity of 1200 U/mg, antivasopressor (antagonist) activity, and oxytocic activity of 8 U/mg. Milk-ejecting activity has not been tested.

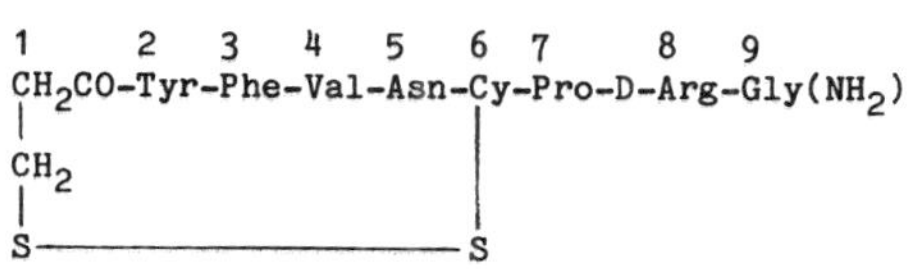

---

2. *Antidiuretic Antagonists*

[1-(β-Mercapto-β,β-cyclopentamethyleneproprionic acid),2-D-leucine,4-valine]arginine vasopressin [d(CH_2)$_5$-DLeu VAVP] has an anti-antidiuretic pA_2 of

7.79, an antivasopressor pA_2 of 6.45, and no anti-CRF-like (antagonist) potency at 50 μ*M*. This analog is 22 times more potent as an antidiuretic antagonist than as a vasopressor antagonist (Manning *et al.*, 1982; Knepel *et al.*, 1984).

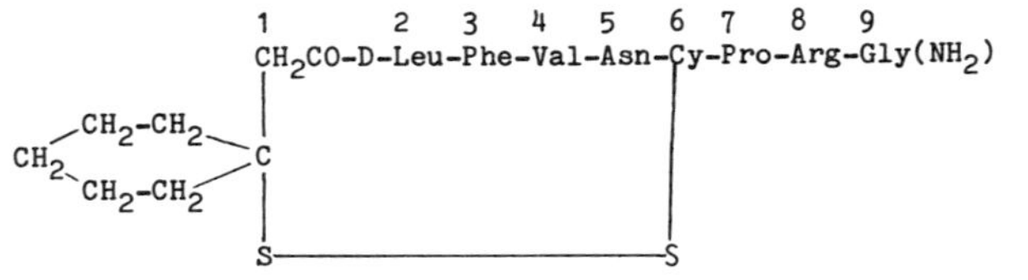

[1-(β-Mercapto-β,β-cyclopentamethylenepropionic acid),2-D-phenylalanine, 4-isoleucine]arginine vasopressin [d(CH_2)$_5$-DPhe[2],Ile[4]AVP] has an anti-antidiuretic pA_2 of 8.24, an antivasopressor pA_2 of 7.86, and an antioxytocic pA_2 of 7.11. Although one of the more potent antidiuretic antagonists, the selectivity of this analog is not great since the effective dose ratio for anti-vasopressor/anti-antidiuretic activity is only 2.2 compared to 22 for d(CH_2)$_5$-DLeu VAVP (Manning *et al.*, 1983).

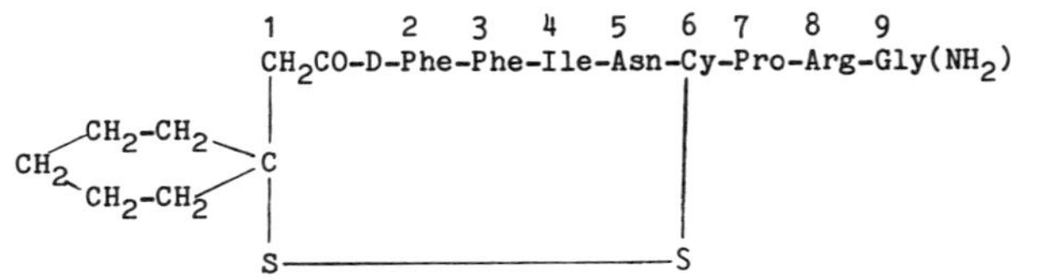

[1-(β-Mercapto-β,β-cyclopentamethylenepropionic acid),2-D-isoleucine,4(α-aminobutyric acid)]arginine vasopressin [d(CH_2)$_5$[DIle[2],Abu[4]]AVP has an anti-antidiuretic pA_2 of 8.22, an antivasopressor pA_2 of 6.73, and an *in vivo* antioxytocic pA_2 of 6.65. The anti-antidiuretic antagonist activity is 29 times more than the antivasopressor activity, so this is a relatively selective analog. By substituting isoleucine or alanine at position 4, d(CH_2)$_5$[DIle[2],Ile[4]]AVP and d(CH_2)$_5$[D Ile[2],Ala[4]]AVP, respectively, the anti-antidiuretic potency is boosted to 39 times more than the antivasopressor potency—that is, these latter analogs are even more selective antidiuretic antagonists (Manning *et al.*, 1983).

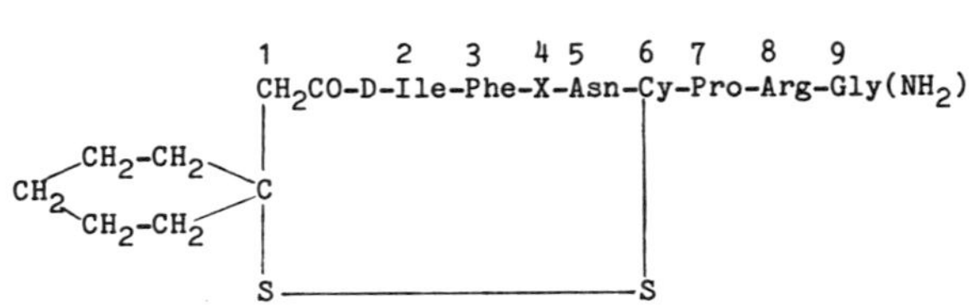

X = the substitute amino groups, Abu, Ile, or Ala.

C. $V_?$ Receptors Mediating Release of ACTH

1. *Unique Structure–Activity Profile of Agonists and Antagonists*

With the large array of agonists and antagonists for V_1 and V_2 receptors, investigators have undertaken studies to determine which receptors might be involved in other actions of vasopressin beside those regulating the circulation and renal free-water clearance. The results of these studies have not been entirely consistent. Aizawa *et al.* (1982) and Rivier *et al.* (1984) reported results consistent with a role of V_1 receptors in CRF-like stimulation of ACTH release by AVP. However, Knepel *et al.* (1984) could not detect any correlation between the ACTH-releasing activity and vasopressor (V_1) or antidiuretic (V_2) activities of AVP and 10 agonist analogs. In addition, one analog with vasopressor antagonist activity, $d(CH_2)_5Tyr(Me)AVP$, blocked release of ACTH by AVP, whereas another vasopressor antagonist with concomitant antidiuretic antagonist activity, $d(CH_2)_5$-DLeu VAVP, did not detectably alter AVP-induced ACTH release. On the basis of these data, Knepel *et al.* (1984) concluded that "results do not support the view that the pituitary vasopressin receptor conforms to the classification of vasopressin receptors in V_1 and V_2 types and, hence, the possibility exists that the pituitary vasopressin receptor represents a third category of vasopressin receptors."

Another possible explanation may be considered. The V_2 receptors mediating antidiuresis are coupled to the second messenger, adenylate cyclase. The V_1 receptors mediating vasoconstriction and glycogenolysis are coupled to phospholipase C and polyphosphoinositide breakdown. It may be that both receptors are present on pituitary cells and that after receptor activation, intracellular processing varies with the different analogs to account for the above results.

VI. SUMMARY

Vasopressin is the prototype neurohypophyseal peptide with actions on the cardiovascular system. These actions are complex. A basis for this complexity of actions has been presented. When receptor subtypes, coupling mechanisms, multiple intracellular mechanisms, and central as well as peripheral effects are considered, it becomes clear that our understanding of this complexity is far from complete and is likely to remain that way for some time into the future.

REFERENCES

Ahmad, Z., DePaoli-Roach, A. A., and Roach, P. J. (1982). *J. Biol. Chem.* **257,** 8348–8355.
Aizawa, T., Yasuda, N., Greer, M., and Sawyer, S. H. (1982). *Endocrinology* **110,** 98–104.

Akatsuka, N., Moran, W. H., Morgan, M. L., and Wilson, M. L. (1977). *J. Physiol. (London)* **266,** 567–588.
Aksoy, M. O., Mras, S., Kamm, K. E., and Murphy, R. A. (1983). *Am. J. Physiol.* **245**(*Cell Physiol.* **14**); C255–C270.
Altura, B. M. (1978). *Microvasc. Res.* **16,** 91–117.
Altura, B. M., and Altura, B. T. (1977). *Fed. Proc.* **36,** 1853–1860.
Aronsen, K. F., and Nylander, G. (1964). *Vasc. Dis.* **1,** 127–133.
Aronsen, K. F., and Nylander, G. (1966). *Acta Radiol.* **4,** 353–364.
Axtelle, T. S., Heesch, C. M., and Abboud, F. M. (1985). *Fed. Proc.* **44,** 1244.
Banks, R. O. (1976). *Proc. Soc. Exp. Biol. Med.* **151,** 547–551.
Barer, G. R. (1961). *J. Physiol. (London)* **156,** 49–66.
Barer, G. R. (1963). *J. Physiol. (London)* **169,** 62–72.
Berridge, M. J., and Fain, J. N. (1979). *Biochem. J.* **178,** 59–69.
Bone, E. A., Fretten, P., Palmer, S., Kirk, C. J., and Michell, R. H. (1984). *Biochem. J.* **221,** 803–811.
Bolton, T. B. (1979). *Physiological Rev.* **56,** 606–718.
Brooks, F. P., and Pickford, M. (1958). *J. Physiol. (London)* **142,** 468–493.
Chalmers, J. P., and Reid, J. L. (1972). *Circ. Res.* **31,** 789–804.
Chatterjee, M., and Murphy, R. A. (1983). *Science* **221,** 464–466.
Chiu, E. K. Y., and McNeill, J. R. (1985). *Am. J. Physiol.* **249,** H193–H197.
Cowley, A. W., Jr. (1982). *Int. Rev. Physiol.* **26,** 289–242.
Cowley, A. W., Monos, E., and Guyton, A. C. (1974). *Circ. Res.* **34,** 506–514.
Cowley, A. W., Merill, D. C., Osborn, J., and Barber, B. J. (1984). *Circ. Res.* **54,** 163–172.
Dawson, R. M. C. (1949). *Biochem. J.* **57,** 237–245.
Delaney, J. P., Goodale, R. L., Cheng, J., and Wang-Ensteen, O. H. (1966). *Surgery* **59,** 397–400.
Elliot, J., West, M. J., and Chalmers, J. (1985). *J. Cardiovasc. Pharmacol.* **7,** 6–11.
Exton, J. H. (1980). *Am. J. Physiol.* **238,** E3–E12.
Exton, J. H. (1981). *Mol. Cell. Endocr.* **23,** 233–264.
Fisher, R. D., Grunfeld, J. P., and Barger, A. C. (1970). *Am. J. Physiol.* **219,** 1348–1358.
Folch, J. (1949). *J. Biol. Chem.* **177,** 497–504.
Gardier, R. W., and Abreu, B. E. (1958). *Fed. Proc.* **17,** 370.
Garrison, J. C., Johnsen, D. E., and Campanile, C. P. (1984). *J. Biol. Chem.* **259,** 3283–3292.
Gerthoffer, W. T., and Murphy, R. A. (1983). *Am. J. Physiol.* **245,** (Cell Physiol **14**), C271–C277.
Gothlin, J. (1976). *Radiol. Diagn.* **17,** 763–772.
Guo, G. B., Sharabi, F. M., Abboud, F. M. and Schmid, P. G. (1982). *Circulation* **66,** II–34.
Guo, G. B., Schmid, P. G., and Abboud, F. M. (1983). *Clin. Res.* **31,** 189A.
Guo, G. B., Schmid, P. G., and Abboud, F. M. (1986). *Am. J. Physiol.* (in press)
Hanley, M. R., Benton, H. P., Lightman, S. L., Todd, K., Bone, E. A., Fretten, P., Palmer, S., Kirk, C. J., and Michell, R. H. (1984). *Nature (London)* **309,** 258–261.
Hausler, H. (1930). *J. Physiol. (London)* **68,** 324–332.
Hoffman, W. E. (1980). *Endocrinology* **107,** 334–41.
Imai, Y., Nolan, P. L., and Johnston, C. I. (1983). *Circ. Res.* **53,** 140–149.
Imaizumi, T., and Thames, M. D. (1986). *Circ. Res.* **58,** 17–25.
Jard, S., and Bockaert, J. (1975). *Physiol. Rev.* **55,** 489–536.
Johnson, M. D., Park, C. S., and Malvin, R. L. (1977). *Am. J. Physiol.* **232,** F111–F116.
Knepel, W., Homolka, L., Vlaskovska, M., and Nutto, D. (1984). *Endocrinology* **114,** 1797–1804.
Kruszynski, M., Lammek, B., and Manning, M. (1980). *J. Med. Chem.* **23,** 364–368.
Lassoff, S., and Altura, B. M. (1980). *Brain Res.* **196,** 266–269.
Liard, J. F. (1986). *Circ. Res.* **58,** 631–640.
Liard, J. F., and Spadone, J. C. (1984). *J. Cardiovasc. Pharmacol.* **6,** 713–719.

Liard, J. F., Deriaz, O., Tschopp, M., and Schoun, J. (1981). *Clin. Sci.* **61,** 345–347.

Liard, J. F., Deriaz, O., Schelling, P., and Thibonnier, M. (1982). *Am. J. Physiol.* **243,** H663–H669.

Lohmeier, T. E., Smith, M. J., Cowley, A. W., Jr., Manning, R. D., and Guyton, A. C. (1981). *Hypertension* **3,** 416–425.

Longo, L. D., Morris, J. A., Smith, R. W., Beck, R., and Assali, N. S. (1964). *Proc. Soc. Exp. Biol. Med.* **115,** 766–770.

Loutzenhizer, R., and Van Breeman, C. (1981). *Am. J. Physiol.* **241** (Cell Physiol. **10**), C243–C249.

Lowbridge, J., and Manning, M. (1978). *J. Med. Chem.* **21,** 313–315.

Lu, F. C. (1951). *J. Pharmacol. Exp. Ther.* **102,** 145–149.

Mancia, G., and Zanchetti, A. (1981). "Handbook of the Hypothalamus," (P. J. Morgane and J. Panksepp, eds.), Vol. 3, Part B, pp. 147–202. Marcol Dekker Inc., New York.

Manning, M., and Sawyer, W. H. (1983). *Prog. Brain Res.* **60,** 367–382.

Manning, M., Olma, A., Klis, W. A., Seto, J., and Sawyer, W. H. (1983). *J. Med. Chem.* **26,** 1607–1613.

Manning, M., and Sawyer, W. H. (1984). *Trends Neurosci.* **7,** 6–9.

Manning, M., Lowbridge, J., Stier, J., Haldar, J., and Sawyer, W. H. (1977). *J. Med. Chem.* **20,** 1228–1230.

Manning, M., Lammek, B., Kolodziejczyk, A., Seto, J., and Sawyer, W. H. (1981). *J. Med. Chem.* **24,** 701–706.

Manning, M., Klis, W. A., Olma, A., Seto, J., and Sawyer, W. H. (1982). *J. Med. Chem.* **25** 414–419.

Matsuguchi, H., and Schmid, P. G. (1982a). *Am. J. Physiol.* **242,** H37–H43.

Matsuguchi, H., and Schmid, P. G. (1982b). *Am. J. Physiol.* **242,** H44–H49.

Mellander, S., and Johansson, B. (1968). *Pharmacol. Rev.* **20,** 117–196.

Melville, K. I. (1936). *J. Am. Med. Assoc.* **106,** 102–105.

Michell, R. H. (1975). *Biochim. Biophys. Acta* **415,** 81–147.

Mohring, J. K., Glanzer, J. A., Maciel, R., Dusing, H., Kramer, J., Arbogast, R., and Koch-Weser, J. (1980). *J. Cardiovasc. Pharmacol.* **2,** 367–376.

Monos, E., Cox, R. H., and Robertson, L. A. (1978). *Am. J. Physiol.* **234,** H167–H172.

Montani, J. P., Liard, J. F., Schoun, J., and Mohring, J. (1980). *Circ. Res.* **47,** 346–355.

Morgan, J. P., and Morgan, K. C. (1983). *Pfluegers Arch. Gesamte Physiol. Menschen Tiere* **395,** 75–77.

Nakano, J. (1967). *J. Pharmacol. Exp. Ther.* **157,** 19–31.

Nakano, J. (1974). *Am. Physiol. Soc.* **4,** 395–442.

Nasjletti, A., and Malik, K. U. (1981). *Med. Clin. North Am.* **65,** 881–889.

Nash, C. B. (1965). *Arch. Int. Pharmacodyn. Ther.* **155,** 90–95.

Payne, M. E., Schworer, C. M., and Soderling, T. R. (1983). *J. Biol. Chem.* **258,** 2376–2382.

Penit, J., Faure, M., and Serge, J. (1983). *Am. J. Physiol.* **244,** E72–E82.

Rasmussen, H., and Barrett, P. Q. (1984). *Physiol. Rev.* **64,** 938–985.

Rivier, C., Rivier, J., Mormede, P., and Vale, W. (1984). *Endocrinology* **115,** 882–886.

Rocha de Silva, M., and Rosenberg, M. (1969). *J. Physiol. (London)* **202,** 535–557.

Saameli, K. (1968). "Handbook of Experimental Pharmacology, Neurohypophyseal Hormones and Similar Polypeptides," (B. Berde, ed.), Vol. 23, pp. 545–612. Springer-Verlag, Berlin and New York.

Sawyer, W. H. (1961). *Methods Med. Res.* **9,** 210–219.

Schmid, P. G., Abboud, F. M., Wendling, M. G., Ramberg, E. S., Mark, A. L., Heistad, D. D., and Eckstein, J. W. (1974). *Am. J. Physiol.* **227,** 998–1004.

Schmid, P. G., Sharabi, F. M., Guo, G. B., Abboud, F. M., and Thames, M. D. (1984). *Fed. Proc.* **43,** 97–102.

Schmid, P. G., Guo, G. B., and Abboud, F. M. (1985a). *Fed. Proc.* **44,** 2388–2392.

Schmid, P. G., Guo, G. B., Axtelle, T. S., and Abboud, F. M. (1985b). *In* "Vasopressin," (R. W. Schrier, ed.), pp. 11–20. Raven, New York.

Schwartz, J., Liard, J. F., Ott, C., and Cowley, A. (1985). *Am. J. Physiol.* **249,** H1001–H1008.

Schwartz, S. I., Bales, H. W., Emmerson, G. L., and Mahoney, E. B. (1959). *Surgery* **45,** 72–80.

Shaldon, S. W., Dolle, L., Guevara, L., Iber, F. L., and Sherlock, S. (1961). *Circulation* **24,** 797–807.

Sharabi, F. M., and Schmid, P. G. (1982). *Am. J. Nephrol.* **3,** 164–171.

Sharabi, F. M., Guo, G. B., Abboud, F. M., Thames, M. D., and Schmid. P. G. (1985). *Am. J. Physiol.* **249,** H922–H928.

Slotnik, I. L., and Teigland, J. D. (1951). *J. Am. Med. Assoc.* **146,** 1126–1129.

Somlyo, A. P., and Somlyo, A. V. (1970). *Pharmacol. Rev.* **22,** 249–353.

Sunahara, F., Duncanson, D., and Edholm, O. G. (1949). *Fed. Proc.* **8,** 152.

Swanson, L. W., and Sawchenko, P. E. (1983). *Annu. Rev. Neurosci.* **6,** 269–324.

Takenawa, T. (1982). *Cell Calcium* **3,** 359–368.

Tanaka, M., de Kloe, E. R., de Wied, D., and Versteeg, D. H. G. (1979). *Life Sci.* **20,** 1799–1808.

Troyer, D. A., Kreisberg, J. I., Schwertz, D. W., and Venkatachalam, M. A. (1985). *Am. J. Physiol.* **249,** F139–F147.

Undesser, K. P., Hasser, E. M., Haywood, J. R., Johnson, A. K., and Bishop, V. S. (1985). *Circ. Res.* **56,** 410–417.

Walker, A. M., Schmidt, C. F., Elson, K. A., and Johnston, C. G. (1937). *Am. J. Physiol.* **118,** 95–110.

Wali, F. A. (1984). *Pharmacol. Res. Commun.* **16,** 55–62.

Webb, R. C., and Bohr, D. F. (1981). *Prog. Cardiovasc. Dis.* **44,** 213–242.

Williamson, J. R., Cooper, R. H., and Hoek, J. B. (1981). *Biochim. Biophys. Acta* **639,** 243–295.

Williamson, J. R., Cooper, R. H., Joseph, S. K., and Thomas, A. P. (1985). *Am. J. Physiol.* **248,** C203–C216.

Index

R

S

T

U

V

X